SpringerWienNewYork

Wolfgang R. Hess, Anita Marchfelder

Regulatory RNAs in Prokaryotes

SpringerWienNewYork

Professor Dr. Wolfgang R. Hess
University Freiburg, Institute of Biology III
Freiburg, Germany

Professor Dr. Anita Marchfelder
Ulm University, Institute of Biology II
Ulm, Germany

Springer-Verlag Wien New York is part of
Springer Science + Business Media
springer.at

Typesetting: le-tex publishing services GmbH, 04229 Leipzig, Germany
Printing: Holzhausen Druck GmbH, 1140 Vienna, Austria

Printed on acid-free and chlorine-free bleached paper
SPIN: 12778393

With 46 (partly coloured) Figures

Library of Congress Control Number: 2011934264

ISBN 978-3-7091-0217-6 SpringerWienNewYork

Editorial

Regulatory RNAs in Prokaryotes

RNA molecules play a central role in gene regulation in all three domains of life. Regulatory RNAs were originally discovered in prokaryotes as early as 1967. Fundamental mechanisms of how these molecules exert their functions were first analyzed in bacteria long before small RNAs were discovered as regulatory molecules in eukaryotes. Research on regulatory RNA in prokaryotes occurred in three major phases.

The first phase started in 1967, when Hindley (1967) identified an RNA species, later named 6S RNA, as a distinct and abundant RNA species in *E. coli*. In pioneering work four years later, its sequence and putative secondary structure were published (Brownlee, 1971). However, several decades passed before 6S RNA function in regulating RNA polymerase activity was determined (Wassarman and Storz 2000). Another enterobacterial regulatory RNA reported early on was the Spot 42 (spf) RNA (Ikemura and Dahlberg 1973). Discovering that the *spf* gene is regulated by the cAMP–CRP system (Sahagan and Dahlberg 1979) and the phenotypic consequences of its overexpression (Rice and Dahlberg 1982) suggested its functional relevance. However, a biological role was determined almost 40 years later, when its significant complementarity to the region around the start codon of the *galK* gene was noticed and its role in discoordinating gene expression of the *galETKM* galactose operon became unraveled (Møller *et al.* 2002).

About the same time the first *trans*-acting regulatory RNAs were discovered, the first regulatory *cis*-antisense RNAs were identified in bacteria. These *cis*-antisense RNAs initially appeared to be a hallmark of extrachromosomal genetic elements, bacteriophages, transposons, and plasmids, controlling their life cycle or copy number. The first of these findings was the identification of antisense transcripts for the gene *cro* in bacteriophage λ (Spiegelman *et al.* 1972). This type of transcription was confirmed for bacteriophage λ when observing that overexpression of the 77 nt OOP antisense transcript leads to its codegradation with the cII mRNA (Krinke and Wulff 1987; Krinke and Wulff 1990; Krinke *et al.* 1991). By studying the plasmid-borne RNA I, another extrachromosomally located *cis*-antisense RNA, many fundamental insights were gained early on. Among those discoveries was that RNA

I regulates maturation of the ColE1 primer for DNA replication (Stougaard *et al.* 1981; Tomizawa *et al.* 1981) and is involved in the control of plasmid incompatibility of ColE1-type plasmids (Tomizawa and Itoh 1981).

In the following two decades, a small number of additional regulatory RNAs were found fortuitously. Although important regulators were discovered, such as the chromosomally encoded small RNA MicF (Mizuno *et al.* 1983, 1984), DicF (Faubladier *et al.* 1990) and OxyS (Altuvia *et al.* 1997), the fundamental importance and broad consequences of all these findings for gene regulation were not initially appreciated. In early 2001, only 12 small RNAs (including the 6S RNA, tmRNA, RNase P RNA and 4.5 S RNA) had been identified in *E. coli* (Argamann *et al.* 2001).

A new phase started in 2001 when computational searches were introduced for more complex and systematic screening. Pioneering studies on small RNA prediction in enterobacteria employed comparative genome analysis of closely related species (Wassarman *et al.* 2001), included a search for transcriptional signals in intergenic regions (Argaman *et al.* 2001), or scored the conservation of predicted RNA secondary structure rather than of primary sequence (Rivas *et al.* 2001). However, the most significant advancement was to integrate these predictions with systematic experimental screens. As result of these seminal studies, several dozens of new *trans*-acting RNAs were identified (Argaman *et al.* 2001; Rivas *et al.* 2001; Wassarmann *et al.* 2001), yielding data for their detailed functional characterization for many years.

A third phase of prokaryotic RNA research began more recently with the advent of RNA-seq technology, triggering a wave of new studies, which have been setting new standards in this field by accelerating the identification of transcripts and transcriptional start sites. Together with progress in RNA bioinformatics and experimental structure determination, new research groups entering this exciting field of research and focusing on the biochemistry, metabolism and molecular biology of RNA, spectacular new insights into the world of prokaryotic regulatory RNAs have been obtained at an unprecedented speed and resolution. To highlight these advancements, this book focuses exclusively on prokaryotic regulatory RNAs.

Current research on regulatory RNA in prokaryotes is presented here by first providing an in depth overview of *trans*- and *cis*-acting small RNAs in various groups of bacteria and archaea and their established mechanisms of action, including the effects mediated by Hfq, an interacting protein with a pivotal role in many bacteria. These chapters are followed by reviews on regulatory mechanisms involving distinct types of RNA (e.g., 6S RNA), control of bacterial heat shock and virulence genes by RNA thermometers, and functions of *cis*-acting metabolite-sensing riboswitches. One chapter is devoted to the major recent discovery of an RNA-based prokaryotic immune system. The two last chapters provide an overview on available computational approaches to predict prokaryotic regulatory RNAs and their targets based on sequence information.

In all, this book is written by leading experts in the field and presents a timely introduction that covers all aspects of prokaryotic regulatory RNAs and their functional mechanisms.

Freiburg and Ulm in May 2011 Anita Marchfelder and Wolfgang R. Hess

References

Altuvia S, Weinstein-Fischer D, Zhang A, Postow L, Storz G (1997) A small, stable RNA induced by oxidative stress: role as a pleiotropic regulator and antimutator. Cell 90: 43–53

Argaman L, Hershberg R, Vogel J, Bejerano G, Wagner EG, Margalit H, *et al.* (2001) Novel small RNA-encoding genes in the intergenic regions of *Escherichia coli*. Curr Biol 11: 941–950

Brownlee GG (1971) Sequence of 6S RNA of *E. coli*. Nature New Biol 229: 147–149

Faubladier M, Cam K, Bouche JP (1990) *Escherichia coli* cell division inhibitor DicF-RNA of the *dicB* operon. Evidence for its generation *in vivo* by transcription termination and by RNase III and RNase E-dependent processing. J Mol Biol 212: 461–471

Hindley J (1967) Fractionation of ^{32}P-labelled ribonucleic acids on polyacrylamide gels and their characterization by fingerprinting. J Mol Biol 30, 125–136

Ikemura T and Dahlberg JE (1973) Small ribonucleic acids of *Escherichia coli*. I. Characterization by polyacrylamide gel electrophoresis and fingerprint analysis. J Biol Chem 248: 5024–5032

Krinke L, Mahoney M, Wulff DL (1991) The role of the OOP antisense RNA in coliphage λ development. Mol Microbiol 5: 1265–1272

Krinke L, Wulff DL (1987) OOP RNA, produced from multicopy plasmids, inhibits lambda cII gene expression through an RNase III-dependent mechanism. Genes Dev 1: 1005–1013

Krinke L, Wulff DL (1990) RNase III-dependent hydrolysis of lambda cII-O gene mRNA mediated by lambda OOP antisense RNA. Genes Dev 4: 2223

Mizuno T, Chou MY, Inouye M (1983) Regulation of gene expression by a small RNA transcript (Micrna) in *Escherichia-coli*-K-12. Proc Japan Acad Ser B Phys Biol Sci 59: 335–338

Mizuno T, Chou MY, Inouye M (1984) A unique mechanism regulating gene expression: translational inhibition by a complementary RNA transcript (micRNA). Proc Natl Acad Sci USA 81: 1966–1970

Møller T, Franch T, Udesen C, Gerdes K, Valentin-Hansen P (2002) Spot 42 RNA mediates discoordinate expression of the *E. coli* galactose operon. Genes Dev 16: 1696–1706

Rice PW, Dahlberg JE (1982) A gene between *polA* and *glnA* retards growth of *Escherichia coli* when present in multiple copies: Physiological effects of the gene for Spot 42 RNA. J Bacteriol 152: 1196–1210

Rivas E, Klein RJ, Jones TA, Eddy SR (2001) Computational identification of noncoding RNAs in *E. coli* by comparative genomics. Curr Biol 11: 1369–1373

Sahagan BG, Dahlberg JE (1979) A small, unstable RNA molecule of *Escherichia coli*: spot 42 RNA. I. Nucleotide sequence analysis. J Mol Biol 131: 573–579

Spiegelman WG, Reichardt LF, Yaniv M, Heinemann SF, Kaiser AD, Eisen H (1972) Bidirectional transcription and the regulation of Phage lambda repressor synthesis. Proc Natl Acad Sci USA 69: 3156–3160

Stougaard P, Molin S, NordstrOm K. (1981) RNAs involved in copy-number control and incompatibility of plasmid R1. Proc Natl Acad Sci USA 78: 6008–6012

Tomizawa J, Itoh T (1981) Plasmid incompatibility determined by interaction of RNAI with primer transcript. Proc Natl Acad Sci USA 78: 6096–6100

Tomizawa J, Itoh T, Seizer G, Som T (1981) Inhibition of ColEI RNA primer formation by a plasmid-specified small RNA. Proc Natl Acad Sci USA 78: 1421–1425

Wassarman KM, Storz G (2000) 6S RNA regulates *E. coli* RNA polymerase activity Cell 101: 613–623

Wassarman KM, Repoila F, Rosenow C, Storz G, Gottesman S (2001) Identification of novel small RNAs using comparative genomics and microarrays. Genes Dev 15: 1637–1651

Contents

Chapter 1
Small RNAs with a Role in the Oxidative Stress Response of Bacteria

Bork Berghoff and Gabriele Klug*

1 Introduction

Most bacteria have to cope with frequent changes in their environment, which generate unfavourable conditions for growth and survival. They have evolved successful strategies as a response to these stresses. Oxidative stress is a stress factor, which is critical in most bacterial habitats and has been defined as an imbalance between pro-oxidants and anti-oxidants in the cell (Storz and Zheng, 2000). Pro-oxidants are mostly reactive oxygen species (ROS) that oxidize proteins, nucleic acids and lipids and thus lead to harmful damage to the cell (Imlay, 2003). Anti-oxidants are cellular components countering these damaging effects: i) enzymes or molecules which remove ROS like peroxidases, superoxide dismutase, thioredoxin or glutathione, ii) proteins that repair the damages like endo- and exonucleases or photolyases, and iii) sensors and regulators necessary to mount the response to oxidative stress like OxyR or SoxRS of *E. coli*. ROS are generated from the ground state (triplet state) of molecular oxygen when less than four electrons are transferred to one O_2 molecule resulting in partially reduced forms of oxygen (Imlay, 2003). Such reactions are e. g. catalyzed by respiratory enzymes and lead to the accumulation of hydrogen peroxide (H_2O_2), superoxide (O_2^-) and hydroxyl radicals ($OH^•$). But ROS are also produced by exposure of cells to metals, redox-active drugs or radiation. Plants as well as animals produce ROS as a defence against pathogens. In addition to electron transfer reactions, a spin conversion of one electron of the oxygen molecule can generate the reactive singlet oxygen ($^1O_2$). This reaction occurs in the combined presence of light and a photosensitizer. In nature, porphyrins (chlorophylls or protoporphyrin) and humic acids can function as photosensitizers.

The oxidative stress response of many model bacteria has been extensively studied in the past and many regulatory proteins and protein based signalling path-

* Institute for Microbiology and Molecular Biology, University of Giessen, Heinrich-Buff-Ring 26–32, 35392 Giessen, Germany

Table 1. Small RNAs with a potential role in the oxidative stress response of bacteria.

sRNA	Bacterium[a]	Length [nt]	Stress conditions	Target mRNA	Mode of regulation	Hfq interaction	References
OxyS	E.c./S.t.	109	H_2O_2	*fhlA, rpoS*	translational repression	yes	Altuvia *et al.*, 1997
RyhB	E.c./S.t. P.a./E.ch.	90	iron limitation	*fur, bfr, ftnA, sodB, iscRSUA*[b]	mRNA destabilization	yes	Massé and Gottesman, 2002
RSs0019	R.s.	298	$^1O_2$	n.k.[c]	n.k.	no[d]	Berghoff *et al.*, 2009
RSs0682	R.s.	206, (130)[e]	$^1O_2$	n.k.	n.k.	yes[d]	Berghoff *et al.*, 2009
RSs0680a RSs1543 RSs2461	R.s.	73 83 116, (75)[e]	$^1O_2$, $O_2^{\cdot-}$, heat	n.k.	n.k.	yes[d]	Berghoff *et al.*, 2009 Nuss *et al.*, 2010
MicF	E.c./S.t.	93	$O_2^{\cdot-}$, membrane perturbation[f]	*ompF*	translational repression	yes	Blanchard *et al.*, 2007 Vogel and Papenfort, 2006
MicC	E.c./S.t.	109	$O_2^{\cdot-}$, membrane perturbation[f]	*ompC, ompD*	translational repression	yes	Blanchard *et al.*, 2007 Vogel and Papenfort, 2006 Pfeiffer *et al.*, 2009
RydB	E.c.	68	$O_2^{\cdot-}$	n.k.	n.k.	n.k.[g]	Blanchard *et al.*, 2007
CyaR	E.c./S.t.	86	$O_2^{\cdot-}$	*ompX*	translational repression	yes	Blanchard *et al.*, 2007 Papenfort *et al.*, 2008
ArcZ	E.c./S.t.	120–130, (~50)[e]	n.k.	*tpx*[h]	translational repression	yes	Papenfort *et al.*, 2009
RgsA	P.f.	~120	H_2O_2	n.k.	n.k.	n.k.	Gonzalez *et al.*, 2008
Yfr1	S.e.	65	$O_2^{\cdot-}$, salt	*sbtA*	mRNA destabilization	n.k.	Nakamura *et al.*, 2007
IsrR	S.sp.	177	iron limitation, H_2O_2	*isiA*	mRNA destabilization	n.k.	Dühring *et al.*, 2006
B11 B55 F6 ASpks	M.t.	93 61 58, 102 78, (~200)[i]	H_2O_2	n.k.	n.k.	n.k.	Arnvig and Young, 2009

Explications for Table 1 *see next page*

◄ **Explications for Table 1**
a E.c.: *Escherichia coli*; E.ch.: *Erwinia chrysanthemi*; M.t.: *Mycobacterium tuberculosis*; P.a.: *Pseudomonas aeruginosa*; P.f.: *Pseudomonas fluorescens*; R.s.: *Rhodobacter sphaeroides*; S.e.: *Synechococcus elongatus*; S.sp.; *Synechocystis* sp. PCC6803; S.t.: *Salmonella typhimurium*
b several other targets like *acnA*, *fumA*, and *sdhCDAB*: at least 18 transcripts, encoding 56 proteins (Massé *et al.*, 2005)
c n.k.: not known
d unpublished data (Hfq co-immunoprecipitation experiments)
e length of processed fragment is shown in brackets
f Transcription of MicF is induced and MicC is repressed by the EnvZ-OmpR system
g putative RydB homolog of *Salmonella typhimurium* interacts with Hfq (Sittka *et al.*, 2008)
h other targets: *sdaCB* and STM3216
i 200-nt fragment only detectable under stress

ways have been elucidated (e.g.: Storz and Imlay, 1999; Storz and Zheng, 2000; Mongkolsuk and Helmann, 2002; Imlay, 2008). The view emerged that the components of oxidative stress response systems overlap with components of other stress response systems, e.g. the heat shock response. It is now widely accepted that we cannot assign strictly defined regulatory systems to a single stress. Instead several components contribute to the response against different stresses and only a few components are specific to a certain stress response. In this review we will focus on responses against ROS or responses affecting genes with a clear function during oxidative stress.

Considering recent advances in the knowledge of the important regulatory roles of small RNAs (sRNAs) in bacteria, it is not surprising to find that they are also part of the oxidative stress response systems. OxyS of *E. coli* was among the first sRNAs to be discovered and analyzed in detail. It links the oxidative stress response to more global responses including other stress resistances, carbon metabolism or cell morphology. In the same organism, the sRNA RyhB plays an important role in linking the response to iron to the oxidative stress response. This review will summarize our current knowledge on the biological function of these two sRNAs and the underlying mechanisms of regulation. In the case of several other sRNAs, changed levels in response to oxidative stress have been reported or they were shown to affect the resistance to ROS (see Table 1). However, their exact function and their mechanisms of action need further elucidation. We attempt to give an overview of those sRNAs and their putative functions.

2 OxyS and the Oxidative Stress Response in Enterobacteria

When studies on the oxidative stress response in enteric bacteria were initiated, the *oxyR* gene was discovered in a screen for *Salmonella* mutants that were hyperresistant to H_2O_2 (Christman *et al.*, 1985). The OxyR protein was shown to function as a redox sensor, which is oxidized at elevated levels of H_2O_2. The oxidized protein binds to DNA target sequences and subsequently activates a small subset of genes (Storz *et al.*, 1990). One of these genes encodes catalase that quickly removes H_2O_2 from the cytoplasm. While following OxyR mRNA levels in *E. coli* by Northern blot hybridization using a probe which in addition to the *oxyR* sequence comprised 200 bp of the upstream region, a strong signal for an sRNA, OxyS was discovered (Altuvia *et al.*, 1997). OxyS is transcribed in opposite direction to OxyR from a promoter that overlaps the promoter for OxyR and is activated by OxyR. Expression of OxyS is quickly and strongly induced upon H_2O_2 addition, while other stress factors only weakly induce OxyS (Altuvia *et al.*, 1997). Deletion of OxyS results in two-fold higher levels of intracellular H_2O_2 (Gonzalez-Flecha and Demple, 1999). Using a genetic screen, eight genes were originally found to be regulated by OxyS (Altuvia *et al.*, 1997), among them the *rpoS* gene for an alternative sigma factor and *fhlA*, a transcriptional activator of formate metabolism. While OxyR-like regulators are found in many bacteria, OxyS seems to be restricted to enteric bacteria.

The mechanism of regulation by OxyS has been best analyzed for the *fhlA* target. Altuvia *et al.* (1998) showed that OxyS represses *fhlA* translation by blocking

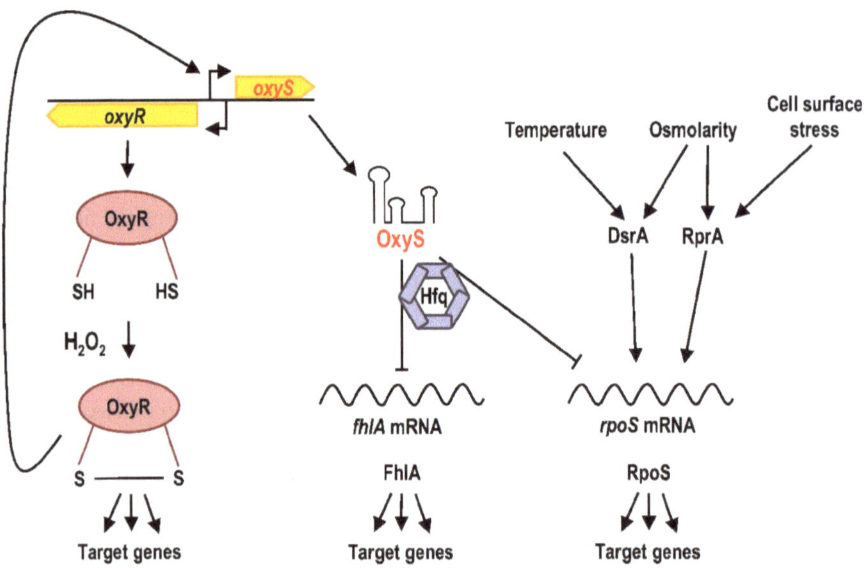

Fig. 1. OxyS is induced by H_2O_2 and acts as a negative riboregulator.
The *oxyS* gene is transcribed divergently from the *oxyR* gene, whereas the promoters are overlapping. OxyR is a transcriptional activator, which is oxidized by H_2O_2 at specific cysteine residues. Oxidized OxyR is active and induces transcription of stress-related genes. Transcription of OxyS sRNA is also induced by OxyR. Together with Hfq, OxyS negatively influences the translation of its target mRNAs, *fhlA* and *rpoS*. FhlA is a transcriptional activator and RpoS is an alternative sigma factor known to regulate gene expression during stationary phase. Translation of *rpoS* mRNA is additionally controlled by the sRNAs DsrA and RprA in a positive manner. DsrA and RprA are induced under cellular stress conditions like changes in temperature, osmolarity or cell surface stress

the ribosome-binding site. Later, the formation of a kissing complex between OxyS and *fhlA* RNAs was demonstrated (Argaman and Altuvia, 2000). Repression of *fhlA* and *rpoS* translation both depend on the RNA chaperone Hfq, since Hfq increases OxyS interaction with its target RNAs (Zhang *et al.*, 2002). Figure 1 illustrates the induction of OxyS by OxyR and its role in post-transcriptional regulation.

Several of the OxyS-regulated genes are also regulated by RpoS, an alternative sigma factor of *E. coli*. OxyS was shown to repress RpoS at post-transcriptional level, most likely by repressing translation. The A-rich single-stranded linker region between the stable OxyS hairpin loop structures is important for this repression (Zhang *et al.*, 2002). Recently it was demonstrated that growth-phase also affects stability of OxyS but altered OxyS stability does not contribute to growth-phase-dependent *rpoS* regulation (Basineni *et al.*, 2009). RpoS was considered as "stationary phase" sigma factor due to its accumulation in stationary phase (Lange and Hengge-Aronis, 1991). It is now well accepted that RpoS is not just a regulator of stationary phase but has a more general role and its target genes are involved in functions such as stress resistance (UV, osmolarity, oxidative and temperature stress), cell envelope composition, cell morphology, and carbon metabolism (Hengge-Aronis, 2002). The exact mechanism by which OxyS affects RpoS levels is less well understood than the OxyS/*fhlA* interaction. The two sRNAs, RprA and DsrA, activate *rpoS* translation in response to changes in osmolarity or temperature, respectively (Figure 1). They bind to the *rpoS* untranslated region and disrupt the formation of a hairpin that masks the ribosome-binding site (Majdalani *et al.*, 1998; 2002). In contrast to RprA and DsrA, OxyS represses *rpoS* translation, but its exact mode of action has not been explained.

The different sRNAs acting on RpoS can be present in the cell simultaneously and may compete for binding to *rpoS* mRNA. The interplay of different sRNAs thus contributes to complex regulatory networks.

3 The Link Between Iron Levels and Oxidative Stress, and the Role of RyhB

3.1 How Iron Can Cause Oxidative Stress

Iron is the most important micronutrient used by bacteria and is essential for cellular processes like respiration, photosynthesis, and nitrogen fixation. It acts as a cofactor for many enzymes and is indispensable for the biogenesis of iron-sulphur (Fe-S) clusters (Wackett *et al.*, 1989; Ayala-Castro *et al.*, 2008). However, iron acquisition and usage have to be tightly controlled in bacteria because high concentrations of free iron favour the generation of hydroxyl radicals ($OH^{\bullet}$) in a process called the Fenton reaction. In this reaction ferrous iron [Fe(II)] catalyzes the conversion of hydrogen peroxide (H_2O_2) to hydroxide ions (OH^-) and $OH^{\bullet}$. The Fe(II) is oxidized to ferric iron [Fe(III)] during this conversion.

$$\mathrm{Fe(II)} + H_2O_2 \rightarrow \mathrm{Fe(III)} + OH^- + OH^{\bullet} \qquad \text{(Fenton reaction)}$$

Accordingly, a less deleterious ROS (H_2O_2) is converted into a highly reactive ROS ($OH^{\bullet}$) by the action of free iron. In addition, H_2O_2 and superoxide (O_2^-) increase the free iron concentration by damaging Fe-S clusters and thereby accelerating the Fenton reaction (Touati, 2000; Varghese *et al.*, 2003). Since iron is a cofactor of proteins involved in defence against ROS (e.g. iron superoxide dismutase, SodB), iron limitation can also lead to elevated oxidative stress. It does not come as a surprise that the iron metabolism is in part coupled to the oxidative stress response. Here we give a review on the RyhB sRNA, which contributes considerably to iron availability and to the avoidance of oxidative stress in *E. coli*.

3.2 Mechanisms of RyhB Regulation

The 90-nt RyhB sRNA was first identified in a genome-wide screen for sRNAs using comparative genomics and microarrays in *E. coli* (Wassarman *et al.*, 2001). Cells overproducing RyhB showed only poor growth on media containing succinate as carbon source. Only one year after its identification, Massé and Gottesman (2002) demonstrated that RyhB negatively regulates a set of six iron-storage and iron-using proteins when iron is limited. Transcription of RyhB itself, is repressed by the global regulator Fur (*F*erric *u*ptake *r*egulator). Besides *ryhB*, essentially all genes involved in iron acquisition are Fur-regulated. In addition, several genes for general metabolism, pathogenicity, and defence against oxidative and acid stresses are also regulated by Fur (Escolar *et al.*, 1999). Since Fur acts as a repressor of transcription under high iron concentrations using Fe(II) as a cofactor (Bagg and Neilands, 1987), repression of *ryhB* explains the earlier observed activation of gene expression by Fur. Fur positively regulates the transcription of *sodB*, *acnA*, *fumA*, and *sdhCDAB*, all encoding iron-containing proteins. *bfr* and *ftnA*, encoding iron-storage proteins, are also activated by Fur. The existence of RyhB provides a nice explanation for this phenomenon and demonstrates that positive regulation by Fur is indirect and needs RyhB (Massé and Gottesman, 2002). RyhB-dependent repression of the *sdhCDAB* operon, encoding the Fe-S cluster containing succinate dehydrogenase, also explains the succinate defective growth of cells overproducing RyhB.

Figure 2 illustrates how the expression and action of RyhB is connected to iron metabolism and, in part, to the oxidative stress response. Under high iron conditions, Fur is active and represses transcription of siderophores and iron-siderophore transporters in order to avoid a further increase of iron concentrations. The *ryhB* gene is also repressed by Fur. As a consequence, destabilization of target mRNAs by RyhB is not possible and regular translation occurs. Translation of the *bfr* and *ftnA* mRNAs remains undisturbed and excess iron is stored by the corresponding proteins bacterioferritin and ferritin. Storage of iron will also lead to a consumption of molecular oxygen (O_2) and H_2O_2 when Fe(II) is oxidized to Fe(III) by the ferroxidase activity of these proteins (Zhao *et al.*, 2002; Bou-Abdallah *et al.*, 2002; Ceci *et al.*, 2003). Therefore, H_2O_2 is detoxified and the Fenton reaction is avoided by keeping free iron concentrations low.

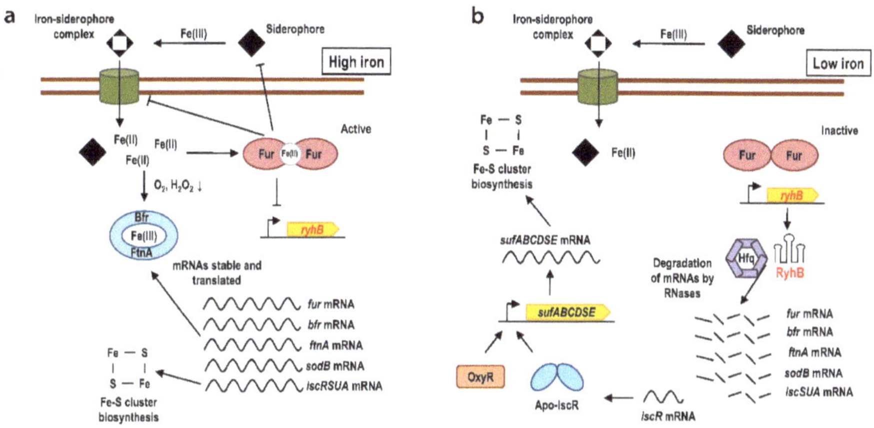

Fig. 2. RyhB links the oxidative stress response to the cellular iron concentration.
a) Under high iron conditions further iron-uptake, by iron-siderophore transport systems, is downregulated by the global iron regulator Fur. Excess iron is stored by iron-storage proteins (bacterioferritin, Bfr, and ferritin, FtnA), which is accompanied by consumption of O_2 and H_2O_2. Both processes lead to a reduced formation of OH˙ during Fe(II)-mediated Fenton reaction. *sodB* mRNA is stable and translated, leading to constant SodB levels. Fe-S cluster biosynthesis is accomplished by genes of the *isc* operon.
b) Under low iron conditions Fur is no longer able to repress gene expression, which also de-represses transcription of the RyhB sRNA. With the help of Hfq, RyhB binding to targets leads to degradation of the respective mRNAs by RNases, thereby inhibiting translation. Bfr, FtnA and SodB are no longer synthesized, and selective destabilization of the *iscRSUA* mRNA results in a shift of Fe-S cluster biosynthesis. Apo-IscR and the oxidative stress activated OxyR induce the *suf* operon, which is then responsible for Fe-S cluster formation. RyhB also destabilizes *fur* mRNA, thereby avoiding high levels of inactive Fur under low iron conditions

Upon iron starvation (low iron), Fur becomes inactive and RyhB is de-repressed leading to RyhB-dependent degradation of *bfr* and *ftnA* mRNAs, thereby circumventing translation (Massé and Gottesman, 2002). RyhB does not only control iron storage but also impairs *sodB* translation, leading to dropping levels of the iron superoxide dismutase. In the well-studied RyhB/*sodB* interaction, several protein partners like Hfq and the RNases E and III play important roles. It was shown that Hfq binds RyhB (Wassarman *et al.*, 2001; Zhang *et al.*, 2003) and that RyhB transcripts are unstable when Hfq is absent (Massé *et al.*, 2003). Since RNase E is involved in cleavage of both RyhB sRNA and *sodB* mRNA (Massé *et al.*, 2003; Afonyushkin *et al.*, 2005), stabilization by Hfq is believed to be due to blocking an AU-rich region within RyhB sRNA, which is also recognized by RNase E. RyhB was also shown to be initially cleaved by RNase III when it is bound to *sodB* mRNA. Accordingly, RyhB contains an intrinsic regulatory mechanism for its own decay, which leads to dropping RyhB levels when its regulatory action is achieved.

Interestingly, RyhB also regulates *fur* expression in a negative feedback loop. Under low iron conditions Fur is inactive, a situation that supports excess *fur* transcription, since active Fur also represses its own gene. However, there is no need for increased Fur synthesis under low iron conditions. It was shown that RyhB also destabilizes *fur* mRNA to ensure balanced synthesis of the iron-responsive repressor (Vecerek *et al.*, 2007).

RyhB also influences the assembly of Fe-S clusters, which are sensitive to high oxygen concentrations because they can be decomposed by ROS (for review see Imlay, 2006). RyhB binds the polycistronic *iscRSUA* mRNA, which encodes the regular machinery for biosynthesis of Fe-S clusters under high iron conditions. Desnoyers *et al.* (2009) showed that binding of RyhB to the second cistron of the polycistronic mRNA under low iron conditions promotes the cleavage of the *isc-SUA* transcript. T he remaining 5'-fragment encodes IscR, which acts as repressor of the *isc* operon when loaded with Fe-S clusters (Holo-IscR; Schwartz *et al.*, 2001). Under low iron, IscR remains as Apo-IscR, which is believed to activate the *suf* operon, encoding an alternative machinery for the Fe-S cluster assembly (Giel *et al.*, 2006; Yeo *et al.*, 2006; Lee *et al.*, 2008). Therefore, RyhB is responsible for shifting Fe-S cluster assembly from the *isc* operon to the *suf* operon. In addition, the *suf* operon was shown to be activated by OxyR under oxidative stress conditions (Outten *et al.*, 2004), showing again the tight connection between iron regulation and oxidative stress responses.

The example of RyhB nicely demonstrates that a single sRNA can link ROS depletion and iron homeostasis by multiple targeting of mRNAs. Accordingly, several regulatory pathways are connected by a single sRNA, which enables them to work together in concert.

3.3 RyhB Homologues in Other Bacteria

RyhB and other sRNAs are best studied in *E. coli*. However, RyhB homologues are also present in other bacteria, e.g. in *Pseudomonas aeruginosa* and *Erwinia chrysanthemi*.

In *E. chrysanthemi*, a 120-nt RyhB homologue was identified that controls expression of the *ftnA* gene, encoding the iron-storage protein ferritin, in a Fur-dependent manner (Boughammoura *et al.*, 2008). Mutants of *E. chrysanthemi*, which lack *ftnA*, are more sensitive to oxidative stress. Like in *E. coli*, RyhB mediates regulation of genes responsible for iron homeostasis and oxidative stress defence.

Wilderman *et al.* (2004) identified two functional homologues of RyhB in *P. aeruginosa*, named PrrF1 and PrrF2. These sRNAs are >95% identical to each other, appear in a tandem duplication in the chromosome and seem to have overlapping roles in the negative regulation of genes involved in diverse functions including iron storage, defence against oxidative stress, and intermediary metabolism. Like RyhB, they are transcribed under low iron conditions in a Fur-dependent manner. It was demonstrated that *sodB* and *katA* mRNAs are regulated by PrrF RNAs and are therefore involved in the detoxification of ROS. Why *P. aeruginosa* has a need for two RyhB-like RNAs is still an open question that needs to be addressed in the future.

4 Photooxidative Stress-Induced sRNAs in Photosynthetic Alpha-Proteobacteria

In bacteria, respiratory enzymes and exposure to metals, like iron, are the main sources of ROS that are generated by unspecific electron transfer. The term "oxidative stress" summarizes the generation of such ROS (H_2O_2, O_2^- and $OH^\bullet$). In contrast, the generation of highly toxic singlet oxygen ($^1O_2$) depends on light energy, which is absorbed by photosensitizers and then transferred to molecular oxygen (triplet oxygen; $^3O_2$). In this case the term "photooxidative stress" is used because light is pivotal for the generation of $^1O_2$. Since bacteriochlorophyll molecules and their precursors act as naturally occurring photosensitizers in the presence of light, it is obvious that photosynthetic bacteria have to cope with photooxidative stress when oxygen is present during photosynthesis. In the group of alpha-proteobacteria, there are several species that are capable of photosynthetic growth. One of the best-studied model organisms, in regard to the regulation of photosynthesis genes, is *Rhodobacter sphaeroides*, which performs anoxygenic photosynthesis in a light-dependent and oxygen-dependent manner (Gregor and Klug, 1999; Zeilstra-Ryalls and Kaplan, 2004). *R. sphaeroides* is an established model organism for studying the $^1O_2$ stress response (Anthony *et al.*, 2005; Glaeser and Klug, 2005), and recently sRNAs have been identified in a genome-wide search by pyrosequencing of cDNA (Berghoff *et al.*, 2009). Among the newly identified sRNAs, four sRNAs were found to have a putative role in the photooxidative stress response. Two of them, RSs0019 and RSs0682, are specific for $^1O_2$. RSs0019 is induced in an RpoE-dependent manner. RpoE is an alternative sigma factor, which is a major regulator in the photooxidative stress response of *R. sphaeroides* (Anthony *et al.*, 2005; Glaeser *et al.*, 2007). RSs0682 is processed after prolonged $^1O_2$ exposure and processing seems to be Hfq-dependent. It is an interesting question whether the $^1O_2$-dependent processing implies an RNA-dependent sensing mechanism for $^1O_2$, especially when taking into account the fact that no direct sensing mechanism for $^1O_2$ is known to date. Two other sRNAs, RSs0680a and RSs2461, are co-transcribed with their upstream genes and induced by photooxidative as well as oxidative stress (Berghoff *et al.*, 2009). Both sRNAs are preceded by an $RpoH_I/RpoH_{II}$-dependent promoter. The work of Nuss *et al.* (2009 and 2010) showed that the alternative sigma factor $RpoH_{II}$ is mainly responsible for the $^1O_2$, and $RpoH_I$ for the heat shock response, although overlapping regulons of the two factors exist in *R. sphaeroides*. It was verified that RSs0680a and RSs2461 really depend on both RpoH sigma factors and can also be induced by heat shock (Nuss *et al.*, 2010). In this study a third sRNA, RSs1543, was presented, which is under direct control of an $RpoH_I/RpoH_{II}$-dependent promoter. Interestingly, RSs1543 is a homologue of RSs2461 and both sRNAs genes are associated with an *ompR/lysR*-like gene, encoding transcriptional regulators. The question as to whether the two sRNAs interact with these regulators needs to be addressed in the future.

The studies on photooxidative stress-responsive sRNAs in *R. sphaeroides* demonstrated that sRNAs can be specific to a single stress, but most likely are induced by several stresses. Consequently, sRNAs enable a connective network of different stress responses, as has already been shown for OxyS and RyhB.

5 Other sRNAs Involved in Oxidative Stress Responses

Some reports present evidence for the involvement of more sRNAs in oxidative stress responses of various bacteria but exactly how they function needs further elucidation. The overview of such sRNAs as given in this chapter may not be complete and does not include all putative sRNAs, which have been found to respond to oxidative stress in global transcriptome analyses.

One important system of *E. coli* in its response to oxidative stress, in particular to superoxide stress, is SoxRS. SoxR contains a [2Fe-2S] cluster that is oxidized by superoxide and subsequently activates transcription of SoxS, an AraC family protein (Ding *et al.*, 1996). SoxS binds to its target promoters and activates genes which encode e. g. superoxide dismutase, DNA repair enzymes and enzymes of the carbon metabolism (Pomposiello and Demple, 2002). More recently, transcriptome studies have identified more protein coding genes and, in addition, sRNAs in *E. coli* that change their expression in response to superoxide. Among those sRNAs are OxyS and RyhB, which we described in previous chapters, as well as MicF, MicC, RydB, and CyaR (formerly RyeE) (Blanchard *et al.*, 2007). MicF, MicC and CyaR regulate the expression of porins (Omp: outer membrane proteins) in enterobacteria, thus linking the oxidative stress response to the outer membrane composition. MicF and MicC act by an antisense mechanism, while CyaR inhibits translation of *ompX* mRNA by sequestering the Shine-Dalgarno sequence (Papenfort *et al.*, 2008). The expression of MicF, MicC, and RydB is SoxR-dependent, whereas expression of CyaR is SoxR-independent (Blanchard *et al.*, 2007).

ArcZ is an abundant enterobacterial sRNA associated with the Hfq protein. It was shown to repress translation of several mRNAs in *Salmonella*, including the *tpx* mRNA for a periplasmic thioredoxin-like thiol peroxidase, an enzyme of the oxidative stress defence (Papenfort *et al.*, 2009). In *E. coli,* Tpx is involved in resistance to diverse oxidative stress compounds (Cha *et al.*, 1995). ArcZ binds *tpx* mRNA within the coding sequence, downstream of known translational control elements (Papenfort *et al.*, 2009). The physiological role of ArcZ in the oxidative stress response has not been analyzed to date.

In *Pseudomonas fluorescens* CHA0, transcription of the three sRNAs RsmY, RsmZ, and RsmX is controlled by the GacS/GacA two-component system (Heeb *et al.*, 2002). These sRNAs contain multiple GGA motifs and when present in high amounts titrate the RNA binding protein, RsmA, and its homologue, RsmE, which leads to increased translation of mRNAs involved in virulence and resistance to oxidative stress (Heeb *et al.*, 2005; Valverde *et al.*, 2003). Recently a novel sRNA, RgsA, was identified in *P. fluorescens* CHA0, which is also under positive control of GacA and the stress sigma factor RpoS and contains a single GGA motif. RgsA contributes to the resistance to hydrogen peroxide (Gonzalez *et al.*, 2008). It is unable to sequester RsmA and RsmE and its mode of action is unknown.

Numerous sRNAs, especially antisense RNAs, have also been identified in cyanobacteria (Georg *et al.*, 2009). The trans-encoded sRNA Yfr1 is highly conserved among cyanobacterial lineages and deletion of the *yfr1* gene results in reduced growth of *Synechococcus elongatus* PCC6301 under different stress conditions, including oxidative stress, and leads to accumulation of the *sbtA* mRNA (Nakamura *et al.*, 2007). SbtA is a sodium-dependent bicarbonate transporter (Shibata *et al.*, 2002). Yfr1 is located between the *guaB* (required for synthesis of GMP) and *trxA* (encoding thioredoxin A) genes in most cyanobacteria (Nakamura *et al.*, 2007). Presently available data rather hint at an indirect effect of Yfr1 in the oxidative stress response.

In the case of the cyanobacterium *Synechocystis* sp. PCC6803, it was shown that the mRNA of *isiA* is under negative control of the antisense RNA IsrR (Dühring *et al.*, 2006). IsiA is the iron stress-induced protein A, which forms a giant ring structure around photosystem I under iron-limiting conditions (see also chapter 5 for additional details). Furthermore, IsiA dissipates excess light energy under high light and oxidative stress. Under iron-replete conditions, transcription of *isiA* is repressed by Fur and residual *isiA* mRNA is bound by its antisense regulator IsrR and degraded. When subject to iron limitation or oxidative stress (H_2O_2), *isiA* mRNA levels increase and exceed IsrR levels. As a consequence, negative control by IsrR is overcome and IsiA is synthesized under conditions where it is needed. This example demonstrates that an antisense RNA is responsible for tight control of a stress-responsive component involved in photosynthesis.

As for pathogenic bacteria, an efficient defence against oxidative stress can be crucial to escaping the host defence. This applies in particular to mycobacteria, which are able to survive and multiply in macrophages. Recently nine sRNAs were identified in *Mycobacterium tuberculosis*, four cis- and five trans-encoded (Arnvig and Young, 2009). Of those nine sRNAs, four (B11, B55, F6 and ASpks) were induced upon oxidative stress applied by hydrogen peroxide treatment. Overexpression of B11 sRNA resulted in poor growth and elongated cells of *M. smegmatis*. The question as to whether sRNAs make a major contribution to the oxidative stress response of mycobacteria needs to be elucidated in future studies.

6 Concluding Remarks

Based on our current knowledge, it emerges that sRNAs have a main function in linking different regulatory networks. This is also the case for OxyS that links the response to oxidative stress to other stress responses via RpoS, for RyhB that links regulation of iron metabolism to the oxidative stress response, and for MicF, MicC, and CyaR that are under control of the oxidative stress responsive SoxRS system and participate in regulation of the composition of the outer membrane. With an increasing number of sRNAs still being identified in bacteria and characterized in regard to their biological function, we can expect to learn much more about their role in the oxidative stress response in the future.

References

Afonyushkin T, Vecerek B, Moll I, Bläsi U, Kaberdin VR (2005) Both RNase E and RNase III control the stability of *sodB* mRNA upon translational inhibition by the small regulatory RNA RyhB. Nucleic Acids Res 33: 1678–1689

Altuvia S, Weinstein-Fischer D, Zhang A, Postow L, Storz G (1997) A small, stable RNA induced by oxidative stress: Role as a pleiotropic regulator and anti-mutator. Cell 90: 43–53

Altuvia S, Zhang A, Argaman L, Tiwari A, Storz G (1998) The *Escherichia coli* OxyS regulatory RNA represses *fhlA* translation by blocking ribosome binding. EMBO J 17: 6069–6075

Anthony JR, Warczak KL, Donohue TJ (2005) A transcriptional response to singlet oxygen, a toxic byproduct of photosynthesis. Proc Natl Acad Sci U S A 102: 6502–6507

Argaman L, Altuvia S (2000) *fhlA* repression by OxyS RNA: Kissing complex formation at two sites results in a stable antisense-target RNA complex. J Mol Biol 300: 1101–1112

Arnvig KB, Young DB (2009) Identification of small RNAs in *Mycobacterium tuberculosis*. Mol Microbiol 73: 397–408

Ayala-Castro C, Saini A, Outten FW (2008) Fe-S cluster assembly pathways in bacteria. Microbiol Mol Biol Rev 72: 110–125

Bagg A, Neilands JB (1987) Ferric uptake regulation protein acts as a repressor, employing iron (II) as a cofactor to bind the operator of an iron transport operon in *Escherichia coli*. Biochemistry 26: 5471–5477

Basineni SR, Madhugiri R, Kolmsee T, Hengge R, Klug G (2009) The influence of Hfq and ribonucleases on the stability of the small non-coding RNA OxyS and its target *rpoS* in *E. coli* is growth phase dependent. RNA Biol 6: 584–594

Berghoff BA, Glaeser J, Sharma CM, Vogel J, Klug G (2009) Photooxidative stress-induced and abundant small RNAs in *Rhodobacter sphaeroides*. Mol Microbiol 74: 1497–1512

Blanchard JL, Wholey WY, Conlon EM, Pomposiello PJ (2007) Rapid changes in gene expression dynamics in response to superoxide reveal SoxRS-dependent and independent transcriptional networks. PLoS One 2: e1186

Bou-Abdallah F, Lewin AC, Le Brun NE, Moore GR, Chasteen ND (2002) Iron detoxification properties of *Escherichia coli* bacterioferritin. Attenuation of oxyradical chemistry. J Biol Chem 277: 37064–37069

Boughammoura A, Matzanke BF, Bottger L, Reverchon S, Lesuisse E, Expert D, Franza T (2008) Differential role of ferritins in iron metabolism and virulence of the plant-pathogenic bacterium *Erwinia chrysanthemi* 3937. J Bacteriol 190: 1518–1530

Ceci P, Ilari A, Falvo E, Chiancone E (2003) The Dps protein of *Agrobacterium tumefaciens* does not bind to DNA but protects it toward oxidative cleavage: X-ray crystal structure, iron binding, and hydroxyl-radical scavenging properties. J Biol Chem 278: 20319–20326

Cha MK, Kim HK, Kim IH (1995) Thioredoxin-linked "thiol peroxidase" from periplasmic space of *Escherichia coli*. J Biol Chem 270: 28635–28641

Christman MF, Morgan RW, Jacobson FS, Ames BN (1985) Positive control of a regulon for defenses against oxidative stress and some heat-shock proteins in *Salmonella typhimurium*. Cell 41: 753–762

Desnoyers G, Morissette A, Prevost K, Massé E (2009) Small RNA-induced differential degradation of the polycistronic mRNA *iscRSUA*. EMBO J 28: 1551–1561

Ding H, Hidalgo E, Demple B (1996) The redox state of the [2Fe-2S] clusters in SoxR protein regulates its activity as a transcription factor. J Biol Chem 271: 33173–33175

Dühring U, Axmann IM, Hess WR, Wilde A (2006) An internal antisense RNA regulates expression of the photosynthesis gene IsiA. Proc Natl Acad Sci U S A 103: 7054–7058

Escolar L, Perez-Martin J, de Lorenzo V (1999) Opening the iron box: Transcriptional metalloregulation by the Fur protein. J Bacteriol 181: 6223–6229

Georg J, Voss B, Scholz I, Mitschke J, Wilde A, Hess WR (2009) Evidence for a major role of antisense RNAs in cyanobacterial gene regulation. Mol Syst Biol 5: 305

Giel JL, Rodionov D, Liu M, Blattner FR, Kiley PJ (2006) IscR-dependent gene expression links iron-sulphur cluster assembly to the control of O_2-regulated genes in *Escherichia coli*. Mol Microbiol 60: 1058–1075

Glaeser J, Klug G (2005) Photo-oxidative stress in *Rhodobacter sphaeroides*: Protective role of carotenoids and expression of selected genes. Microbiology 151: 1927–1938

Glaeser J, Zobawa M, Lottspeich F, Klug G (2007) Protein synthesis patterns reveal a complex regulatory response to singlet oxygen in *Rhodobacter*. J Proteome Res 6: 2460–2471

Gonzalez N, Heeb S, Valverde C, Kay E, Reimmann C, Junier T, Haas D (2008) Genome-wide search reveals a novel GacA-regulated small RNA in *Pseudomonas* species. BMC Genomics 9: 167

Gonzalez-Flecha B, Demple B (1999) Role for the OxyS gene in regulation of intracellular hydrogen peroxide in *Escherichia coli*. J Bacteriol 181: 3833–3836

Gregor J, Klug G (1999) Regulation of bacterial photosynthesis genes by oxygen and light. FEMS Microbiol Lett 179: 1–9

Heeb S, Blumer C, Haas D (2002) Regulatory RNA as mediator in GacA/RsmA-dependent global control of exoproduct formation in *Pseudomonas fluorescens* CHA0. J Bacteriol 184: 1046–1056

Heeb S, Valverde C, Gigot-Bonnefoy C, Haas D (2005) Role of the stress sigma factor RpoS in GacA/RsmA-controlled secondary metabolism and resistance to oxidative stress in *Pseudomonas fluorescens* CHA0. FEMS Microbiol Lett 243: 251–258

Hengge-Aronis R (2002) Signal transduction and regulatory mechanisms involved in control of the sigma(S) (RpoS) subunit of RNA polymerase. Microbiol Mol Biol Rev 66: 373–395

Imlay JA (2003) Pathways of oxidative damage. Annu Rev Microbiol 57: 395–418

Imlay JA (2006) Iron-sulphur clusters and the problem with oxygen. Mol Microbiol 59: 1073–1082

Imlay JA (2008) Cellular defenses against superoxide and hydrogen peroxide. Annu Rev Biochem 77: 755–776

Lange R, Hengge-Aronis R (1991) Identification of a central regulator of stationary-phase gene expression in *Escherichia coli*. Mol Microbiol 5: 49–59

Lee KC, Yeo WS, Roe JH (2008) Oxidant-responsive induction of the *suf* operon, encoding a Fe-S assembly system, through Fur and IscR in *Escherichia coli*. J Bacteriol 190: 8244–8247

Majdalani N, Cunning C, Sledjeski D, Elliott T, Gottesman S (1998) DsrA RNA regulates translation of RpoS message by an anti-antisense mechanism, independent of its action as an antisilencer of transcription. Proc Natl Acad Sci U S A 95: 12462–12467

Majdalani N, Hernandez D, Gottesman S (2002) Regulation and mode of action of the second small RNA activator of RpoS translation, RprA. Mol Microbiol 46: 813–826

Massé E, Escorcia FE, Gottesman S (2003) Coupled degradation of a small regulatory RNA and its mRNA targets in *Escherichia coli*. Genes Dev 17: 2374–2383

Massé E, Gottesman S (2002) A small RNA regulates the expression of genes involved in iron metabolism in *Escherichia coli*. Proc Natl Acad Sci U S A 99: 4620–4625

Massé E, Vanderpool CK, Gottesman S (2005) Effect of RyhB small RNA on global iron use in *Escherichia coli*. J Bacteriol 187: 6962–6971

Mongkolsuk S, Helmann JD (2002) Regulation of inducible peroxide stress responses. Mol Microbiol 45: 9–15

Nakamura T, Naito K, Yokota N, Sugita C, Sugita M (2007) A cyanobacterial non-coding RNA, Yfr1, is required for growth under multiple stress conditions. Plant Cell Physiol 48: 1309–1318

Nuss AM, Glaeser J, Berghoff BA, Klug G (2010) Overlapping alternative sigma factor regulons in the response to singlet oxygen in *Rhodobacter sphaeroides*. J Bacteriol 192: 2613–2623

Nuss AM, Glaeser J, Klug G (2009) RpoH(II) activates oxidative-stress defense systems and is controlled by RpoE in the singlet oxygen-dependent response in *Rhodobacter sphaeroides*. J Bacteriol 191: 220–230

Outten FW, Djaman O, Storz G (2004) A *suf* operon requirement for Fe-S cluster assembly during iron starvation in *Escherichia coli*. Mol Microbiol 52: 861–872

Papenfort K, Pfeiffer V, Lucchini S, Sonawane A, Hinton JC, Vogel J (2008) Systematic deletion of *Salmonella* small RNA genes identifies CyaR, a conserved Crp-dependent riboregulator of OmpX synthesis. Mol Microbiol 68: 890–906

Papenfort K, Said N, Welsink T, Lucchini S, Hinton JC, Vogel J (2009) Specific and pleiotropic patterns of mRNA regulation by ArcZ, a conserved, Hfq-dependent small RNA. Mol Microbiol 74: 139–158

Pfeiffer V, Papenfort K, Lucchini S, Hinton JC, Vogel J (2009) Coding sequence targeting by MicC RNA reveals bacterial mRNA silencing downstream of translational initiation. Nat Struct Mol Biol 16: 840–846

Pomposiello PJ, Demple B (2002) Global adjustment of microbial physiology during free radical stress. Adv Microb Physiol 46: 319–341

Schwartz CJ, Giel JL, Patschkowski T, Luther C, Ruzicka FJ, Beinert H, Kiley PJ (2001) IscR, an Fe-S cluster-containing transcription factor, represses expression of *Escherichia coli* genes encoding Fe-S cluster assembly proteins. Proc Natl Acad Sci U S A 98: 14895–14900

Shibata M, Katoh H, Sonoda M, Ohkawa H, Shimoyama M, Fukuzawa H, Kaplan A, Ogawa T (2002) Genes essential to sodium-dependent bicarbonate transport in cyanobacteria: Function and phylogenetic analysis. J Biol Chem 277: 18658–18664

Sittka A, Lucchini S, Papenfort K, Sharma CM, Rolle K, Binnewies TT, Hinton JC, Vogel J (2008) Deep sequencing analysis of small noncoding RNA and mRNA targets of the global post-transcriptional regulator, Hfq. PLoS Genet 4: e1000163

Storz G, Imlay JA (1999) Oxidative stress. Curr Opin Microbiol 2: 188–194

Storz G, Tartaglia LA, Ames BN (1990) Transcriptional regulator of oxidative stress-inducible genes: Direct activation by oxidation. Science 248: 189–194

Storz G, Zheng M (2000) Oxidative stress. In: Storz G, Hengge-Aronis R (Hrsg) Bacterial stress responses. American Society for Microbiology, Washington, S 47–59

Touati D (2000) Iron and oxidative stress in bacteria. Arch Biochem Biophys 373: 1–6

Valverde C, Heeb S, Keel C, Haas D (2003) RsmY, a small regulatory RNA, is required in concert with RsmZ for GacA-dependent expression of biocontrol traits in *Pseudomonas fluorescens* CHA0. Mol Microbiol 50: 1361–1379

Varghese S, Tang Y, Imlay JA (2003) Contrasting sensitivities of *Escherichia coli* aconitases A and B to oxidation and iron depletion. J Bacteriol 185: 221–230

Vecerek B, Moll I, Bläsi U (2007) Control of Fur synthesis by the non-coding RNA RyhB and iron-responsive decoding. EMBO J 26: 965–975

Vogel J, Papenfort K (2006) Small non-coding RNAs and the bacterial outer membrane. Curr Opin Microbiol 9: 605–611

Wackett LP, Orme-Johnson WH, Walsh CT (1989) Transition metal enzymes in bacterial metabolism. In: Beveridge TJ, Doyle RJ (Hrsg) Metal ions and bacteria. John Wiley und Sons, Inc., New York, S 165–206

Wassarman KM, Repoila F, Rosenow C, Storz G, Gottesman S (2001) Identification of novel small RNAs using comparative genomics and microarrays. Genes Dev 15: 1637–1651

Wilderman PJ, Sowa NA, FitzGerald DJ, FitzGerald PC, Gottesman S, Ochsner UA, Vasil ML (2004) Identification of tandem duplicate regulatory small RNAs in *Pseudomonas aeruginosa* involved in iron homeostasis. Proc Natl Acad Sci U S A 101: 9792–9797

Yeo WS, Lee JH, Lee KC, Roe JH (2006) IscR acts as an activator in response to oxidative stress for the *suf* operon encoding Fe-S assembly proteins. Mol Microbiol 61: 206–218

Zeilstra-Ryalls JH, Kaplan S (2004) Oxygen intervention in the regulation of gene expression: The photosynthetic bacterial paradigm. Cell Mol Life Sci 61: 417–436

Zhang A, Wassarman KM, Ortega J, Steven AC, Storz G (2002) The Sm-like Hfq protein increases OxyS RNA interaction with target mRNAs. Mol Cell 9: 11–22

Zhang A, Wassarman KM, Rosenow C, Tjaden BC, Storz G, Gottesman S (2003) Global analysis of small RNA and mRNA targets of Hfq. Mol Microbiol 50: 1111–1124

Zhao G, Ceci P, Ilari A, Giangiacomo L, Laue TM, Chiancone E, Chasteen ND (2002) Iron and hydrogen peroxide detoxification properties of DNA-binding protein from starved cells. A ferritin-like DNA-binding protein of *Escherichia coli*. J Biol Chem 277: 27689–27696

Chapter 2
Hfq-associated Regulatory Small RNAs

Colin P. Corcoran, Kai Papenfort, and Jörg Vogel*

1 Introduction

Regulatory RNAs in bacteria represent a diverse group of molecules that act through a variety of intricate mechanisms to modulate a broad range of physiological responses. Some regulatory RNAs act to modulate protein function, often by acting as molecular mimics of other nucleic acids that are regulatory targets of the protein and thus capturing the protein in an unproductive complex (Babitzke & Romeo, 2007). The majority of regulatory RNAs characterized to date act by base-pairing with target messenger RNAs (mRNAs) to modulate their stability and/or translation (Papenfort & Vogel, 2010, Waters & Storz, 2009). Base-pairing regulatory RNAs can be divided into *cis*- and *trans*-acting RNAs. Well characterized groups of *cis*-acting regulatory RNAs include riboswitches, which are regulatory elements encoded within an mRNA that can control translation by adopting alternative structures in response to signals such as temperature or the presence of small molecules specifically bound by the riboswitch (Grundy & Henkin, 2006). Other *cis*-acting RNAs include antisense encoded RNAs, which are transcribed from the opposite strand within or near the target coding sequence and therefore contain perfect sequence complementarity to the target mRNA (Thomason & Storz, 2010).

Trans-acting regulatory RNAs are the best characterized class of RNA regulators (Waters & Storz, 2009). A recently discovered class of *trans*-acting regulatory RNAs are the CRISPR (clustered regularly interspaced short palindromic repeats) RNAs. CRISPR RNAs act as an adaptive immune response to confer resistance to bacteriophage and conjugative plasmids through specific base-pairing with foreign nucleic acid, which targets the duplex for destruction by the conserved CRISPR associated (CAS) proteins (Deveau *et al.,* 2010).

* RNA Biology Group, Institute for Molecular Infection Biology, University of Würzburg, Josef-Schneider-Strasse 2, D-97080 Würzburg, Germany

1.1 *Trans*-acting sRNAs and the Role for Hfq

Clearly the best studied and largest group of *trans*-acting RNAs are those small regulatory RNAs (sRNAs) that modulate mRNA translation by base-pairing with limited sequence homology to the target mRNA. There are very few commonalities among sRNAs, which vary dramatically in size (from 50–500 nt), sequence and secondary structure (Table 1; Papenfort & Vogel, 2010, Waters & Storz, 2009). One defining factor that links many of these sRNAs is their association with the RNA chaperone Hfq, which is a bacterial member of the extensive Sm/LSm family of RNA-binding proteins that assemble into a ring-like multimeric quaternary structure, such as a doughnut-shaped homohexamer in the case of Hfq (see also chapter 8 of this book). The Hfq protein has been known in *E. coli* since the late 1960's as host factor of RNA phage Qbeta (hence the name), but began to emerge as a global regulator of gene expression only rather recently. Hfq binds tightly to many sRNAs and is required for both intracellular stability and target mRNA pairing of sRNAs in many bacteria (Aiba, 2007, Brennan & Link, 2007, Chao & Vogel, 2010, Valentin-Hansen *et al.*, 2004). The molecular basis for this stabilizing effect involves protection from degradation by a major component of the bacterial degradosome, RNase E. The binding sites for both Hfq and RNase E are remarkably similar; specifically, A/U rich regions flanked by a stem-loop structure. Thus, in general, when Hfq binds to an sRNA it occludes the RNase E binding site and thus increases stability of the sRNA (Moll *et al.*, 2003, Folichon *et al.*, 2003). Hfq also acts as a chaperone in promoting the interaction of many sRNAs with their target mRNAs (Geissmann & Touati, 2004) and in some cases, facilitates mRNA degradation through the formation of a specific ternary complex between the sRNA, Hfq and RNase E (Afonyushkin *et al.*, 2005, Kawamoto *et al.*, 2005).

Although the exact concentration of Hfq has yet to be conclusively defined, it is clear that Hfq is limiting in the cell relative to the amount of Hfq targets (Hussein & Lim, 2011, Papenfort *et al.*, 2009). On the other hand, the rapid acquisition of Hfq by an induced sRNA is essential for its function. This is likely facilitated by the active cycling of RNA on Hfq, which results in the concentration dependent acquisition of Hfq (Fender *et al.,* 2010). Active cycling results in the rapid equilibriation of the Hfq binding profile upon sRNA induction and allows for the rapid action of sRNAs (Fender *et al.,* 2010).

Hfq has been identified and phenotypically characterized in many diverse bacterial species. Moreover, global approaches using microarrays, RNA deep sequencing and proteomics have been taken to understand how many and which genes are regulated by Hfq *in vivo* (Sharma & Vogel, 2009). The sum of such studies in *E. coli* and *Salmonella* suggests that close 100 sRNAs might target several hundred mRNAs, arguing that post-transcriptional regulation by Hfq and sRNAs affects almost 20% of all genes in these Gram-negative model bacteria (Ansong *et al.*, 2009, Guisbert *et al.*, 2007, Sittka *et al.*, 2007; 2008; 2009, Zhang *et al.*, 2003). This book chapter aims to highlight generic principles of gene regulation by Hfq-associated sRNAs. Rather than giving a complete overview of all gene circuits in which these regulators have been discovered to have functional roles, we have selected a limited number of well-investigated sRNAs of *E. coli*, *Salmonella* and *Vibrio* species to illustrate emerging features of this class.

2 Regulatory Mechanisms Employed by Hfq-associated sRNAs

2.1 Translational Control Near the SD Sequence and AUG Start Codon

In bacteria, translation starts with the formation of an initiation complex consisting of 30S ribosomal subunits bound to the ribosomal binding site (RBS) of the mRNA, fMET-tRNA, and initiation factors (Laursen *et al.,* 2005). Subsequently, 50S ribosomes bind and form the translationally active 70S ribosome. The Shine-Dalgarno sequence (SD), located several nucleotides upstream of the mRNA start codon (Shine & Dalgarno, 1974), forms an integral component of the RBS and plays a key role in the initial capture of 30S ribosomes. The archetypal mechanism of an sRNA regulating a *trans*-encoded target was established by the discovery of MicF (Mizuno *et al.,* 1984). MicF, which associates with Hfq *in vivo* (Sittka *et al.*, 2008, Zhang *et al.*, 2003), is divergently transcribed from the *ompC* gene. MicF represses translation of the major outer-membrane protein OmpF by base-pairing with the SD and start codon of the *ompF* mRNA (Mizuno *et al.*, 1984, Schmidt *et al.*, 1995). This canonical mechanism for translational repression has since been supported in numerous studies showing that the association of 30S with the RBS (a prerequisite for translation initiation) can be inhibited by sRNAs *in vitro* (e.g. (Argaman & Altuvia, 2000, Bouvier *et al.*, 2008, Geissmann & Touati, 2004, Møller *et al.*, 2002b, Pfeiffer *et al.*, 2009, Sharma *et al.*, 2007, Udekwu *et al.*, 2005).

2.2 Primary Role for sRNAs in Translational Silencing

Early studies on MicF interaction with *ompF* also highlighted the rapid degradation of the *ompF* mRNA (Andersen *et al.*, 1989, Mizuno *et al.*, 1984). Thus, the primary mechanism by which protein synthesis is inhibited by sRNAs has been an area of intensive research. Two mechanisms are possible: firstly that the sRNA acts primarily to inhibit translation, which subsequently results in increased susceptibility of the mRNA to degradation due to reduced protection from ribosomes. A second possibility is that the primary function of the sRNA is not to inhibit translation but specifically to target the mRNA for degradation, which will clearly abrogate protein synthesis (Fig. 1a). The two events are intimately coupled. mRNA degradation will certainly cause loss of translation, whereas inhibition of translation can destabilize an mRNA by exposing it to RNases (Deana & Belasco, 2005, Carpousis, 2007). Thus, the relative role of direct translational repression versus that of mRNA destabilization in sRNA mediated translational silencing is a chicken-or-egg question: which is the primary event?

A seminal study which showed the RNase E-mediated coupled degradation of RyhB and its mRNA target *sodB* (Massé *et al.*, 2003) was closely followed by evidence of a tripartite complex of RyhB, Hfq and RNase E which promoted the decay of the *sodB* mRNA (Kawamoto *et al.*, 2005, Afonyushkin *et al.*, 2005). This tripartite complex was distinct from the RNA degradosome which is the usual functional

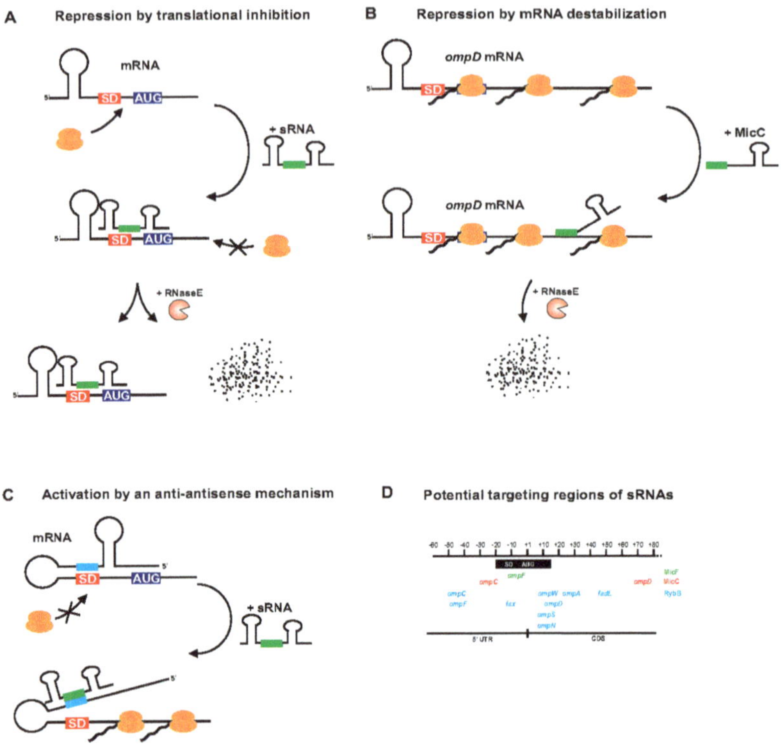

Fig. 1. Mechanisms for translational control by Hfq-dependent sRNAs
(**A**) sRNA targeting of the SD/AUG region prevents ribosome (yellow ovals) binding and thus inhibits translation. Translational repression can result in silenced mRNA (intact but not active) or the coupled degradation of the sRNA/mRNA complex by RNase E. (**B**) MicC binding within the CDS of *ompD* does not affect ribosome binding or elongation but instead targets the mRNA for degradation by RNase E. (**C**) The most common mechanism of sRNA-mediated translational activation involves an anti-antisense mechanism. In this case, the SD/AUG region of an mRNA is sequestered in an inhibitory secondary structure that prevents association of ribosomes. Binding of the sRNA to the upstream region disrupts the inhibitory secondary structure allows ribosome binding and efficient translation. (**D**) The known binding regions of sRNAs on target mRNAs has expanded significantly beyond the canonical targeting of the SD/AUG region, as exemplified by MicF (green). For example, MicC (red) binds deep within the CDS to cause destabilization of the *ompD* mRNA (see **B**) while RybB (blue) targets multiple mRNAs outside of the SD/AUG region. Targets are aligned with a scale indicating distance from the transcriptional start site. The SD and AUG regions are indicated within a black box, which represents the region whereby sRNA targeting can directly antagonize binding of the 30S initiating ribosome.

form of RNase E established by C-terminal binding of PNPase, helicase RhlB and enolase, none of which co-purified with Hfq (Kawamoto *et al.*, 2005). Collectively these results supported the exciting prospect that the primary function of Hfq-dependent sRNAs was to target mRNAs for destruction by programmed RNA decay machines.

In contrast, there has also been significant evidence for translation initiation as the primary mechanism of sRNA-mediated translational silencing. For example, Spot42 is responsible for the discoordinate expression of the galactose operon, which contains four genes involved in carbon metabolism encoded in a single polycistronic mRNA transcript (*galETKM*) (Møller *et al.*, 2002a). In the absence of glucose, D-galactose induces transcription of the operon and all four genes accumulate in an equimolar fashion, which is required for catabolism of the sugar (Lee *et al.*, 2008, Ullmann *et al.*, 1979). In contrast, when cells grow in the presence of glucose, the first two genes of the operon, which are necessary for providing substrates for biosynthetic glycosylation reactions, are expressed in higher concentrations than the downstream encoded enzymes GalK and GalM (Lee *et al.*, 2008, Ullmann *et al.*, 1979). Spot42 targets the *galK* RBS, which interferes with the formation of *galK*-30S initiation complexes and thus represses translation of this cistron. Hfq is required for both stabilization of Spot42 and the annealing of Spot42 to *galK* mRNA (Møller *et al.*, 2002a). Notably, the repression of *galK* only weakly effects the expression of *galE* and *galT* indicating that the integrity of the *gal* mRNA is not affected by binding of Spot42 and/or Hfq. This example of translational silencing independent of mRNA degradation supported translation inhibition as the primary mechanism for sRNA-mediated translational silencing (Møller *et al.*, 2002a). However, it was recently reported that over-expression of Spot 42 does result in destabilization of the *galK* mRNA (Beisel & Storz, 2011). Whether this interesting observation is a physiologically relevant component of the regulation of the *gal* operon or a result of Spot 42 over-expression is as yet unclear.

Irrespective of this, further evidence for translation inhibition as the primary mechanism of sRNA-mediated translational inhibition was supplied in a number of studies using the SgrS-*ptsG* model system. SgrS sRNA is induced under glucose phosphate stress in *E. coli* and represses translation of the mRNA of the glucose transporter, PtsG. Similar to RyhB, SgrS has been proposed to form a specific ribonucleoprotein complexes with RNase E through Hfq, which results in translational repression and rapid degradation of *ptsG* mRNA (Kawamoto *et al.,* 2005). Using an *in vitro* system, Maki *et al.* (2008) defined SgrS and Hfq as the minimum components required for translational silencing of *ptsG* that faithfully reproduced the *in vivo* reaction. Hfq stimulates the base pairing between SgrS and *ptsG* mRNA, which correlated strongly with translational repression. This confirmed the earlier reports that translational repression of target mRNAs can occur in the absence of the rapid mRNA destabilization (see above) and supported the model whereby inhibition of translation initiation is the primary event in sRNA actions. It appears therefore that RNase E-mediated mRNA degradation is an added step that functions to make translational silencing irreversible.

Similar to SgrS, RyhB was shown to destabilize *sodB* translation through formation of a ternary of complex involving Hfq and RNase E (Morita *et al.,* 2005). However, RyhB is also known to down-regulate some targets without causing mRNA degradation (Massé & Gottesman, 2002, Desnoyers *et al.*, 2009) indicating that the formation of these special mRNA decay machines depends on the unique features of each sRNA-mRNA-Hfq-RNase E interaction and is not a general mechanism for a subset of sRNAs to induce irreversible translational silencing.

Interestingly, while RyhB can block translation without inducing mRNA destablization, a recent study has revealed that RyhB can induce mRNA destabilization even in the absence of translation on the mRNA target (Prévost *et al.*, 2011). While introduction of a premature stop codon (thus blocking translation independently of RyhB) resulted in reduced *sodB* levels, the effect was not sufficient to account for RyhB-induced mRNA degradation. Induction of RyhB increased degradation of the *sodB* mRNA containing the premature stop codon, suggesting that RhyB can actively induce *sodB* degradation independently of the translational block mechanism. Collectively these studies, rather than defining a single universal mechanism for translational repression by sRNAs, have highlighted that sRNAs can silence translation by multiple mechanisms.

2.3 Non-canonical Repression of Translation Initiation

While many of the characterized sRNA-mRNA interactions involve the canonical mechanism of sequestration of the immediate translation start site, recent studies have unearthed numerous examples of repression by sRNA pairing outside this narrow region (Fig. 1). While some of these findings were by serendipity, others came through searches that were motivated by the question whether the recognition of SD and AUG sequences would permit sufficient selectivity in target selection.

GcvB, a global regulator of amino-acid biosynthesis, is one such example of an sRNA with non-canonical target sites. For example, GcvB binds the *gltI* mRNA ~50 nucleotides upstream of the AUG start codon which is clearly outside the region covered by a ribosome during the initiation step (Sharma *et al.*, 2007). Nonetheless, 30S toeprinting and 70S *in vitro* translation showed that GcvB binding at this upstream position inhibited translation of the *gltI* mRNA. This was suprising since the previously defined footprint for the 30S subunit covered a ~40 nucleotide region extending from residue -20 in the 5' UTR to residue +19 in the coding sequence (CDS) (Hüttenhofer & Noller, 1994, Beyer *et al.*, 1994). Further investigation revealed that the region targeted by GcvB coincided with a C/A-rich upstream enhancer element required for translation initiation.

A second novel mechanism involving repression of translation intiation was deciphered for the RybB interaction with *ompN* mRNA (Bouvier *et al.*, 2008). RybB was shown to interact within the first codons of *ompN*, downstream of AUG. This raised the possibility that sRNA base-pairing within the early CDS could sterically hinder 30S ribosome binding without directly interfering with the SD sequence. Systematic analysis using 30S footprinting assays showed conclusively that RybB represses ribosome binding to *ompN* even though SD and AUG are in principle accessible. This mechanism of regulation is effective only within a "5 codon window" after which sRNA binding fails to inhibit 30S binding and translation is unaffected. Notably, further investigation may reveal other examples of RybB targeting within the 5 codon window. In fact, many of the RybB repressed mRNAs are bound downstream of the AUG (Fig.1d; Balbontin *et al.*, 2010, Papenfort *et al.*, 2010).

2.4 Control of Protein Synthesis Through Regulation of mRNA Decay

While Bouvier *et al.* (2008) provided a novel mechanism of CDS targeting that affected translation initiation, it was still assumed that sRNAs could not repress targets in the deeper CDS owing to the efficient RNA helicase activity of elongating 70S ribosomes, which would prevent stable sRNA-mRNA pairing (Takyar *et al.,* 2005). This assumption was supported by the recurring discovery of new sRNAs that paired to SD/AUG or sensitive regions in the 5' UTR. Recently, however, Pfeiffer *et al.*, (2009) identified a novel target for MicC sRNA deep within the CDS of *ompD* mRNA (Fig. 1b).

Translational GFP fusions designed to localize MicC binding within *ompD* showed no effect of MicC over-expression on constructs containing the 5' UTR of *ompD* or the 5' UTR and the first 5 codons. In contrast, similar constructs using the well characterized MicC target, *ompC* (Chen *et al.,* 2004), as a target mRNA were efficiently down-regulated. MicC regulation of *ompD* was only restored when the first 26 codons of *ompD* were included in the fusion. Structure probing of MicC-*ompD* interactions showed that the 5' region of MicC forms a 12-bp RNA duplex with *ompD* codons 23–26 but did not generate consistent changes elsewhere in *ompD* mRNA, such as the 5' UTR. Binding of MicC within the CDS had no effect on translation initiation and failed to prevent translation of *ompD* in a 70S ribosome translation system system indicating that MicC binding within the *ompD* CDS cannot directly interfere with translation initiation or elongation. Instead, MicC induces RNase E-mediated target cleavage downstream of the MicC-*ompD* duplex. This post-transcriptional repression involving endonucleolytic degradation in the absence of translation inhibition represents a new mechanism for bacterial sRNAs.

The signal which targets the MicC-*ompD* complex for degradation is as yet unclear. The available crystal structure for RNase E shows no binding pocket for double stranded RNA which argues against the possibility that the enzyme senses the MicC-*ompD* RNA duplex directly (Callaghan *et al.,* 2005). However, *in vitro* results for reconstituted MicC-induced decay of the *ompD* mRNA have suggested that the sRNA might indeed program RNase E for decay of specific targets (unpublished results from the J. Vogel and B.F. Luisi laboratories).

sRNA-mediated mRNA decay can also be used to promote the discoordinate expression of operons. RyhB binds in the *iscRSUA* operon overlapping the RBS of *iscS* and promotes degradation of the *iscSUA* section. The upstream mRNA containing the *iscR* transcript contains a strong secondary structure that protects it from degradation, thus resulting in sustained translation of *iscR* (Desnoyers *et al.*, 2009). Degradation of the downstream message requires both the RNA degradosome and Hfq, which is reminiscent of the mechanism of RyhB mediated destabilization of the *sodB* mRNA. Unusually, degradation of the *iscSUA* mRNA results in a feedback loop that increases *iscR* transcription. Therefore, transcription of *ryhB* and its target *iscRSUA* both increase in response to low iron conditions (Desnoyers *et al.*, 2009). RyhB also promotes the discoordinate expression of a second operon *sdhCDAB*, and similar to *iscRSUA*, promotes degradation of the transcript downstream of the first gene, although direct pairing remains to be confirmed (Massé & Gottesman, 2002, Desnoyers *et al.*, 2009).

2.5 Multiple Target Control by sRNAs

The characterization of MicF, which established the paradigm for targeting of the SD/AUG as a general mechanism of translational repression also supported a model whereby sRNAs were dedicated regulators of single target mRNA. This was supported by the significant homology with which MicF targets *ompF*, which involves 44 nt spanning the SD and 5' CDS (Mizuno *et al.,* 1984). The physiological role for MicF in downregulation of OmpF and its genetic location beside the *ompC* locus all suggested that this sRNA functioned in a small regulatory circuit with a single target (Mizuno *et al.,* 1984). The 'one-sRNA one-target' dogma was supported by other early examples of seemingly dedicated sRNA-target pairs including Spot 42-*galK* (Møller *et al.*, 2002a), MicC-*ompC* (Chen *et al.,* 2004) and MicA-*ompA* (Udekwu *et al.*, 2005, Rasmussen *et al.*, 2005).

The Hfq-associated **DsrA** sRNA was one of the first antisense regulators that was shown to target multiple mRNAs sequence specifically. DsrA controls the expression of two global transcriptional regulators H-NS, a major histone-like protein responsible for silencing of a large number of bacterial genes, and the stationary phase σ factor RpoS. Although DsrA uses an antisense mechanism for both up-regulation of *rpoS* and down-regulation of *hns*, different regions of DsrA are involved in the regulation of each target (Lease *et al.*, 1998, Majdalani *et al.*, 1998, Sledjeski *et al.*, 1996). Translation of *rpoS* is inhibited by formation of repressive secondary structure in the 5' UTR that sequesters the SD sequence (Brown & Elliott, 1997). DsrA binding in the *rpoS* 5' UTR, ~100 nt upstream of SD, relieves the inhibitory pairing and thus allows increased translation of *rpoS* (Lease *et al.*, 1998, Lease & Belfort, 2000, Majdalani *et al.*, 1998, Sledjeski *et al.*, 2001). This 'anti-antisense' mechanism is the most common mechanism for direct positive regulation by sRNAs (Fig.1c; Frohlich & Vogel, 2009).

Transcriptional profiling using microarrays later revealed that over-expression of DsrA results in pleiotropic changes in gene expression including induction of several operons (*hdeAB*, *gadAX*, and *gadBC*) involved in acid resistance (Lease *et al.,* 2004). Since these findings were based on constituitive over-expression of DsrA, it is likely that many of the changes were downstream effects from DsrA altering H-NS and RpoS levels. Nevertheless, this study highlighted the position of DsrA at the apex of two global regulons.

A number of putative target sites for DsrA have also been identified on mRNAs such as *argR*, *ilvIH* and *rbsD.* Interestingly, the regions predicted to interact with these targets were independent of the *hns* and *rpoS* targeting sites (Lease *et al.*, 1998, Lease & Belfort, 2000). In this respect, DsrA appeared chimeric in structure with individual targeting sites for each mRNA. Importantly, since there was no core region that targeted the sRNA to multiple mRNAs, the number of targets would be limited by the size of the sRNA.

OxyS is another sRNA that acts as a pleiotropic regulator of gene expression. OxyS is induced in response to oxidative stress and alters expression of a number of genes required for protection against DNA damage (Altuvia *et al.,* 1997). OxyS accumulates rapidly, within a few minutes after treatment with hydrogen peroxide, to high intracellular levels (~4500 copies per cell). OxyS is highly stable with a half-life of >10 min and in contrast to many sRNAs, does not require Hfq for stability *in vivo* (Altuvia *et al.,* 1997).

OxyS represses two transcriptional regulators FhlA (a transcriptional activator for formate metabolism) and the aforementioned stress factor RpoS, indicating that OxyS functions to integrate the response to hydrogen peroxide with other regulatory networks (Altuvia *et al.,* 1997). OxyS uses different base-pairing regions to target *flhA* and *rpoS.* OxyS inhibition of *fhlA* involves a unique 'kissing complex' that forms between the loop structures of two extended hairpins of OxyS and similar structures on *fhlA*, which overlap the RBS (Altuvia *et al.*, 1998, Argaman & Altuvia, 2000). In contrast, translational inhibition of *rpoS* mRNA by OxyS appears to occur indirectly via association with Hfq and does not involve the stem-loop regions required for *fhlA* repression (Zhang *et al.*, 1998, Altuvia *et al.*, 1997). In fact, OxyS does not base-pair directly with *rpoS* mRNA. Instead, it is predicted that OxyS sequesters Hfq and thus the factor required to melt the inhibitory secondary structure in the *rpoS* mRNA (Zhang *et al.,* 1998). Therefore, OxyS might inhibit *rpoS* translation by modulating the intracellular availability of Hfq.

In a change that revolutionized sRNA target discovery, the simulataneous identification of multiple sRNA targets was achieved by global transcriptomic analysis following brief ectopic expression of the sRNA of interest (Sharma & Vogel, 2009). This allowed the identification of multiple target mRNAs independent of target predictions and resulted in a dramatic expansion of known sRNA targets. Contrasting the earlier 'one sRNA, one target view, many sRNAs have now been shown to regulate multiple target genes simultaneously and it is now believed that that multiple target regulation by sRNAs constitutes the rule rather than an exception (Table 1).

RyhB was originally identified in *E. coli* in two independent systematic sRNA screens (Argaman *et al.*, 2001, Wassarman *et al.*, 2001). This sRNAs accumulates in response to growth in minimal medium and when over-expressed causes slow growth in media containing succinate as a sole carbon source (Wassarman *et al.,* 2001). Further investigation showed that the *ryhB* gene was strictly controlled by the ferric uptake repressor (Fur), which is the master regulator of genes involved in iron uptake and metabolism in many bacteria (Massé & Gottesman, 2002). Repression of the *ryhB* promoter is alleviated under iron limiting conditions. Induction of RyhB causes the rapid downregulation of multiple mRNAs of non-essential iron-using proteins (Massé & Gottesman, 2002, Massé *et al.*, 2005). Transcriptional profiling following pulse over-expression of RyhB expanded this target repertoire to 18 target mRNAs encoding for 56 proteins, most of which with documented functions in iron metabolism (Massé *et al.,* 2005). Thus far, base-pairing interactions have been experimentally validated for five targets, i.e. the *sodB*, *uof, icsS, cysE* and *shiA* mRNAs (Desnoyers *et al.*, 2009, Massé & Gottesman, 2002, Prévost *et al.*, 2007, Vecerek *et al.*, 2007, Salvail *et al.*, 2010). The down-regulation of multiple non-essential iron binding proteins coupled with the up-regulation of proteins involved in iron scavenging by RyhB results in the accumulation of intracellular iron, which can be used by essential iron-binding proteins to maintain growth under low-iron conditions and sustain viability during iron starvation (Jacques *et al.*, 2006, Salvail *et al.*, 2010).

Table 1. Physiological roles of Hfq-dependent sRNAs in *E. coli* and *Salmonella* (listed by genomic location in *Salmonella*).

Name	Alternative IDs	Adjacent genes in *Salmonella*	Length	Important features	References
ArcZ	*sraH/ryhA*	*yhbL/arcB*	120	Induced under anaerobic conditions. Represses *sdaC*, STM3216 and *tpx* mRNAs. Activates *rpoS* mRNA. Processed rapidly *in vivo*.	(Papenfort *et al.*, 2009; Mandin & Gottesman 2010; Soper *et al.*, 2010)
CyaR	*ryeE*	*yegQ*/STM2137	87	Induced by low glucose. Regulated by CRP. Represses *ompX* and *nadE* synthesis.	(Johansen *et al.*, 2008; Papenfort *et al.*, 2008; De Lay & Gottesman 2009)
DsrA	-	yodD/yedP	87	Expression increases at low temperature. Represses *hns* mRNA, activates *rpoS* mRNA. Implicated in acid resistance response.	(Sledjeski *et al.*, 1996; Lease *et al.*, 1998; Majdalani *et al.*, 1998; Lease & Belfort 2000; Lease *et al.*, 2004; Jones *et al.*, 2006).
FnrS	*STnc580*	*dbpA/zntB*	113	Contains multiple independent targeting domains. Induced under anaerobic conditions. Represses synthesizes enzymes required during aerobic growth.	(Boysen *et al.*, 2010; Durand & Storz, 2010)
GadY	IS183	*gadX/gadW*	105	Induced in stationary phase. Stabilizes *gadX* mRNA, which encodes a transcriptional regulator of the acid stress response.	(Opdyke *et al.*, 2004; Tramonti *et al.*, 2008)
GcvB	*IS145*	*gcvA/ygdI*	209	Expressed in nutrient rich conditions. Targets upstream activator regions of multiple mRNAs encoding proteins involved in amino acid uptake (e.g. *dppA* and *gltI*) using a conserved G/U rich targeting domain.	(Urbanowski *et al.*, 2000; Sharma *et al.*, 2007; Pulvermacher *et al.*, 2008, 2009)

Table 1. (*Continued*) Physiological roles of Hfq-dependent sRNAs in *E. coli* and *Salmonella* (listed by genomic location in *Salmonella*).

Name	Alternative IDs	Adjacent genes in *Salmonella*	Length	Important features	References
GlmY	*tke1, sroF*	*yfhK/purG*	183	Induced in response to decreasing GlcN-6-P. Stabilizes the homologous sRNA, GlmZ.	(Kalamorz *et al.*, 2007; Urban *et al.*, 2007; Reichenbach *et al.*, 2008; Urban & Vogel 2008)
GlmZ	*k19, ryiA, sraJ*	*yifK/hemY*	207	Activated by the homologous GlmY sRNA in response to decreasing GlcN-6-P. Causes the discoordinate regulation of *glmUS* by stabilizing the *glmS* mRNA and thus increasing levels of GlcN-6-P synthase.	(Kalamorz *et al.*, 2007; Urban *et al.*, 2007; Reichenbach *et al.*, 2008; Urban & Vogel 2008)
InvR	*STnc270*	*inv*H/STM2901	80	Induced by HilD under conditions that favour invasion. Represses *ompD* mRNA.	(Pfeiffer *et al.*, 2007)
MgrR	*STnc560*	*ydeI/ydeE*	99	Induced by low magnesium through PhoPQ. Increases resistance against antimicrobial peptides through repression of *etpB*, which is involved in LPS modifcation. Also targets *ygdQ*.	(Sittka *et al.*, 2008; Moon & Gottesman 2009)
MicA	*sraD*	*luxS/gshA*	72	Activated in response to envelope stress through σE. Represses synthesis of the *ompA* and *lamB* porins and *phoP*.	(Rasmussen *et al.*, 2005; Udekwu *et al.*, 2005; Papenfort *et al.*, 2006)
MicC	*IS063, tke8*	*nifJ/ynaF*	109	Induced during conditions that promote virulence. Repression of porin synthesis (*ompC* and *ompD*). Targets *ompD* deep within the coding sequence to induce degradation.	(Chen *et al.*, 2004; Pfeiffer *et al.*, 2009)
MicF	-	*ompC/yojN*	93	First *trans*-acting RNA identified. Represses *ompF* mRNA.	(Mizuno *et al.*, 1984; Hartog *et al.*, 2008)

Table 1. (*Continued*) Physiological roles of Hfq-dependent sRNAs in *E. coli* and *Salmonella* (listed by genomic location in *Salmonella*).

Name	Alternative IDs	Adjacent genes in *Salmonella*	Length	Important features	References
MicM	*rybC/sroB/ chiX*	*ybaK/ybaP*	85	Constitutively expressed. Represses *ybfM* (a.k.a *chiP*) mRNA, encoding a transporter for chitosugar uptake. ChiP activity is antagonized by a novel mechanism involving *chbBC* mRNA, which acts as bait to promote MicM degradation.	(Vogel *et al.*, 2003; Figueroa-Bossi *et al.*, 2009; Overgaard *et al.*, 2009; Rasmussen *et al.*, 2009)
OmrA/B	*t59,rygA/B, sraE*	*aas/galR*	88/82	Homologous sRNAs. OmpR regulated, induced by high osmolarity. Represses multiple omp mRNAs (*ompT*, *cirA*, *fecA* and *fepA*) and *ompR*. Also regulates *csgD*, a transcriptional regulator of curli synthesis.	(Guillier & Gottesman 2006, 2008; Holmqvist *et al.*, 2010)
OxyS	-	*argH/oxyR*	109	Induced in response to oxidative stress. Represses translation of *fhlA* and *rpoS*. Hfq required for function but not stability.	(Altuvia *et al.*, 1998; Zhang *et al.*, 1998; Argaman & Altuvia 2000; Zhang *et al.*, 2002)
RprA	*IS083*	*ydiL/ydiK*	106	Induced by cell surface stress. Activates *rpoS* mRNA. Processed rapidly *in vivo*.	(Majdalani *et al.*, 2001; Majdalani *et al.*, 2002; Jones *et al.*, 2006; Soper *et al.*, 2010)
RseX	-	STM1994/*ompS*	91	Required for normal σE activation. Repression of porin synthesis (*ompA* and *ompC*).	(Douchin *et al.*, 2006)
RybB	*p25*	*STM0869/STM0870*	80	Activated in response to envelope stress through σE. Repressor of multiple *omp* mRNAs (e.g. *ompF*, *ompC* and *ompA*).	(Johansen *et al.*, 2006; Papenfort *et al.*, 2006; Thompson *et al.*, 2007; Papenfort *et al.*, 2010)
RydC	*IS067*	STM1638/*cybB*	62	Forms a pseudo-knot. Represses expression of a putative ABC transporter (*yejABEF*).	(Zhang *et al.*, 2003; Antal *et al.*, 2005)

Table 1. (*Continued*) Physiological roles of Hfq-dependent sRNAs in *E. coli* and *Salmonella* (listed by genomic location in *Salmonella*).

Name	Alternative IDs	Adjacent genes in *Salmonella*	Length	Important features	References
RyhB	*sraI, IS176, RfrA*	*yhhX/yhhY*	90	Induced under iron limiting conditions. Represses multiple mRNAs encoding non-essential iron utilizing proteins (e.g. *sodB*, *uof*, *icsS*, *cysE* and *shiA*).	(Massé & Gottesman 2002; Massé *et al.*, 2005; Jacques *et al.*, 2006; Prévost *et al.*, 2007; Vecerek *et al.*, 2007; Desnoyers *et al.*, 2009; Salvail *et al.*, 2010)
SdsR	*ryeB, tpke79*	*STM1871/STM1872*	104	Hotspot for integration of foreign DNA sequences. Encoded antisense to SraC. Implicated in targeting SdsR for processing by RNase III.	(Vogel *et al.*, 2003; Balbontin *et al.*, 2008)
SgrS	*ryaA*	*yabN/leuD*	227	Targets *ptsG* and *manX* to inhibit glucose uptake in response to glucose phosphate stress. Encodes a small peptide, SgrT. SgrS-*ptsG* is the model system for role the role of translational block in mRNA degradation.	(Vanderpool & Gottesman 2004; Wadler & Vanderpool 2007; Maki *et al.*, 2008; Rice & Vanderpool 2011)
Spot42	*spf*	*polA/yihA*	109	Abundant in the presence of glucose. Regulated by CRP. Targets multiple genes involved in the uptake and catabolism of non-prefered carbon sources (e.g. *fucI*, *fucK*: L-fuctose; *srlA*:D-sorbitol). Contains multiple independent targeting domains.	(Sahagan & Dahlberg 1979; Møller *et al.*, 2002a; Beisel & Storz 2011)
SraC	*ryeA*	*STM1871/STM1872*	270	Hotspot for integration of foreign DNA sequences. Encoded antisense to SdsR.	(Vogel *et al.*, 2003)
SraF	*tpk1, IS160*	*ygjR/ygjT*	185	pH responsive *cis* acting riboregulator. May also act in *trans*.	(Nechooshtan *et al.*, 2009)

The role of Hfq in regulation by RyhB has been particularly well characterized. Firstly, Hfq is required for RyhB stability *in vivo* (Massé *et al.*, 2003). Hfq binding to RyhB masks an RNase E recognition site and thus increases its stability (Moll *et al.*, 2003). Hfq also acts as a chaperone for RyhB targeting of the superoxide dismutase gene, *sodB* (Geissmann & Touati, 2004). Binding of Hfq to an A/U rich segment on the *sodB* 5' UTR results in a conformational change in *sodB* that allows RyhB binding and inhibition of translation (Geissmann & Touati, 2004). As mentioned earlier, Hfq also acts as a scaffold to allow recruitment of RNase E, which causes the coupled degradation of the RyhB-*sodB* duplex (Massé *et al.*, 2003, Morita *et al.*, 2005). Interestingly, RyhB binding at the RBS of *sodB* does not induce cleavage locally within the mRNA but instead induces cleavage at a distal site (>350 nt) downstream of the RBS (Prévost *et al.*, 2011). RyhB induced mRNA cleavage in this region could be disrupted without altering RyhB association with the RBS but instead by altering the sequence of the cleavage site indicating that the structural features of the target mRNA are an important component defining its susceptibility to sRNA induced degradation.

RyhB is a versatile post-transcriptional regulator and acts by a number of mechanisms. For example, RyhB controls the expression of its own transcriptional regulator (Fur) by binding far upstream of the *fur* start codon (Vecerek *et al.*, 2007). Bläsi and co-workers showed that the *fur* gene is co-transcribed with a small upstream ORF (*uof*; upstream of fur), which is not annotated in the *E. coli* genome sequence (Vecerek *et al.*, 2007). RyhB targets translation initiation of the *uof* mRNA by RBS occlusion. Since *uof* and *fur* are translationally coupled, RyhB association with *uof* represses translation of *fur*. Physiologically, the negative control of Fur by RyhB probably establishes a feed-forward loop that amplifies production of RyhB when iron is scarce (Vecerek *et al.*, 2007).

RyhB also acts through an anti-antisense mechanism in the upregulation of *shiA* mRNA mRNA, which encodes a permease of shikimate that is believed to function in the biosynthesis of siderophores that acquire extracellular iron (Prévost *et al.*, 2007). More generally, similar sRNAs whose expression is Fur-controlled have now been identified and studied in a great variety of bacteria, e.g., *Pseudomonas*, *Vibrio* and *Neisseria* suggesting that this functional mechanism is widely conserved. Interestingly, the dependence on Hfq varies between species (Wilderman *et al.*, 2004, Metruccio *et al.*, 2009, Davis *et al.*, 2005). Lastly, RyhB has also been identified as a regulator required for virulence in *Shigella dysenteriae*, the causative agent of bacillary dysentery in humans (Murphy & Payne, 2007).

The intergenic space between *aas* and *galR* contains two tandemly orientated loci encoding the nearly identical **OmrA/B** sRNAs that are widely conserved in many enterobacterial species (Argaman *et al.*, 2001, Vogel *et al.*, 2003, Wassarman *et al.*, 2001). Transcription of these loci, *omrA* and *omrB* (OmpR regulated sRNAs A and B), is induced by high osmolarity and other stresses that perturb the outer membrane, through the OmpR-EnvZ two component system (Guillier & Gottesman, 2006). Detailed investigations by Guillier and Gottesman (Guillier & Gottesman, 2006, Guillier & Gottesman, 2008) revealed that OmrA and OmrB control overlapping regulons, repressing multiple OMP mRNAs (Fig. 2). Interestingly, unlike the well-chacterized sRNAs RybB, MicA, MicC, or MicF that regu-

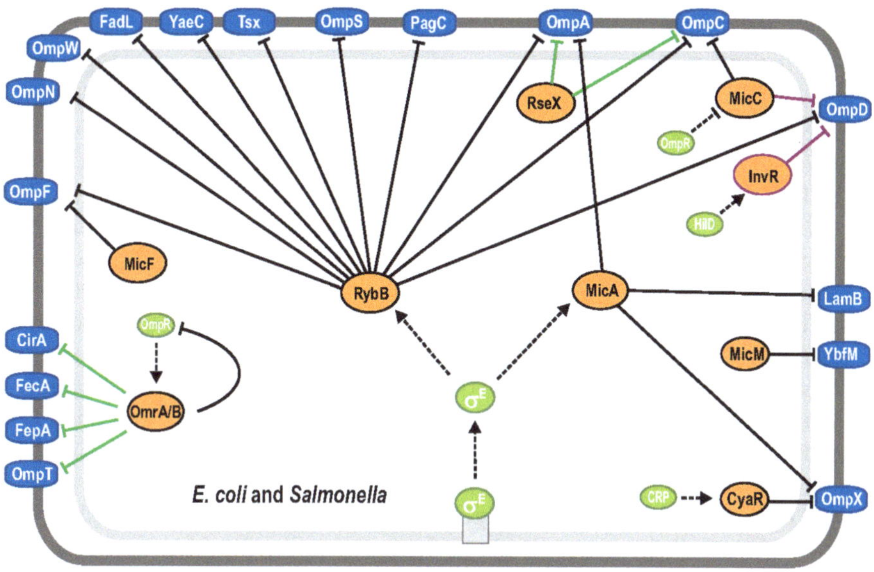

Fig. 2. OMP regulating Hfq-dependent sRNAs in *E. coli* and ***Salmonella***
The sRNAs (orange bubbles) and corresponding OMP targets (blue bubbles) are shown within a schematic drawing of a cell (the thick dark gray or light gray lines indicate outer or inner membranes, respectively). sRNAs and OMPs common to both *E. coli* and *Salmonella* are indicated by black lines. *E. coli* or *Salmonella* specific interactions are indicated by dark green and purple lines, respectively. Key transcriptional regulators are identified in lime green bubbles.

late, with some overlap, the expression of the major OmpA, OmpC, OmpD, and OmpF porins in *E. coli* and *Salmonella*, the OmrA/B sRNAs control a distinct set of OMPs, e.g. the outer membrane protease, OmpT, and the specific gated channels for iron-siderophore complexes CirA, FecA and FepA (Guillier & Gottesman, 2006, Guillier & Gottesman, 2008). More recently OmrA/B have been shown to regulate *csgD,* a transcriptional regulator of curli fimbriae expression (Holmqvist *et al.,* 2010). The two ~80 nt RNAs contain the highest regions of homology at the 5' and 3' ends. In fact, the first 20 nt of each sRNA, which are required for targeting of multiple mRNAs including *cirA*, *csgD*, *ompT* and *ompR*, are identical.

While the OmrA/B target sites are consistent with repression through targeting of the translation initiation step near the SD (*cirA*, *ompR*) or within the 5 codon window (*ompT*), RNase E is required for OmrA/B mediated repression of gene expression (Guillier & Gottesman, 2008). Regulation of OmpR by OmrA/B allows for homeostatic control of OmrA/B levels. Why two homologous sRNAs are employed to seemingly control identical target mRNAs is as yet unclear. It is possible that the heterologous sequences in the central portion of the sRNA may be involved in regulating different targets that were not expressed during experimental conditions and thus have not yet been identified. Systems where the roles for homologous sRNAs have been elucidated (*Qrr* and *GlmY/Z*) are discussed later.

As already indicated, the repression of mRNAs encoding outer membrane proteins (OMPs) has emerged as a common theme amongst Hfq-associated sRNAs (Fig. 2); almost one third of sRNAs characterized in enterobacteria target *omp* mRNAs (Guillier & Gottesman, 2006, Vogel & Papenfort, 2006). The importance of Hfq in this control is clear since deletion of *hfq* results in misregulation of major OMPs and chronic induction of the σE controlled envelope stress response (Ding *et al.*, 2004, Figueroa-Bossi *et al.*, 2006, Guisbert *et al.*, 2007, Sittka *et al.*, 2007, Vytvytska *et al.*, 1998; 2000). The *rpoE*-encoded σE protein, which is widely conserved among pathogenic and non-pathogenic bacteria (Rowley *et al.,* 2006) is activated upon accumulation of misfolded OMPs in the periplasmic space or through perturbations in the membrane caused by external stresses (Mecsas *et al.*, 1993, Missiakas *et al.*, 1996, Raivio *et al.*, 1999) resulting in the rapid disappearance of multiple *omp* mRNAs (Rhodius *et al.,* 2006). This rapid down-regulation of OMP production is controlled predominantly by the **RybB** sRNA, transcription of which is induced by σE (Thompson *et al.*, 2007, Papenfort *et al.*, 2006, Johansen *et al.*, 2006). Upon induction RybB acts post-transcriptionally to simultaneously repress translation of all the major OMPs in *Salmonella* (Fig. 2), dramatically reducing half-lives of the normally stable mRNAs. For example, over-expression of RybB reduced the half-lives of *ompC* and *ompD* mRNAs from >10 min to ~1 min (Papenfort *et al.*, 2006). Collectively, these observations have led to a model that, whenever envelope perturbations and misfolded OMPs induce the σE-regulon, RybB is expressed to immediately limit OMP synthesis on a global scale (Thompson *et al.*, 2007, Papenfort *et al.*, 2006, Johansen *et al.*, 2006). This response is dependent on Hfq, which is required for the stability of RybB *in vivo* and also facilitates base-pairing with some RybB targets *in vitro* (Papenfort *et al.*, 2006, Bouvier *et al.*, 2008). A highly conserved ~16 nucleotide core region located at the 5' end of RybB is essential and sufficient to for full regulation of the RybB regulon (Balbontin *et al.*, 2010, Papenfort *et al.*, 2010).

Besides playing a role in the stress response RybB also functions as a surveillance system in unstressed cells, where the constant flux of OMPs to the membrane will naturally result in some accumulation of misfolded proteins. The role for RybB in unstressed cells is clear since deletion of the *rybB* gene results in over-production of OMP mRNAs (Papenfort *et al.*, 2006), which in turn chronically activates σE (Papenfort *et al.*, 2006, Thompson *et al.*, 2007). That both RybB and its σE dependent promoter are widely conserved in the gamma-proteobacteria (Papenfort *et al.*, 2006) suggest that this Hfq-dependent mechanism of envelope homeostasis using post-transcriptional regulators plays an important role in diverse range of environments.

A second σE induced sRNA, **MicA** (Papenfort *et al.*, 2006, Udekwu & Wagner, 2007, Johansen *et al.*, 2006), directly represses expression of at least three OMPs, OmpA, OmpX and LamB (Fig.2; Rasmussen *et al.*, 2005, Udekwu *et al.*, 2005, Bossi & Figueroa-Bossi, 2007, Johansen *et al.*, 2008). MicA also regulates expression of the pleiotropic two component system, PhoPQ. MicA repression of PhoPQ results in activation of a PhoPQ-regulated sRNA, **MgrR** (Coornaert *et al.,* 2010). MgrR is a regulator of LPS modification and modulates resistance to antimicrobial peptides (Moon & Gottesman, 2009). Therefore MicA functions as a link between the σE dependent envelope stress response and the PhoPQ regulon, which results in increased resistance to antimicrobial peptides and facilitates membrane stabilization.

The view that the global control of multiple mRNAs within a defined physiological response may be a common function of sRNAs is further supported by the characterization of **GcvB**, which regulates many targets with functions in amino acid biosynthesis. Expression of this ~200nt sRNA is tightly controlled by the glycine responsive regulators, GcvA and GcvR, which results in the increased expression of GcvB during rapid growth in nutrient-rich media (Argaman *et al.*, 2001, Sharma *et al.*, 2007, Urbanowski *et al.*, 2000). Inactivation of GcvB causes the accumulation of the periplasmic components of major ABC transport systems such as OppA and DppA, which are involved in the uptake of amino acids in nutrient poor conditions. GcvB directly regulates multiple targets including *oppA*, *dppA*, *cycA*, *gltI*, *livJ*, *livK*, *argT* and *sstT* which encode components of ABC transport systems (Pulvermacher *et al.*, 2008, Pulvermacher *et al.*, 2009a, Sharma *et al.*, 2007, Urbanowski *et al.*, 2000). This suggested a central role for GcvB in repression of amino acid uptake systems during growth in rich media (Sharma *et al.*, 2007, Pulvermacher *et al.*, 2009a, Guisbert *et al.*, 2007, Sittka *et al.*, 2007).

Hfq associates with GcvB *in vivo* (Zhang *et al* 2003, Sittka *et al* 2008) and is required for GcvB stability (Urban and Vogel, 2007; Sharma *et al*, 2007). However, even in the presence of Hfq GcvB is relatively unstable (half-life 2–3 min) (Vogel *et al.*, 2003; Sharma *et al.*, 2007), perhaps ensuring that upon nutrient depletion, GcvB concentrations will decrease rapidly and allow up-regulation of ABC transporters required for cell growth. GcvB targeting of *cycA,* which encodes a key component of the major glycine transporter, suggests a homeostatic mechanism based on the accumulation of glycine. Accumulation of glycine results in increased levels of GcvB, which will in turn lead to repression of *cycA* translation and thus repression of glycine uptake. This coupled with the central role for GcvB in controlling ABC transporter biosynthesis at first suggested a role for GcvB in improving the economy of nutrient uptake in fast-growing cells (Sharma *et al.*, 2007).

Detailed investigation of the molecular basis for GcvB targeting revealed the existence of a hyper-conserved ~30 nt G/U-rich region (R1) that mediates translational repression of seven ABC transporter mRNAs through targeting of a C/A-rich upstream activator sequence common to each of the mRNAs (Sharma *et al.*, 2007). Intriguingly, the target sites on each of the mRNAs were not conserved for sequence, and were linked simply by their high C/A content. Multimers of the dinucleotide CA are known to stimulate translation through increasing the association of 30S subunits (Martin-Farmer & Janssen, 1999). Thus it appears that GcvB evolved rapidly as a global regulator by targeting a conserved element common to mRNAs encoding periplasmic transporters. Pulse over-expression and microarray analysis of GcvB variants lacking either the R1 or a second hyper-conserved domain (R2) were used to further define the roles for each conserved segment and confirmed R1 as the primary targeting region of GcvB (Sharma C.M. *et al.*, in revision). Intringuingly, deletion of both the R1 and R2 domains failed to relieve repression of *cycA*. GcvB represses translational initiation by binding to a C/A rich element located just upstream of the *cycA* RBS. Further mutational analysis and *in vitro* 30S footprinting assays revealed an incredible functional redundancy within GcvB that allowed efficient translational repression of *cycA*. Importantly, transcriptional profiling and *in silico* analysis utilizing a C/A rich motif generated from known GcvB targets was used to identify novel GcvB targets, many of which were confirmed using a *gfp* based reporter system (Urban & Vogel, 2007). The identification of multiple new targets has challenged the model whereby GcvB acts simply to repress expression of multiple ABC transporters during growth in a nutrient rich environment (Sharma C.M. *et al.*, in revision). Using branched change amino acids (leucine, isoleucine and valine) as an example, it appears that GcvB concomitantly represses expression of both ABC transporters for import (*brnQ*, *livJ* and *livK*) and enzymes for *de novo* synthesis of the same amino acids. Thus GcvB can shut down the processes that provide the cell with essential amino acids.

Quorum-sensing (QS) is a process whereby bacteria assess population density through secretion and detection of molecules called autoinducers (Ng & Bassler, 2009). QS in the human pathogen *Vibrio cholera* and the closely related marine bacterium *Vibrio harveyi* involves Hfq and multiple homologous quorum regulated RNAs known as **Qrr** (Lenz *et al.*, 2004). At low cell density, the transcription of the *qrr* genes is activated by the quorum-sensing response regulator LuxO. Transcription of the Qrr sRNAs blocks production of the master regulators for quorum regulated processes, i.e. LuxR (bioluminescence of *V. harveyi*) and HapR (virulence gene expression in *V. cholera*). This occurs through targeting of the SD sequences of the respective mRNAs using highly conserved regions, which represses translation and promotes mRNA degradation (Lenz *et al.*, 2004, Tu & Bassler, 2007). At high cell density, transcription of Qrr sRNAs decreases resulting in higher levels of LuxR and HapR and reduced levels of bioluminescence and virulence, respectively (Lenz *et al.*, 2004, Tu & Bassler, 2007).

V. cholerae contains 4 Qrr sRNAs while *V. harveyi* contains 5. In both species the *qrr* genes are expressed at different levels with *qrr4* expressed at the highest levels during low cell density (Lenz *et al.*, 2004, Tu & Bassler, 2007). Although the sequences required for targeting of LuxR and HapR are conserved between all of the Qrr sRNAs, Qrr activity differs between the two *Vibrio* species. The key difference between the two systems is that while in *V. cholera* the 4 Qrr RNAs function redundantly, i.e. only deletion of all 4 Qrr sRNAs results in altered quorum phentoypes (Lenz *et al.*, 2004), in *V. harveyi* the sRNAs act co-operatively, with deletion of a single Qrr sRNA resulting in altered quorum regulation (Tu & Bassler, 2007). The functional redundancy in the *V. cholera* Qrr system stems from two negative feed-back loops involving two Qrr targets, *luxO* and *hapR*. Deletion of a *V. cholera* Qrr sRNA results in increased levels of LuxO and HapR, which increases transcription of the remaining *qrr* genes and restores homeostasis in the system (Svenningsen *et al.*, 2008; 2009). Although similar feedback loops exist in *V. harveyi*, deletion of a Qrr sRNA is not fully compensated by increased expression of the other *qrr* genes (Tu & Bassler, 2007, Tu *et al.*, 2008; 2010). Since the *qrr* genes are expressed at different levels during low cell density (Qrr4 > Qrr2 > Qrr3 > Qrr1 > Qrr5), deletion of each *qrr* gene has different effects on quorum phenotype which relates to the levels to which the gene is normally expressed. For example, expression of only *qrr4* results in almost wild-type QS behavior whereas expression of only *qrr5* results in dramatically reduced QS behavior (Tu & Bassler, 2007). The additive effect of Qrr sRNAs in *V. harvevi* allows for the incorporation of additional environmental signals into the quorum sensing regulon through the differential expression of the *qrr* genes and the resulting rheostatic control of LuxR levels (Tu & Bassler, 2007). The negative feed-back loops also function to increase the sensitivity of the QS system to autoinducer levels, facillitating rapid alterations in gene expression during transition between low and high cell density (Svenningsen *et al.*, 2008, Tu *et al.*, 2008; 2010).

Intringuingly, some pandemic strains of *V. cholerae* contain nonsense mutations in *hapR* (Joelsson *et al.,* 2006) but still exhibit quorum sensing behavior (Hammer & Bassler, 2007). In these strains the Qrr RNAs promote *hapR* independent quorum controlled gene expression by directly targeting multiple mRNAs. For example, the Qrr sRNAs target *vca0939* mRNA (which encodes a putative guanylate cyclase) using an anti-antisense mechanism that relieves a repressive loop and allows efficient translation (Hammer & Bassler, 2007). The increased conservation of the Qrr sRNAs relative to HapR, which was previously thought to be essential for QS, highlights their essential roles in controlling quorum sensing in *Vibrio*.

2.6 Small RNAs with Multiple Conserved Targeting Regions

Similar to proteins, most sRNAs display regions of conservation marking them as functionally relevant residues. Such conservation may be further classified into structural components, *e.g.* transcriptional terminators, Hfq binding sites or mRNA targetting regions. While most characterized Hfq-dependent sRNAs utilize a single conserved region to control multiple target mRNAs, there is a growing number of regulators that contain multiple distinct regions that act independently to regulate multiple targets.

Expression of the highly conserved sRNA, **FnrS**, is induced upon a shift from aerobic to anaerobic conditions primarily through the global regulator FNR (fumarate and nitrate reduction) (Boysen *et al.*, 2010, Durand & Storz, 2010), which is a transcriptional regulator that directly senses oxygen availability. Induction of FnrS is also affected by the ArcA response regulator, which is active under low oxygen conditions, and CRP, a global regulator of carbon metabolism. Over-expression of FnrS resulted in depletion of 32 mRNAs encoding a variety of metabolic enzymes that are dispensable during growth in low oxygen, such as those involved in aerobic respiration (Boysen *et al.*, 2010, Durand & Storz, 2010).

FnrS interacts with different targets using two core regions: firstly, using a single stranded region between the first and second stem-loops and secondly, using a partially open stem at the 5' end. The dual targeting regions of FnrS were novel since they are truly distinct (Durand & Storz, 2010). While DsrA interacts with multiple targets using different regions, these regions are closely linked and probably overlap, since some mutations affect the regulation of both *hns* and *rpoS* (Lease *et al.*, 1998, Majdalani *et al.*, 1998). Equally, GcvB has two highly conserved regions,

R1 and R2 (Sharma *et al.*, 2007). However, while R1 targets many mRNAs independently of R2 (Sharma *et al.*, 2007), predicted targets for R2 have not yet been confirmed by mutational analysis (Pulvermacher *et al.*, 2009b).

Perhaps the most interesting aspect of FnrS is that each region regulates a set of genes with different physiological roles. Namely, the 5' region regulates genes linked with oxidative stress as well as folate and methionine metabolism while the central targeting region regulates genes involved in central metabolism promoting the idea that perhaps FnrS is a chimera of what were originally two separate sRNAs (Durand & Storz, 2010). Although FnrS is known to associate with Hfq *in vivo* (Sittka *et al.*, 2008) the effect of Hfq on the stability and function of FnrS has not been investigated. Predicted FnrS targets are predominantly located near the SD, indicating that this novel sRNA primarily utilizes the canonical method of SD/AUG targeting for translational repression (Durand & Storz, 2010).

Spot 42 is best characterized for its role in the discoordinate gene expression of the *galETKM* operon (discussed earlier). The recent study by Beisel and Storz (2011) describes a new global role for Spot42 in the regulation of catabolite respression, which ensures that preferred sugars such as glucose are selectively utilized before non-preferred, low energy, sugars. CRP is a transcriptional regulator involved in catabolite repression (Görke & Stülke, 2008). CRP is only active the presence of cyclic AMP (cAMP), which accumulates in response to low levels of glucose. CRP-cAMP upregulates the transcription of multiple genes involved in metabolism of non-preferred carbon sources (Reichenbach *et al.,* 2008). Pulse over-expression of Spot42 revealed 16 targets, including the known *galK* target (Beisel & Storz, 2011). Spot42 is abundant in the presence of glucose (Sahagan & Dahlberg, 1979a, Sahagan & Dahlberg, 1979b) and consistent with this, many of the targets identified are involved in the uptake and catabolism of non-preferred carbon sources (e.g. *fucI*, *fucK*: L-fuctose; *srlA*: D-sorbitol). A striking element is that Spot 42 seems to use 3 independent targeting domains to regulate the multiple mRNAs. Two of the three elements flank the base of a stem-loop structure while the third motif is located apically within the stem-loop structure (Beisel & Storz, 2011, Møller *et al.*, 2002b,). Although more than one motif is involved in targeting some mRNAs (*fucI* regulation involves all of them) each of the three elements controls expression of at least one target independently of the other regions (Beisel & Storz, 2011). One of these targetting regions is located at the 5' end and targets multiple mRNAs (*gltA*, *nanC*), further supporting the increasingly strong evidence for a conserved role of 5' located motifs in the regulation of multiple targets (Papenfort *et al.,* 2010).

Consistent with the involvement of multiple regions in mRNA targeting, the entire Spot 42 sRNA is highly conserved in enterobacteria (Møller *et al.*, 2002a). Similar to FnrS, since most of its targets seem to be bound near the RBS (Beisel & Storz, 2011) it appears that the Spot 42 primarily represses translation by the canonical method of RBS occlusion. In physiological terms, the incorporation of Spot 42 into the catabolite repression regulon results in reduced leakiness of CRP-controlled promoters when CRP is not active. Since CRP is a regulator of Spot42, this system represents a coherent feedforward loop that results in delayed accumulation of CRP activated proteins (SrlA and FucI). However, Spot42 also increases the rate the disappearance of SrlA and FucI upon CRP inactivation, highlighting the potential benefits of post-transcriptional control elements within feedforward regulatory loops (Beisel & Storz, 2011).

2.7 Unusually Complex Mechanisms

While the majority of characterized sRNA-mRNA interactions occur through the canonical RBS targeting mechanism, some sRNAs have evolved unusually complex mechanisms to control target mRNA levels or even adjust the activity of other sRNAs.

In *E. coli* the *glmS* gene encoding the glucosamine 6-phosphate (GlcN-6-P) synthase is feedback regulated by GlcN-6-P concentration in a pathway that involves two homologous sRNAs, **GlmY** and **GlmZ**. The mechanism of action of these sRNAs was described in a number of studies from the Görke and Vogel groups (Kalamorz *et al.*, 2007, Reichenbach *et al.*, 2008, Urban *et al.*, 2007, Urban & Vogel, 2008). The GlmS enzyme is encoded within the bicistronic *glmUS* operon. *glmUS* mRNA undergoes RNase E-mediated processing to accumulate a monocistronic *glmS* mRNA species. GlmZ is required, in conjunction with Hfq, to unlock an inhibitory structure in the 5' UTR of *glmS* mRNA that sequesters the SD and prevents efficient translation. GlmZ is subject to endonucleolytic cleavage that removes most of the *glmS* targeting region and thus the processed GlmZ is incapable of activating *glmS*. The unusual aspect of this system is the mechanism for preventing GlmZ processing, which involves a second sRNA (GlmY) that contains extensive homology to GlmZ. The homology between GlmY and GlmZ is restricted to the 5' end of each molecule and does not extend to the *glmS* targeting site, which is only contained in GlmZ. GlmY suppresses cleavage of GlmZ, through an as-yet undefined mechanism but which may involve sequestration of YhbJ, a protein which is essential for GlmZ processing. That GlmY is not required for GlmZ stability in the absence of YhbJ supports this indirect mechanism for control of GlmZ processing by GlmY.

GlmY expression is tightly regulated both transcriptionally and post-trancriptionally; it is induced in response to decreasing GlcN-6-P and thus functions as a post-transcriptional mechanism to link GlmS synthesis with GlcN-6-P levels. The RNA is unstable due to 3‘ end polyadenylation, which reduces the GlmY half-life from ~7 to ~1 min. GlmS expression is therefore tightly regulated by the rapid processing of both sRNAs required for its translation. The interaction of GlmY and GlmZ provides the first example of multiple sRNAs acting in a hierarchical cascade to control a single mRNA.

Studies from the Valentin-Hansen and Bossi labs involving the conserved Hfq-dependent sRNA, **MicM** (a.k.a. SroB, RybC or ChiX) provide the next example of an unexpectedly complex mechanism in enterobactetrial post-transcriptional regulation. MicM represses the synthesis of a conserved porin YbfM (a.k.a. ChiP; Fig.2) that is involved in chitosugar uptake (Overgaard *et al.*, 2009, Figueroa-Bossi *et al.*, 2009, Rasmussen *et al.*, 2009). MicM exhibits a number of traits uncharacteristic of sRNAs. Firstly, unlike many stress responsive sRNAs which are conditionally expressed, MicM is constitutively expressed (Vogel *et al.*, 2003). Secondly, MicM targeting of *ybfM* mRNA results in the degradation of *ybfM* but not of MicM, which is refractory to RNase E-mediated cleavage and recycled from the *ybfM*-MicM duplex (Overgaard *et al.*, 2009, Figueroa-Bossi *et al.*, 2009). Cumulatively, the constant presence of MicM and its catalytic activity strongly bias the system towards *ybfM* degradation. A novel and simple mechanism has evolved to overcome this bias and allow YbfM expression in the presence of the chitosugar, chitobiose. The *chbBCARPG* operon, which encodes enzymes required for the utilization of chitobiose, is upregulated in the presence of chitobiose (Figueroa-Bossi *et al.*, 2009). The resulting mRNA contains a sequence in the intergenic region between *chbB* and *chbC* that is an ideal MicM target site almost identical to the site in *ybfM*, but with a key difference. The complementarity between MicM and the *chbBC* intergenic region extends +2 nt compared to MicM-*ybfM*. These extra interactions are predicted to disrupt a stem-loop required for the stability of MicM, which makes it susceptible to RNase E-mediated cleavage (Figueroa-Bossi *et al.*, 2009, Overgaard *et al.*, 2009). The *chbBC* target therefore not only acts as to sequester MicM, but causes its degradation. The induction of *chbBC* and consequent reduction in MicM levels allows for expression of YbfM that will result in a positive feed-forward loop dependent on the presence of extracellular chitosugars. Upon depletion of chitosugars, the resulting depletion of the *chbBC* mRNA will result in rapid restoration of the tight repression of *ybfM* by MicM.

2.8 5' Regions as a Conserved Mechanism for Targeting Multiple mRNAs

While a number of internally pairing sRNAs have been described that regulate multiple targets (e.g., RyhB, GcvB, or CyaR) the location of the targeting sequence in the 5' region is emerging as a common theme among sRNAs which have multiple targets. The best characterized example to-date is RybB, which contains a 5' conserved region that functions as an autonomous targeting system (Fig. 3). Transfer of this short conserved region to an unrelated RNA backbone confers the full regulatory functions of RybB (Bouvier *et al.*, 2008, Papenfort *et al.*, 2010). A similarly autonomous targeting domain is present in the 5' region of MicC, which confers full regulation of *ompD* and *ompC* when fused to an unrelated sRNA backbone (Chen *et al.*, 2004, Pfeiffer *et al.*, 2009). The homologous sRNAs OmrA and OmrB also target multiple mRNAs using a conserved 5' region (Guillier & Gottesman, 2008) while one of the two conserved core regions of FnrS is contained at the 5' end (Durand & Storz, 2010). In eukaryotes, the discovery that 5'-terminal seed sequences in the ~22-nucleotide microRNAs provide the base-pairing specificity to target mRNAs greatly advanced target predictions (Lewis *et al.*, 2003, Friedman *et al.*, 2009). Recognition of the importance of 5' targeting regions has already proved fruitful in identifying new targets, such as OmrA/B regulation of *ompR*

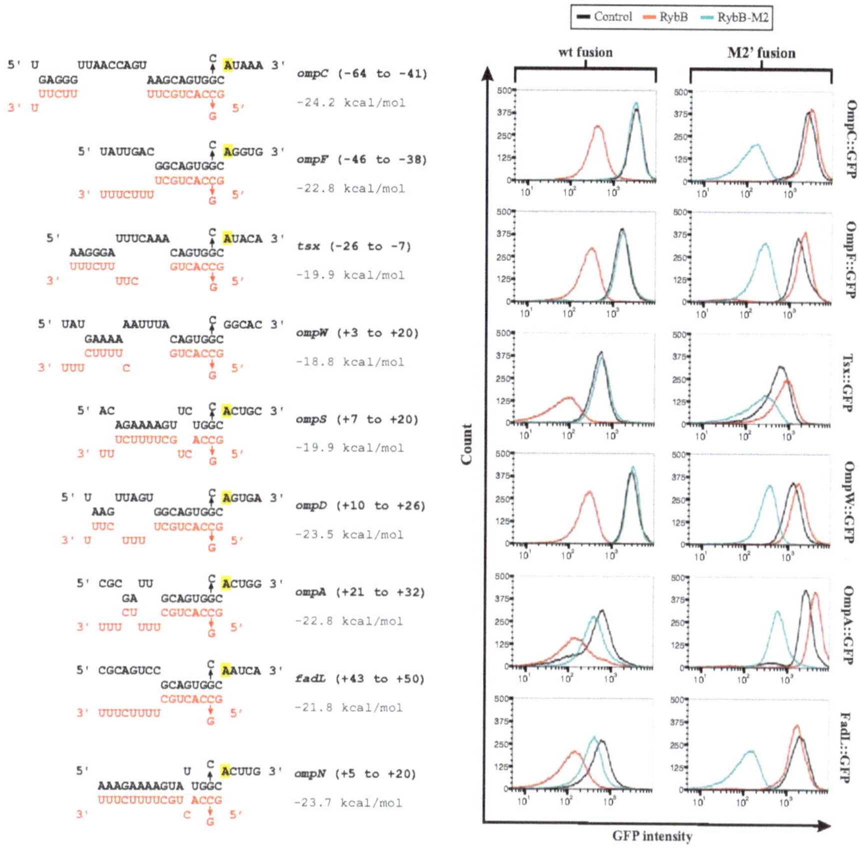

Fig. 3. Multiple targeting by RybB seed pairing
Left: Base-pairing of RybB with multiple target mRNAs. Vertical arrows denote the C_2->G mutation in RybB (RybB-M2), and the compensatory G->C change in the mRNA target alleles (M2'). The 3' adenosine flanking most RybB sites is shadowed. The predicted change in free energy of an interaction is shown in grey. **Right:** Validation of selected target gene interactions. Stationary-phase *Salmonella* Δ*rybB* carrying *gfp*-target fusions in combination with RybB expression or control plasmids were analyzed by FACS. Results are plotted in fluorescence histograms (FH) derived from 30,000 events per sample. Cellular fluorescence is given in arbitrary units (GFP intensity). Histograms on the same line represent the wild-type or M2' mutant fusions of the same target. Each FH represent triplicate results from pPl-RybB (red), pPl-RybB-M2 (blue) or control (black) samples.

(Guillier & Gottesman, 2008). The functional importance of locating a targeting region at the 5' end of an sRNA has yet to be identified. One possibility is that targeting of mRNAs using a 5' end may facilitate stronger interactions than some internally pairing regions, which may be sterically hindered by flanking non-targetting RNA-elements. The reduced steric hindrance would allow reduced complementrarity to achieve sufficient interaction for stable binding. Alternatively, the length of the region 5' to the base-paired sequence may be an important determinant of the rate of RNase E-mediated degradation of the duplex RNA (J. Vogel and B.F. Luisi labs, unpublished).

2.9 Maturation of Small RNAs

Considering that a number of sRNAs undergo endonucleolytic processesing, it is an exciting possibility that processing of an sRNA could re-locate the target sequence to the 5' end and alter the function of the sRNA. For example, the highly conserved **ArcZ** sRNA involved in the anaerobic/aerobic stress response (Mandin & Gottesman, 2010) is rapidly processed from 121 nt into a short (56 nt) stable sRNA. Processing exposes a 5' targeting region which regulates multiple mRNAs such as *sdaC*, *tpx*, STM3216, and *rpoS* (Mandin & Gottesman, 2010, Papenfort *et al.*, 2009). Due to the inherent instability of the full-length transcript, it is as yet unclear whether full-length ArcZ containing the targeting region internally would have the same activity as the processed ArcZ. It appears, however, that processing is required for the activation of *rpoS* translation since only processed ArcZ and not full-length ArcZ binds *rpoS* mRNA efficiently *in vitro* (Soper *et al.*, 2010).

RprA is another well studied sRNA that undergoes processing, resulting in two detectable species (Argaman *et al.*, 2001). The primary 105-nt RprA was originally identified as a sRNA that allowed activation of *rpoS* translation in the absence of DsrA (Majdalani *et al.*, 2001). Transcription of the *rprA* gene increases during stationary phase (Argaman *et al.*, 2001, Wassarman *et al.*, 2001) and is dependent on RcsB, the response regulator of the capsular polysaccharide genes (Majdalani *et al.*, 2002).

Although DsrA and RprA share only limited homology (Majdalani *et al.*, 2001), they act by similar mechanisms to upregulate translation of *rpoS* (Majdalani *et al.*, 2002, Majdalani *et al.*, 2001). Similar to ArcZ, RprA is processed into a short stable 3' fragment. What is unusual about RprA is that the sequence required for targeting of *rpoS* is located in the 5' end of the molecule. Full-length RprA is sufficiently stable (half-life ~7 min) to efficiently regulate *rpoS* (Majdalani *et al.*, 2001). A second highly conserved targeting domain is located within the 3' domain of RprA allows targeting of multiple mRNAs by RprA (Papenfort *et al.*, unpublished results). Similar to ArcZ, RNase E-dependent processing of RprA relocates this targeting region to the 5' end of the 3' fragment. Whether this alters the function of the targeting region is unclear. However, since only the full-length molecule is capable of regulating *rpoS*, processing of RprA creates two distinct regulatory molecules from a single sRNA transcript. Stress conditions often results in increased mRNA decay and thus may cause titration of the degradosome complex. It is therefore possible that stress conditions will alter the processing rate of RprA and allow increased accumulation of the full-length transcript, which in turn can activate translation of *rpoS*.

Processing of **MicX** in *Vibrio cholera* produces a stable active sRNA which is the dominant form *in vivo* (Davis & Waldor, 2007). The full-length molecule has some regulatory activity indicating that in this case processing is not essential for activity (Davis & Waldor, 2007). Interestingly, in contrast to the role for Hfq in protecting sRNAs from degradation, processing of MicX is dependent on Hfq (Davis & Waldor, 2007).

An exciting prospect is that sRNA processing may facilitate the more rapid degradation of target mRNA through generation of a 5' monophosphate group that allows more rapid degradation by RNase E (Carpousis, 2007). Thus the generation of a short, stable sRNA by processing may result in more rapid degradation of target mRNAs by RNase E than that of a primary transcript. This is supported by recent work characterizing the mechanism by which MicC induces the degradation of *ompD* (J. Vogel and B. F. Luisi labs, unpublished). MicC is dephosphorylated *in vivo* to produce a 5' monophosphate, which targets the sRNA/mRNA duplex for rapid degradation by RNase E.

2.10 Potential Evolution of New sRNAs from 3' UTRs

Given the requirement for a 3' terminator among Hfq-associated sRNAs, one exciting prospect is that new sRNAs could be generated rapidly from the 3' UTR of mRNAs containing a terminator. Since terminator regions are A/U rich and contain a stem-loop structure they are ideal targets for Hfq-binding. Equally, since the 3' UTR is unlikely to be evolutionarily constrained, targetting domains could evolve by genetic drift. High-throughput sequencing of Hfq-bound sRNAs (Sittka *et al.*, 2008) has uncovered a number of examples of such sRNAs, including STnc820, located immediately downstream of the *dapB* gene. A promoter for STnc820 is located within the 3' end of the *dapB* gene, which allows expression of the sRNA independent from the *dapB* promoter. Intriguingly, STnc820 contains a highly conserved internal G/U rich segment that is almost a perfect match to the R1 targetting domain of GcvB, suggesting a global role in amino acid uptake (Y. Chao & J. Vogel, unpublished results). Although there are no characterized examples, in theory novel regulatory sRNAs could also be generated from 3' UTRs by endonucleolytic cleavage.

3 Perspective

3.1 Overlap of sRNAs and Targets in Regulons

Since the first characterization of multiple target regulation by DsrA and OxyS, numerous examples have been uncovered. In fact, it now appears that multiple target regulation by sRNAs is the rule rather than the exception. This is exemplified by the archetypal sRNA MicF, which for over 25 years has been believed to only target a single mRNA, *ompF*. A recent study utilizing pulse over-expression of MicF followed by microarray analysis revealed multiple previously uncharacterized targets for MicF, including the global transcriptional regulator LRP (J. Vogel and E.G. Wagner labs, unpublished). It is therefore unlikely that *trans*-acting sRNAs function as dedicated regulators of a single mRNA. The role for sRNAs in targeting multiple mRNAs is further supported by deep sequencing analysis of Hfq-associated mRNA in *Salmonella*, which predicts 7-fold more target mRNAs than sRNAs (Sittka *et al.*, 2008). Equally, it is becoming increasing clear that there is significant overlap in many of the sRNA regulons. For example, to-date 3 sRNAs have been shown to directly activate translation of RpoS (DsrA, ArcZ, RprA), while even more significant overlap occurs in the regulation of other mRNAs such as *csgD* (OmrA/B, HbrD, RprA) and *ompD* (RybB, MicC, InvR, SdsR). Whether particular sequences or structures within certain mRNAs makes them susceptible to sRNA targeting or makes other mRNAs refractory to sRNA mediated regulation is as yet unknown.

3.2 Titration of Hfq: Regulation or Side-effect?

While it is clear that the vast majority of sRNAs act sequence specifically to alter translation, it is possible that some sRNAs may act sequence independently to cause pleiotropic changes in gene expression through the sequestration of Hfq.

The number of Hfq hexamers per cell has not yet been defined conclusively. Estimates of Hfq hexamers per cell have varied from 400 (Carmichael *et al.*, 1975) to 5,000–10,000 (Kajitani *et al.*, 1994) making predictions about the amount of sRNA required to cause pleiotropic effects through Hfq titration difficult. A recent study (Papenfort *et al.*, 2009) addressed this issue directly through deep sequencing of Hfq-bound RNA following varying levels of sRNA induction. Amazingly, expression of ArcZ to ~150 molecules per cell (~8 fold more than chromosomal wild-type levels) resulted in pleiotropic effects reminiscent of those associated with deletion of *hfq* (Sittka *et al.*, 2008, Papenfort *et al.*, 2009).

OxyS can reach ~4500 copies per cell in response to oxidative stress (Altuvia *et al.*, 1997), which may cause pleiotropic changes in gene expression through titration of Hfq (Zhang *et al.*, 1998). Moreover, OxyS is unusual among Hfq-dependent sRNAs in that it does not require Hfq for stability (Zhang *et al.*, 2002). This would further bias the acquisition of Hfq towards OxyS since sRNAs that are dependent on Hfq for stability but cannot access Hfq due to OxyS concentration will be rapidly degraded further reducing the competition for OxyS.

Whether high-level induction and subsequent titration of Hfq is a mechanism evolved to aid survival in response to oxidative stress, or whether the direct regulatory action of each of the OxyS molecules is required, with Hfq-titration as a side-effect, is not yet known. Interestingly, it was recently shown that Hfq sequestered by an sRNA can be efficiently recycled through induction of an mRNA target of the bound sRNA (Hussein & Lim, 2011). Thus the potential for a sRNA to act by Hfq sequestration depends not only on the levels of the sRNA itself but on those of its target mRNAs.

Further evidence for Hfq titration as a physiologically relevant regulatory mechanism comes from recent studies from the Gottesman and Woodson groups. Firstly, a library based screen to identify new regulators of *rpoS* translation identified multiple sRNAs (DicF, RybC, CyaR, MicM and OxyS) that negatively regulated translation of *rpoS* (Soper *et al.,* 2010). A direct role for these sRNAs was ruled out through mutational analysis of predicted interaction sites and thus it is believed that each of these sRNAs acts indirectly through titration of Hfq (Soper *et al.,* 2010). Equally, studies utilizing the native (not plasmid-based or over-expressed) system revealed that MicM, which is constitutively expressed, acts indirectly to alter regulation by the PhoP-activated sRNA MgrR. Deletion of MicM led to increased target repression by MgrR consistent with increased association of MgrR with Hfq and thus increased MgrR activity. This study strongly supports Hfq titration as a regulatory mechanism to elicit pleiotropic changes in gene expression in wild-type systems (Moon & Gottesman, 2009).

3.3 Titration of Hfq: Implications for Horizontal Gene Transfer

The fact that Hfq function can be titrated with pleiotropic effects is reminiscent of a similar global regulatory protein H-NS. Similar to Hfq, the H-NS protein binds with low sequence specificity to A/T rich DNA (Dorman, 2007) and A/U rich RNA (Park *et al.*, 2010). H-NS can be titrated by the horizontal acquisition of large A/T rich self-transmissable plasmids resulting in pleiotropic changes in gene expression and a dramatic reduction in fitness (Dillon *et al.*, 2010, Doyle *et al.*, 2007). Acquisition of such plasmids and subsequent transcription of the A/T rich DNA would lead to similar titration of Hfq. Self-transmissable plasmids such as the R-27 plasmid originally isolated from *Salmonella typhimurium* in the 1960s or its *Shigella flexneri* homologue (pSF-R27) carry H-NS homologues which dramatically reduce the fitness cost associated with plasmid acquisition (Doyle *et al.,* 2007). It is possible that a similar mechanism exists to protect the native post-transcriptional regulatory systems. Although convential BLAST searches have not revealed an Hfq homologoue on these plasmids, one may yet be identified by other methods. For example, a recent study identified an Hfq homologue in *Borrelia* that contained only limited sequence homology to the Hfq from *E. coli* but contained the characteristic Sm-like binding motifs and a similar predicted tertiary structure (Lybecker *et al.,* 2010). The identification of RNA binding proteins that facilitate horizontal gene transfer would provide a key insight into the evolution of post-transcriptional networks and specifically, how horizontally acquired sRNAs that are dependent on Hfq are integrated into existing regulatory networks. One characterized example of a horizontally acquired sRNA that has been successfully integrated into the host post-transcriptional regulatory network is the invasion gene-associated RNA (InvR), which is located within the horizonatally acquired *Salmonella* pathogenicity island SPI-1 (Pfeiffer *et al.*, 2007). Unexpectedly, InvR was shown to have little effect on invasion gene expression but instead regulates the ancestral *ompD* gene encoding the most abundant *Salmonella* porin (Lee & Schnaitman, 1980). The post-transcriptional repression of *ompD* by InvR, which is induced under conditions that promote expression of the type 3 secretion system required for invasion by *Salmo-*

nella, may improve virulence by balancing envelope stability or simply by making the cell membrane less permeable to antimicrobial peptides. In either case, InvR provides a key example of the successful incorporation of a sRNA into an existing post-transcriptional regulatory network.

3.4 Design of Synthetic sRNAs

RNA based control switches are rapidly emerging as integral components of synthetic biology (Khalil & Collins, 2010). While this revolution in synthetic system design has focused on *cis*-acting riboswitches there is potential for global control of gene expression using *trans*-acting sRNAs. The diversity of sRNAs is well described, yet more recent studies have highlighted commonalities among sRNAs. Specifically, Hfq-dependent sRNAs are modular in nature consisting of 3 principal components; a targeting domain, Hfq-binding domain and a 3' -located terminator. To date a number of studies (Papenfort *et al.*, 2010, Bouvier *et al.*, 2008, Pfeiffer *et al.*, 2009) have fused 5' targeting regions to unrelated sRNA backbones to confer new regulatory function to the sRNA. Similar to the design of novel regulatory circuits using bio-bricks (Shetty *et al.,* 2008), the simple design of custom sRNAs with the combined characteristics of any three modules from the vast pool of sRNAs may soon be a reality.

The extensive information gained from mutational analysis of sRNAs coupled with new *in silico* models that accurately predict sRNA function (Mitarai *et al.,* 2009) may aid the design of custom sRNAs. These designer sRNAs could be combined with existing software that allows the rapid construction of complex synthetic systems (Cai *et al.,* 2010).

4 Outlook

The rapid expansion of sRNA discovery that began in 2001 has continued unabated through the application of innovative new techniques and the expansion beyond typical model organisms such as *E. coli* and *Salmonella.* Intrincate sRNA mediated post-transcriptional regulons are continually being discovered in organisms previously presumed to be devoid of sRNAs (Sharma *et al.,* 2010). While the list of candidate sRNAs generated by deep sequencing is certain to be expanded over the coming years, a major challenge will be to characterize the physiological role for these sRNAs. The affordability of new high-throughput sequencing technologies will aid discovery of virulence associated sRNAs by identifying those sRNAs that are differentially expressed under virulence and non-virulence conditions. While these conditions are often mimicked through modification in culture media, more precise data will be derived from the deep sequencing of bacteria isolated at various stages of interaction with host cells. Equally, the continued discovery and characterization of Hfq homologs in a vast range of gram positive and gram negative bacteria highlights the importance of these post-transcriptional mechanisms and should provide important information on the evolution of post-transcriptional regulatory networks.

One fundamental aspect of Hfq-dependent sRNA regulation which has not yet been investigated is the spatial organization of sRNAs. The advent of super-resolution microscopy technology (Huang *et al.*, 2010) coupled with developments in RNA localization in bacteria (Montero Llopis *et al.*, 2010, Nevo-Dinur *et al.*, 2011) may soon provide tools to delve into the localization of sRNAs. This would allow visualization of how *trans*-acting sRNAs find their targets and the role for Hfq in this process. One promising candidtate for these studies is the regulation of SgrS with *ptsG*, which is dependent on the association of *ptsG* with the membrane. This indicates that SgrS association with *ptsG* is targeted and not simply achieved by passive diffusion through the cell (Kawamoto *et al.*, 2005) and is surely only the first of many examples of localization dependent regulation by sRNAs.

Excitingly, considering the continuing advances in technology and growth in the field, the vast advance in our understanding of Hfq-dependent sRNAs achieved in the past 10 years may yet be eclipsed by what we will learn in the near future.

References

Afonyushkin, T., B. Vecerek, I. Moll, U. Bläsi & V. R. Kaberdin (2005) Both RNase E and RNase III control the stability of *sodB* mRNA upon translational inhibition by the small regulatory RNA RyhB. *Nucleic Acids Res* **33**: 1678–1689.

Aiba, H. (2007) Mechanism of RNA silencing by Hfq-binding small RNAs. *Curr Opin Microbiol* **10**: 134–139.

Altuvia, S., D. Weinstein-Fischer, A. Zhang, L. Postow & G. Storz (1997) A small, stable RNA induced by oxidative stress: role as a pleiotropic regulator and antimutator. *Cell* **90**: 43–53.

Altuvia, S., A. Zhang, L. Argaman, A. Tiwari & G. Storz (1998) The *Escherichia coli* OxyS regulatory RNA represses *fhlA* translation by blocking ribosome binding. *EMBO J* **17**: 6069–6075.

Andersen, J., S. A. Forst, K. Zhao, M. Inouye & N. Delihas (1989) The function of *micF* RNA. *micF* RNA is a major factor in the thermal regulation of OmpF protein in *Escherichia coli*. *J Biol Chem* **264**: 17961–17970.

Ansong, C., H. Yoon, S. Porwollik, H. Mottaz-Brewer, B. O. Petritis, N. Jaitly, J. N. Adkins, M. McClelland, F. Heffron & R. D. Smith (2009) Global systems-level analysis of Hfq and SmpB deletion mutants in *Salmonella*: implications for virulence and global protein translation. *PLoS One* **4**: e4809.

Antal, M., V. Bordeau, V. Douchin & B. Felden (2005) A small bacterial RNA regulates a putative ABC transporter. *J Biol Chem* **280**: 7901–7908.

Argaman, L. & S. Altuvia (2000) *fhlA* repression by OxyS RNA: kissing complex formation at two sites results in a stable antisense-target RNA complex. *J Mol Biol* **300**: 1101–1112.

Argaman, L., R. Hershberg, J. Vogel, G. Bejerano, E. G. Wagner, H. Margalit & S. Altuvia (2001) Novel small RNA-encoding genes in the intergenic regions of *Escherichia coli*. *Curr Biol* **11**: 941–950.

Babitzke, P. & T. Romeo (2007) CsrB sRNA family: sequestration of RNA-binding regulatory proteins. *Curr Opin Microbiol* **10**: 156–163.

Balbontin, R., N. Figueroa-Bossi, J. Casadesus & L. Bossi (2008) Insertion hot spot for horizontally acquired DNA within a bidirectional small-RNA locus in *Salmonella enterica*. *J Bacteriol* **190**: 4075–4078.

Balbontin, R., F. Fiorini, N. Figueroa-Bossi, J. Casadesus & L. Bossi (2010) Recognition of heptameric seed sequence underlies multi-target regulation by RybB small RNA in *Salmonella enterica. Mol Microbiol* **78**: 380–394.

Beisel, C. L. & G. Storz (2011) The Base-Pairing RNA Spot 42 Participates in a Multioutput Feedforward Loop to Help Enact Catabolite Repression in *Escherichia coli. Mol Cell* **41**: 286–297.

Beyer, D., E. Skripkin, J. Wadzack & K. H. Nierhaus (1994) How the ribosome moves along the mRNA during protein synthesis. *J Biol Chem* **269**: 30713–30717.

Bossi, L. & N. Figueroa-Bossi, (2007) A small RNA downregulates LamB maltoporin in *Salmonella. Mol Microbiol* **65**: 799–810.

Bouvier, M., C. M. Sharma, F. Mika, K. H. Nierhaus & J. Vogel (2008) Small RNA binding to 5' mRNA coding region inhibits translational initiation. *Mol Cell* **32**: 827–837.

Boysen, A., J. Møller-Jensen, B. Kallipolitis, P. Valentin-Hansen & M. Overgaard (2010) Translational regulation of gene expression by an anaerobically induced small non-coding RNA in *Escherichia coli. J Biol Chem* **285**: 10690–10702.

Brennan, R. G. & T. M. Link (2007) Hfq structure, function and ligand binding. *Curr Opin Microbiol* **10**: 125–133.

Brown, L. & T. Elliott (1997) Mutations that increase expression of the rpoS gene and decrease its dependence on *hfq* function in *Salmonella. J Bacteriol* **179**.

Cai, Y., M. L. Wilson & J. Peccoud (2010) GenoCAD for iGEM: a grammatical approach to the design of standard-compliant constructs. *Nucleic Acids Res* **38**: 2637–2644.

Callaghan, A. J., M. J. Marcaida, J. A. Stead, K. J. McDowall, W. G. Scott & B. F. Luisi (2005) Structure of *Escherichia coli* RNase E catalytic domain and implications for RNA turnover. *Nature* **437**: 1187–1191.

Carmichael, G. G., K. Weber, A. Niveleau & A. J. Wahba (1975) The host factor required for RNA phage Qbeta RNA replication *in vitro*. Intracellular location, quantitation, and purification by polyadenylate-cellulose chromatography. *J Biol Chem* **250**: 3607–3612.

Carpousis, A. J. (2007) The RNA degradosome of *Escherichia coli*: an mRNA-degrading machine assembled on RNase E. *Annu Rev Microbiol* **61**: 71–87.

Chao, Y. & J. Vogel (2010) The role of Hfq in bacterial pathogens. *Curr Opin Microbiol* **13**: 24–33.

Chen, S., A. Zhang, L. B. Blyn & G. Storz (2004) MicC, a second small-RNA regulator of Omp protein expression in *Escherichia coli. J Bacteriol* **186**: 6689–6697.

Coornaert, A., A. Lu, P. Mandin, M. Springer, S. Gottesman & M. Guillier (2010) MicA sRNA links the PhoP regulon to cell envelope stress. *Mol Microbiol* **76**: 467–479.

Davis, B. M., M. Quinones, J. Pratt, Y. Ding & M. K. Waldor (2005) Characterization of the small untranslated RNA RyhB and its regulon in *Vibrio cholerae. J Bacteriol* **187**: 4005–4014.

Davis, B. M. & M. K. Waldor (2007) RNase E-dependent processing stabilizes MicX, a *Vibrio cholerae* sRNA. *Mol Microbiol* **65**: 373–385.

De Lay, N. & S. Gottesman (2009) The Crp-activated small noncoding regulatory RNA CyaR (RyeE) links nutritional status to group behavior. *J Bacteriol* **191**: 461–476.

Deana, A. & J. G. Belasco (2005) Lost in translation: the influence of ribosomes on bacterial mRNA decay. *Genes Dev* **19**: 2526–2533.

Desnoyers, G., A. Morissette, K. Prévost & E. Massé (2009) Small RNA-induced differential degradation of the polycistronic mRNA *iscRSUA. EMBO J* **28**: 1551–1561.

Deveau, H., J. E. Garneau & S. Moineau (2010) CRISPR/Cas system and its role in phage-bacteria interactions. *Annu Rev Microbiol* **64**: 475–493.

Dillon, S. C., A. D. S. Cameron, K. Hokamp, S. Lucchini, J. C. Hinton & C. J. Dorman (2010) Genome-wide analysis of the H-NS and Sfh regulatory networks in *Salmonella* Typhimurium identifies a plasmid-encoded transcription silencing mechanism. *Mol Microbiol* **76**: 1250–1265.

Dorman, C.J. (2007). H-NS, the genome sentinel. *Nat Rev Microbiol* **5**(2): 157–61.

Ding, Y., B. M. Davis & M. K. Waldor (2004) Hfq is essential for *Vibrio cholerae* virulence and downregulates sigma expression. *Mol Microbiol* **53**: 345–354.

Douchin, V., C. Bohn & P. Bouloc (2006) Down-regulation of porins by a small RNA bypasses the essentiality of the regulated intramembrane proteolysis protease RseP in *Escherichia coli. J Biol Chem* **281**: 12253–12259.

Doyle, M., M. Fookes, A. Ivens, M. W. Mangan, J. Wain & C. J. Dorman (2007) An H-NS-like stealth protein aids horizontal DNA transmission in bacteria. *Science* **315**: 251–252.

Durand, S. & G. Storz (2010) Reprogramming of anaerobic metabolism by the FnrS small RNA. *Mol Microbiol* **75**: 1215–1231.

Fender, A., J. Elf, K. Hampel, B. Zimmermann & E. G. Wagner (2010) RNAs actively cycle on the Sm-like protein Hfq. *Genes Dev* **24**: 2621–2626.

Figueroa-Bossi, N., S. Lemire, D. Maloriol, R. Balbontín, J. Casadesús & L. Bossi (2006) Loss of Hfq activates the sigmaE-dependent envelope stress response in *Salmonella enterica. Mol Microbiol* **62**: 838–852.

Figueroa-Bossi, N., M. Valentini, L. Malleret, F. Fiorini & L. Bossi (2009) Caught at its own game: regulatory small RNA inactivated by an inducible transcript mimicking its target. *Genes Dev* **23**: 2004–2015.

Folichon, M., V. Arluison, O. Pellegrini, E. Huntzinger, P. Regnier & E. Hajnsdorf (2003) The poly(A) binding protein Hfq protects RNA from RNase E and exoribonucleolytic degradation. *Nucleic Acids Res* **31**: 7302–7310.

Friedman, R. C., K. K. Farh, C. B. Burge & D. P. Bartel (2009) Most mammalian mRNAs are conserved targets of microRNAs. *Genome Res* **19**: 92–105.

Frohlich, K. S. & J. Vogel, (2009) Activation of gene expression by small RNA. *Curr Opin Microbiol* **12**: 674–682.

Geissmann, T. A. & D. Touati (2004) Hfq, a new chaperoning role: binding to messenger RNA determines access for small RNA regulator. *EMBO J* **23**: 396–405.

Görke, B. & J. Stülke (2008) Carbon catabolite repression in bacteria: many ways to make the most out of nutrients. *Nat Rev Microbiol* **6**: 613–624.

Grundy, F. J. & T. M. Henkin (2006) From ribosome to riboswitch: control of gene expression in bacteria by RNA structural rearrangements. *Crit Rev Biochem Mol Biol* **41**: 329–338.

Guillier, M. & S. Gottesman (2006) Remodelling of the *Escherichia coli* outer membrane by two small regulatory RNAs. *Mol Microbiol* **59**: 231–247.

Guillier, M. & S. Gottesman (2008) The 5‘ end of two redundant sRNAs is involved in the regulation of multiple targets, including their own regulator. *Nucleic Acids Res* **36**: 6781–6794.

Guisbert, E., V. A. Rhodius, N. Ahuja, E. Witkin & C. A. Gross (2007) Hfq modulates the sigmaE-mediated envelope stress response and the sigma32-mediated cytoplasmic stress response in *Escherichia coli. J Bacteriol* **189**: 1963–1973.

Hammer, B. K. & B. L. Bassler (2007) Regulatory small RNAs circumvent the conventional quorum sensing pathway in pandemic *Vibrio cholerae. Proc Natl Acad Sci U S A* **104**: 11145–11149.

Hartog, E., L. Ben-Shalom, D. Shachar, K. R. Matthews & S. Yaron (2008) Regulation of *marA, soxS, rob, acrAB* and *micF* in *Salmonella enterica* serovar *Typhimurium. Microbiol Immunol* **52**: 565–574.

Holmqvist, E., J. Reimegard, M. Sterk, N. Grantcharova, U. Romling & E. G. Wagner (2010) Two antisense RNAs target the transcriptional regulator CsgD to inhibit curli synthesis. *EMBO J* **29**: 1840–1850.

Huang, B., H. Babcock & X. Zhuang (2010) Breaking the Diffraction Barrier: Super-Resolution Imaging of *Cells Cell* **143**: 1047–1058.

Hussein, R. & H. N. Lim (2011) Disruption of small RNA signaling caused by competition for Hfq. Proc *Natl Acad Sci U S A* **108**: 1110–1115.

Hüttenhofer, A. & H. F. Noller (1994) Footprinting mRNA-ribosome complexes with chemical probes. *EMBO J* **13**: 3892–3901.

Jacques, J. F., S. Jang, K. Prévost, G. Desnoyers, M. Desmarais, J. Imlay & E. Massé (2006) RyhB small RNA modulates the free intracellular iron pool and is essential for normal growth during iron limitation in *Escherichia coli. Mol Microbiol* **62**: 1181–1190.

Joelsson, A., Z. Liu & J. Zhu (2006) Genetic and phenotypic diversity of quorum-sensing systems in clinical and environmental isolates of *Vibrio cholerae. Infect Immun* **74**: 1141–1147.

Johansen, J., M. Eriksen, B. Kallipolitis & P. Valentin-Hansen (2008) Down-regulation of outer membrane proteins by noncoding RNAs: unraveling the cAMP-CRP- and sigmaE-dependent CyaR-*ompX* regulatory case. *J Mol Biol* **383**: 1–9.

Johansen, J., A. A. Rasmussen, M. Overgaard & P. Valentin-Hansen (2006) Conserved small non-coding RNAs that belong to the sigmaE regulon: role in down-regulation of outer membrane proteins. *J Mol Biol* **364**: 1–8.

Jones, A. M., A. Goodwill & T. Elliott (2006) Limited role for the DsrA and RprA regulatory RNAs in *rpoS* regulation in *Salmonella enterica. J Bacteriol* **188**: 5077–5088.

Kajitani, M., A. Kato, A. Wada, Y. Inokuchi & A. Ishihama (1994) Regulation of the *Escherichia coli hfq* gene encoding the host factor for phage Q beta. *J Bacteriol* **176**: 531–534.

Kalamorz, F., B. Reichenbach, W. Marz, B. Rak & B. Görke (2007) Feedback control of glucosamine-6-phosphate synthase GlmS expression depends on the small RNA GlmZ and involves the novel protein YhbJ in *Escherichia coli. Mol Microbiol* **65**: 1518–1533.

Kawamoto, H., T. Morita, A. Shimizu, T. Inada & H. Aiba (2005) Implication of membrane localization of target mRNA in the action of a small RNA: mechanism of post-transcriptional regulation of glucose transporter in *Escherichia coli. Genes Dev* **19**: 328–338.

Khalil, A. S. & J. J. Collins (2010) Synthetic biology: applications come of age. *Nat Rev Genet* **11**: 367–379.

Laursen, B. S., H. P. Sorensen, K. K. Mortensen & H. U. Sperling-Petersen (2005) Initiation of protein synthesis in bacteria. *Microbiol Mol Biol Rev* **69**: 101–123.

Lease, R. A., M. E. Cusick & M. Belfort (1998) Riboregulation in Escherichia coli: DsrA acts by RNA:RNA interactions at multiple loci. *Proc Natl Acad Sci U S A* **95**: 12456–12461.

Lease, R. A. & M. Belfort (2000) A trans-acting RNA as a control switch in Escherichia coli: DsrA modulates function by forming alternative structures. *Proc Natl Acad Sci U S A* **97**: 9919–9924.

Lease, R. A., D. Smith, K. McDonough & M. Belfort (2004) The small noncoding DsrA RNA is an acid resistance regulator in *Escherichia coli. J Bacteriol* **186**: 6179–6185.

Lee, D. R. & C. A. Schnaitman (1980) Comparison of outer membrane porin proteins produced by *Escherichia coli and Salmonella typhimurium. J Bacteriol* **142**: 1019–1022.

Lee, H. J., H. J. Jeon, S. C. Ji, S. H. Yun & H. M. Lim (2008) Establishment of an mRNA gradient depends on the promoter: an investigation of polarity in gene expression. *J Mol Biol* **378**: 318–327.

Lenz, D. H., K. C. Mok, B. N. Lilley, R. V. Kulkarni, N. S. Wingreen & B. L. Bassler (2004) The small RNA chaperone Hfq and multiple small RNAs control quorum sensing in *Vibrio harveyi* and *Vibrio cholerae. Cell* **118**: 69–82.

Lewis, B. P., I. H. Shih, M. W. Jones-Rhoades, D. P. Bartel & C. B. Burge (2003) Prediction of mammalian microRNA targets. *Cell* **115**: 787–798.

Lybecker, M. C., C. A. Abel, A. L. Feig & D. S. Samuels (2010) Identification and function of the RNA chaperone Hfq in the Lyme disease spirochete *Borrelia burgdorferi. Mol Microbiol* **78**: 622–635.

Majdalani, N., C. Cunning, D. Sledjeski, T. Elliott & S. Gottesman (1998) DsrA RNA regulates translation of RpoS message by an anti-antisense mechanism independent of its action as an antisilencer of transcription. *Proc Natl Acad Sci U S A* **95**: 12462–12467.

Majdalani, N., S. Chen, J. Murrow, K. St John & S. Gottesman (2001) Regulation of RpoS by a novel small RNA: the characterization of RprA. *Mol Microbiol* **39**: 1382–1394.

Majdalani, N., D. Hernandez & S. Gottesman (2002) Regulation and mode of action of the second small RNA activator of RpoS translation, RprA. *Mol Microbiol* **46**: 813–826.

Maki, K., K. Uno, T. Morita & H. Aiba (2008) RNA, but not protein partners, is directly responsible for silencing by a bacterial Hfq-binding small RNA. *Proc Natl Acad Sci U S A* **105**: 10332–10337.

Mandin, P. & S. Gottesman (2010) Integrating anaerobic/aerobic sensing and the general stress response through the ArcZ small RNA. *EMBO J* **29**: 3094–3107.

Martin-Farmer, J. & G. R. Janssen (1999) A downstream CA repeat sequence increases translation from leadered and unleadered mRNA in *Escherichia coli. Mol Microbiol* **31**: 1025–1038.

Massé, E. & S. Gottesman (2002) A small RNA regulates the expression of genes involved in iron metabolism in *Escherichia coli. Proc Natl Acad Sci U S A* **99**: 4620–4625.

Massé, E., F. E. Escorcia & S. Gottesman (2003) Coupled degradation of a small regulatory RNA and its mRNA targets in *Escherichia coli. Genes Dev* **17**: 2374–2383.

Massé, E., C. K. Vanderpool & S. Gottesman (2005) Effect of RyhB small RNA on global iron use in *Escherichia coli. J Bacteriol* **20**.

Mecsas, J., P. E. Rouviere, J. W. Erickson, T. J. Donohue & C. A. Gross (1993) The activity of sigma E, an *Escherichia coli* heat-inducible sigma-factor, is modulated by expression of outer membrane proteins. *Genes Dev* **7**: 2618–2628.

Metruccio, M. M., L. Fantappie, D. Serruto, A. Muzzi, D. Roncarati, C. Donati, V. Scarlato & I. Delany (2009) The Hfq-dependent small noncoding RNA NrrF directly mediates Fur-dependent positive regulation of succinate dehydrogenase in Neisseria meningitidis. *J Bacteriol* **191**: 1330–1342.

Missiakas, D., J. M. Betton & S. Raina (1996) New components of protein folding in extracytoplasmic compartments of *Escherichia coli* SurA, FkpA and Skp/OmpH. *Mol Microbiol* **21**: 871–884.

Mitarai, N., J. A. M. Benjamin, S. Krishna, S. Semsey, Z. Csiszovszki, E. Massé & K. Sneppen (2009) Dynamic features of gene expression control by small regulatory RNAs. *Proc Natl Acad Sci U S A* **106**: 10655–10659.

Mizuno, T., M. Y. Chou & M. Inouye (1984) A unique mechanism regulating gene expression: translational inhibition by a complementary RNA transcript (micRNA). *Proc Natl Acad Sci U S A* **81**: 1966–1970.

Moll, I., T. Afonyushkin, O. Vytvytska, V. R. Kaberdin & U. Bläsi (2003) Coincident Hfq binding and RNase E cleavage sites on mRNA and small regulatory RNAs. *RNA* **9**: 1308–1314.

Møller, T., T. Franch, C. Udesen, K. Gerdes & P. Valentin-Hansen (2002a) Spot 42 RNA mediates discoordinate expression of the *E. coli* galactose operon. *Genes Dev* **16**: 1696–1706.

Møller, T., T. Franch, P. Hojrup, D. R. Keene, H. P. Bachinger, R. G. Brennan & P. Valentin-Hansen, (2002b) Hfq: a bacterial Sm-like protein that mediates RNA-RNA interaction. *Mol Cell* **9**: 23–30.

Montero Llopis, P. M., A. F. Jackson, O. Sliusarenko, I. Surovtsev, J. Heinritz, T. Emonet & C. Jacobs-Wagner (2010) Spatial organization of the flow of genetic information in bacteria. *Nature* **466**: 77–81.

Moon, K. & S. Gottesman (2009) A PhoQ/P-regulated small RNA regulates sensitivity of *Escherichia coli* to antimicrobial peptides. *Mol Microbiol* **74**: 1314–1330.

Morita, T., K. Maki & H. Aiba (2005) RNase E-based ribonucleoprotein complexes: mechanical basis of mRNA destabilization mediated by bacterial noncoding RNAs. *Genes Dev* **19**: 2176–2186.

Murphy, E. R. & S. M. Payne (2007) RyhB, an iron-responsive small RNA molecule, regulates *Shigella dysenteriae* virulence. *Infect Immun* **75**: 3470–3477.

Nechooshtan, G., M. Elgrably-Weiss, A. Sheaffer, E. Westhof & S. Altuvia (2009) A pH-responsive riboregulator. *Genes Dev* **23**: 2650–2662.

Nevo-Dinur, K., A. Nussbaum-Shochat, S. Ben-Yehuda & O. Amster-Choder (2011) Translation-independent localization of mRNA in *E. coli. Science* **331**: 1081–1084.

Ng, W. L. & B. L. Bassler (2009) Bacterial quorum-sensing network architectures. *Annu Rev Genet* **43**: 197–222.

Opdyke, J. A., J. G. Kang & G. Storz (2004) GadY, a small-RNA regulator of acid response genes in *Escherichia coli. J Bacteriol* **186**: 6698–6705.

Overgaard, M., J. Johansen, J. Møller-Jensen & P. Valentin-Hansen (2009) Switching off small RNA regulation with trap-mRNA. *Mol Microbiol* **73**: 790–800.

Papenfort, K., V. Pfeiffer, F. Mika, S. Lucchini, J. C. Hinton & J. Vogel (2006) Sigma(E)-dependent small RNAs of *Salmonella* respond to membrane stress by accelerating global *omp* mRNA decay. *Mol Microbiol* **62**: 1674–1688.

Papenfort, K., V. Pfeiffer, S. Lucchini, A. Sonawane, J. C. Hinton & J. Vogel (2008) Systematic deletion of *Salmonella* small RNA genes identifies CyaR, a conserved CRP-dependent riboregulator of OmpX synthesis. *Mol Microbiol* **68**: 890–906.

Papenfort, K., N. Said, T. Welsink, S. Lucchini, J. C. Hinton & J. Vogel (2009) Specific and pleiotropic patterns of mRNA regulation by ArcZ, a conserved, Hfq-dependent small RNA. *Mol Microbiol* **74**: 139–158.

Papenfort, K. & J. Vogel (2010) Regulatory RNA in bacterial pathogens. *Cell Host Microbe* **8**: 116–127.

Papenfort, K., M. Bouvier, F. Mika, C. M. Sharma & J. Vogel (2010) Evidence for an autonomous 5' target recognition domain in an Hfq-associated small RNA. *Proc Natl Acad Sci U S A* **107**: 20435–20440.

Park, H.S., Y. Ostberg, J. Johansson, E.G. Wagner & B.E. Uhlin (2010). Novel role for a bacterial nucleoid protein in translation of mRNAs with suboptimal ribosome-binding sites. *Genes Dev* **24**(13): 1345–50.

Pfeiffer, V., A. Sittka, R. Tomer, K. Tedin, V. Brinkmann & J. Vogel (2007) A small non-coding RNA of the invasion gene island (SPI-1) represses outer membrane protein synthesis from the *Salmonella* core genome. *Mol Microbiol* **66**: 1174–1191.

Pfeiffer, V., K. Papenfort, S. Lucchini, J. C. Hinton & J. Vogel (2009) Coding sequence targeting by MicC RNA reveals bacterial mRNA silencing downstream of translational initiation. *Nat Struct Mol Biol* **16**: 840–846.

Prévost, K., H. Salvail, G. Desnoyers, J. F. Jacques, E. Phaneuf & E. Massé (2007) The small RNA RyhB activates the translation of *shiA* mRNA encoding a permease of shikimate, a compound involved in siderophore synthesis. *Mol Microbiol* **64**: 1260–1273.

Prévost, K., G. Desnoyers, J. F. Jacques, F. Lavoie & E. Massé (2011) Small RNA-induced mRNA degradation achieved through both translation block and activated cleavage. *Genes Dev.*

Pulvermacher, S. C., L. T. Stauffer & G. V. Stauffer (2008) The role of the small regulatory RNA GcvB in GcvB/mRNA posttranscriptional regulation of *oppA* and *dppA* in *Escherichia coli. FEMS Microbiol Lett* **281**: 42–50.

Pulvermacher, S. C., L. T. Stauffer & G. V. Stauffer (2009a) Role of the *Escherichia coli* Hfq protein in GcvB regulation of *oppA* and *dppA* mRNAs. *Microbiology* **155**: 115–123.

Pulvermacher, S. C., L. T. Stauffer & G. V. Stauffer (2009b) Role of the sRNA GcvB in regulation of *cycA* in *Escherichia coli. Microbiology* **155**: 106–114.

Raivio, T. L., D. L. Popkin & T. J. Silhavy, (1999) The Cpx envelope stress response is controlled by amplification and feedback inhibition. *J Bacteriol* **181**: 5263–5272.

Rasmussen, A. A., M. Eriksen, K. Gilany, C. Udesen, T. Franch, C. Petersen & P. Valentin-Hansen (2005) Regulation of *ompA* mRNA stability: the role of a small regulatory RNA in growth phase-dependent control. *Mol Microbiol* **58**: 1421–1429.

Rasmussen, A. A., J. Johansen, J. S. Nielsen, M. Overgaard, B. Kallipolitis & P. Valentin-Hansen (2009) A conserved small RNA promotes silencing of the outer membrane protein YbfM. *Mol Microbiol* **72**: 566–577.

Reichenbach, B., A. Maes, F. Kalamorz, E. Hajnsdorf & B. Görke (2008) The small RNA GlmY acts upstream of the sRNA GlmZ in the activation of *glmS* expression and is subject to regulation by polyadenylation in Escherichia coli. *Nucleic Acids Res* **36**: 2570–2580.

Rhodius, V. A., W. C. Suh, G. Nonaka, J. West & C. A. Gross (2006) Conserved and variable functions of the sigmaE stress response in related genomes. *PLoS Biol* **4**: e2.

Rice, J. B. & C. K. Vanderpool (2011) The small RNA SgrS controls sugar-phosphate accumulation by regulating multiple PTS genes. *Nucleic Acids Res. doi: 10.1093/nar/gkq1219*

Rowley, G., M. Spector, J. Kormanec & M. Roberts (2006) Pushing the envelope: extracytoplasmic stress responses in bacterial pathogens. *Nat Rev Microbiol* **4**: 383–394.

Sahagan, B. G. & J. E. Dahlberg (1979a) A small, unstable RNA molecule of *Escherichia coli*: spot 42 RNA. I. Nucleotide sequence analysis. *J Mol Biol* **131**: 573–592.

Sahagan, B. G. & J. E. Dahlberg (1979b) A small, unstable RNA molecule of *Escherichia coli*: spot 42 RNA. II. Accumulation and distribution. *J Mol Biol* **131**: 593–605.

Salvail, H., P. Lanthier-Bourbonnais, J. M. Sobota, M. Caza, J. A. M. Benjamin, M. E. S. Mendieta, F. Lepine, C. M. Dozois, J. Imlay & E. Massé (2010) A small RNA promotes siderophore production through transcriptional and metabolic remodeling. *Proc Natl Acad Sci U S A* **107**: 15223–15228.

Schmidt, M., P. Zheng & N. Delihas (1995) Secondary structures of *Escherichia coli* antisense *micF* RNA, the 5'-end of the target ompF mRNA, and the RNA/RNA duplex. *Biochemistry* **34**: 3621–3631.
Sharma, C. M., F. Darfeuille, T. H. Plantinga & J. Vogel (2007) A small RNA regulates multiple ABC transporter mRNAs by targeting C/A-rich elements inside and upstream of ribosome-binding sites. *Genes Dev* **21**: 2804–2817.
Sharma, C. M. & J. Vogel (2009) Experimental approaches for the discovery and characterization of regulatory small RNA. *Curr Opin Microbiol* **12**: 536–546.
Sharma, C. M., S. Hoffmann, F. Darfeuille, J. Reignier, S. Findeiss, A. Sittka, S. Chabas, K. Reiche, J. Hackermüller, R. Reinhardt, P. F. Stadler & J. Vogel (2010) The primary transcriptome of the major human pathogen *Helicobacter pylori. Nature* **464**: 250–255.
Shetty, R. P., D. Endy & T. F. Knight, Jr. (2008) Engineering BioBrick vectors from BioBrick parts. *J Biol Eng* **2**: 5.
Shine, J. & L. Dalgarno (1974) The 3'-terminal sequence of *Escherichia coli* 16S ribosomal RNA: complementarity to nonsense triplets and ribosome binding sites. *Proc Natl Acad Sci U S A* **71**: 1342–1346.
Sittka, A., V. Pfeiffer, K. Tedin & J. Vogel (2007) The RNA chaperone Hfq is essential for the virulence of *Salmonella typhimurium. Mol Microbiol* **63**: 193–217.
Sittka, A., S. Lucchini, K. Papenfort, C. M. Sharma, K. Rolle, T. T. Binnewies, J. C. Hinton & J. Vogel (2008) Deep sequencing analysis of small noncoding RNA and mRNA targets of the global post-transcriptional regulator, Hfq. *PLoS Genet* **4**: e1000163.
Sittka, A., C. M. Sharma, K. Rolle & J. Vogel (2009) Deep sequencing of *Salmonella* RNA associated with heterologous Hfq proteins in vivo reveals small RNAs as a major target class and identifies RNA processing phenotypes. *RNA Biol* **6**: 266–275.
Sledjeski, D. D., A. Gupta & S. Gottesman (1996) The small RNA, DsrA, is essential for the low temperature expression of RpoS during exponential growth in *Escherichia coli. EMBO J* **15**: 3993–4000.
Sledjeski, D. D., C. Whitman & A. Zhang (2001) Hfq is necessary for regulation by the untranslated RNA DsrA. *J Bacteriol* **183**: 1997–2005.
Soper, T., P. Mandin, N. Majdalani, S. Gottesman & S. A. Woodson (2010) Positive regulation by small RNAs and the role of Hfq. *Proc Natl Acad Sci U S A* **107**: 9602–9607.
Svenningsen, S. L., K. C. Tu & B. L. Bassler (2009) Gene dosage compensation calibrates four regulatory RNAs to control *Vibrio cholerae* quorum sensing. *EMBO J* **28**: 429–439.
Svenningsen, S. L., C. M. Waters & B. L. Bassler (2008) A negative feedback loop involving small RNAs accelerates Vibrio cholerae's transition out of quorum-sensing mode. *Genes Dev* **22**: 226–238.
Takyar, S., R. P. Hickerson & H. F. Noller (2005) mRNA helicase activity of the ribosome. *Cell* **120**: 49–58.
Thomason, M. K. & G. Storz (2010) Bacterial antisense RNAs: how many are there, and what are they doing? *Annu Rev Genet* **44**: 167–188.
Thompson, K. M., V. A. Rhodius & S. Gottesman (2007) SigmaE regulates and is regulated by a small RNA in *Escherichia coli. J Bacteriol* **189**: 4243–4256.
Tramonti, A., M. De Canio & D. De Biase (2008) GadX/GadW-dependent regulation of the *Escherichia coli* acid fitness island: transcriptional control at the *gadY-gadW* divergent promoters and identification of four novel 42 bp GadX/GadW-specific binding sites. *Mol Microbiol* **70**: 965–982.
Tu, K. C. & B. L. Bassler (2007) Multiple small RNAs act additively to integrate sensory information and control quorum sensing in *Vibrio harveyi. Genes Dev* **21**: 221–233.
Tu, K. C., C. M. Waters, S. L. Svenningsen & B. L. Bassler (2008) A small-RNA-mediated negative feedback loop controls quorum-sensing dynamics in Vibrio harveyi. *Mol Microbiol* **70**: 896–907.
Tu, K. C., T. Long, S. L. Svenningsen, N. S. Wingreen & B. L. Bassler (2010) Negative feedback loops involving small regulatory RNAs precisely control the *Vibrio harveyi* quorum-sensing response. *Mol Cell* **37**: 567–579.

Udekwu, K. I., F. Darfeuille, J. Vogel, J. Reimegard, E. Holmqvist & E. G. Wagner (2005) Hfq-dependent regulation of OmpA synthesis is mediated by an antisense RNA. *Genes Dev* **19**: 2355–2366.

Udekwu, K. I. & E. G. Wagner (2007) Sigma E controls biogenesis of the antisense RNA MicA. *Nucleic Acids Res* **35**: 1279–1288.

Ullmann, A., E. Joseph & A. Danchin (1979) Cyclic AMP as a modulator of polarity in polycistronic transcriptional units. *Proc Natl Acad Sci U S A* **76**: 3194–3197.

Urban, J. H., K. Papenfort, J. Thomsen, R. A. Schmitz & J. Vogel (2007) A conserved small RNA promotes discoordinate expression of the *glmUS* operon mRNA to activate GlmS synthesis. *J Mol Biol* **373**: 521–528.

Urban, J. H. & J. Vogel (2007) Translational control and target recognition by *Escherichia coli* small RNAs *in vivo*. *Nucleic Acids Res* **35**: 1018–1037.

Urban, J. H. & J. Vogel (2008) Two seemingly homologous noncoding RNAs act hierarchically to activate *glmS* mRNA translation. *PLoS Biol* **6**: e64.

Urbanowski, M. L., L. T. Stauffer & G. V. Stauffer (2000) The *gcvB* gene encodes a small untranslated RNA involved in expression of the dipeptide and oligopeptide transport systems in *Escherichia coli*. *Mol Microbiol* **37**: 856–868.

Valentin-Hansen, P., M. Eriksen & C. Udesen (2004) The bacterial Sm-like protein Hfq: a key player in RNA transactions. *Mol Microbiol* **51**: 1525–1533.

Vanderpool, C. K. & S. Gottesman (2004) Involvement of a novel transcriptional activator and small RNA in post-transcriptional regulation of the glucose phosphoenolpyruvate phosphotransferase system. *Mol Microbiol* **54**: 1076–1089.

Vecerek, B., I. Moll & U. Bläsi (2007) Control of Fur synthesis by the non-coding RNA RyhB and iron-responsive decoding. *EMBO J* **26**: 965–975.

Vogel, J., V. Bartels, T. H. Tang, G. Churakov, J. G. Slgater-Jäger, A. Huttenhofer & E. G. Wagner (2003) RNomics in *Escherichia coli* detects new sRNA species and indicates parallel transcriptional output in bacteria. *Nucleiceic Acids Res* **31**: 6435–6443.

Vogel, J. & K. Papenfort (2006) Small non-coding RNAs and the bacterial outer membrane. *Curr Opin Microbiol* **9**: 605–611.

Vytvytska, O., J. S. Jakobsen, G. Balcunaite, J. S. Andersen, M. Baccarini & A. von Gabain (1998) Host factor I, Hfq, binds to *Escherichia coli ompA* mRNA in a growth rate-dependent fashion and regulates its stability. *Proc Natl Acad Sci U S A* **95**: 14118–14123.

Vytvytska, O., I. Moll, V. R. Kaberdin, A. von Gabain & U. Bläsi (2000) Hfq (HF1) stimulates *ompA* mRNA decay by interfering with ribosome binding. *Genes Dev* **14**: 1109–1118.

Wadler, C. S. & C. K. Vanderpool (2007) A dual function for a bacterial small RNA: SgrS performs base pairing-dependent regulation and encodes a functional polypeptide. *Proc Natl Acad Sci U S A* **104**: 20454–20459.

Wassarman, K. M., F. Repoila, C. Rosenow, G. Storz & S. Gottesman (2001) Identification of novel small RNAs using comparative genomics and microarrays. *Genes Dev* **15**: 1637–1651.

Waters, L. S. & G. Storz (2009) Regulatory RNAs in bacteria. *Cell* **136**: 615–628.

Wilderman, P. J., N. A. Sowa, D. J. FitzGerald, P. C. FitzGerald, S. Gottesman, U. A. Ochsner & M. L. Vasil (2004) Identification of tandem duplicate regulatory small RNAs in *Pseudomonas aeruginosa* involved in iron homeostasis. *Proc Natl Acad Sci U S A* **101**: 9792–9797.

Zhang, A., S. Altuvia, A. Tiwari, L. Argaman, R. Hengge-Aronis & G. Storz (1998) The OxyS regulatory RNA represses *rpoS* translation and binds the Hfq (HF-1) protein. *EMBO J* **17**: 6061–6068.

Zhang, A., K. M. Wassarman, J. Ortega, A. C. Steven & G. Storz (2002) The Sm-like Hfq protein increases OxyS RNA interaction with target mRNAs. *Mol Cell* **9**: 11–22.

Zhang, A., K. M. Wassarman, C. Rosenow, B. C. Tjaden, G. Storz & S. Gottesman (2003) Global analysis of small RNA and mRNA targets of Hfq. *Mol Microbiol* **50**: 1111–1124.

Chapter 3
A Current Overview of Regulatory RNAs in Staphylococcus Aureus

Isabelle Caldelari*, Pierre Fechter*, Efthimia Lioliou*, Cédric Romilly*, Clément Chevalier*, Christine Gaspin**, and Pascale Romby*

1 Introduction

Staphylococcus aureus is a common commensal bacterial species that is usually found in the nose and on the skin of 30 % of healthy adults. However, the bacterium is a remarkably versatile pathogen that is one of the main causes of community as well as hospital-acquired infections (Cheung *et al.*, 2004; Novick, 2003). *S. aureus* is responsible for systemic infections such as sepsis and endocarditis, which can be difficult to treat due to the acquisition of resistance towards numerous antibiotics in clinical use. *S. aureus* causes a wide spectrum of human diseases in part due to its ability to produce an array of virulence factors, which are mostly encoded by laterally acquired genomic regions, the so-called pathogenicity islands. These factors include surface proteins responsible for the adhesion and invasion of the host, exoproteins required for host immune evasion, and toxins involved in dissemination in host tissues and acquisition of nutrients (Novick, 2003). Redundancies exist to ensure that a productive infection still occurs even though one factor may be lost. In recent decades, many studies have been carried out to understand how *S. aureus* is able to coordinate the expression of a large panel of virulence factors at the appropriate time in order to facilitate successful infections (Novick and Geisinger, 2008). These works offer the possibility for developing anti-virulence therapies as alternative strategies for affecting the bacteria viability, i. e. by inhibiting the expression of the virulence factors that cause host damage or the interaction between the pathogen and the host (Clatworthy *et al.*, 2007). Inhibiting virulence instead of viability may have little impact on human flora and result in weaker selective pressure for the development of antibiotic resistance. Hence, determining the regulatory networks and the dynamics involved in virulence and in fast adaptive responses are of prime importance to combating *S. aureus* infections.

* Architecture et Réactivité de l'ARN, Université de Strasbourg, CNRS, IBMC, 15 rue René Descartes, F-67084 Strasbourg, France.

** INRA, UR 875, F-31320 Castanet-Tolosan, France.

The regulation of virulence gene expression and stress response pathways in *S. aureus* involves complex interactions between multiple *trans*-acting regulators. These factors include two-component systems, regulatory proteins, peptide secondary metabolites, and RNAs (Novick and Geisinger, 2008; Wyatt *et al.*, 2010). The ability of bacteria to use RNA to regulate virulence gene expression is a relatively new concept (Vogel, 2009; Romby and Charpentier, 2010; Toledo-Arana *et al.*, 2007). However, research on *S. aureus* has gained interest since 1993 with the discovery of one of the largest regulatory RNAs (RNAIII), the main intracellular effector of the quorum-sensing system, which controls the temporal expression of numerous virulence genes (Novick *et al.*, 1993). Besides this multifunctional regulatory RNA, recent studies have demonstrated that it is likely that the *S. aureus* genome encodes a high diversity of RNAs. These studies include *in silico* predictions coupled with expression studies (Livny *et al.*, 2008; Marchais *et al.*, 2009; Geissmann *et al.*, 2009a; Pichon and Felden, 2005), microarrays (Anderson *et al.*, 2006; Roberts *et al.*, 2006), and cDNA libraries combined with high throughput sequencing approaches (Beaume *et al.*, 2010; Abu-Qatouseh *et al.*, 2010; Bohn *et al.*, 2010). Taken together, these RNAs[1] include *cis*-acting regulatory regions of mRNAs, *cis*-acting antisense RNAs (asRNA), and small non-coding RNAs (ncRNAs). In addition, RNAs containing small open reading frames (ORF) and overlapping untranslated regions (UTR) were also identified (Beaume *et al.*, 2010).

In this review, we will provide a summary of the newly identified ncRNAs and will give an overview of the functions they regulate. Our particular focus will be on RNAs that are involved in the regulation of metabolism, stress response, DNA maintenance, defense mechanisms against foreign DNA, and virulence.

2 Cis-acting Regulatory Elements in mRNAs

mRNA is now recognized as the main actor of post-transcriptional events. Since the folding of mRNA gradually proceeds during the transcription process, the elongation speed of RNA polymerase and specific pauses by the enzyme influence the mRNA folding, the formation of alternative mRNA structures, and the access of *trans*-acting regulators. Furthermore, transcription-translation synchronization plays an essential role in regulation (Iost and Dreyfus, 1995; Proshkin *et al.*, 2010). This is likely to be the reason why the 5' untranslated regions (5'-UTR) of numerous mRNAs are quite structured and have evolved dedicated regulatory sites. These regions of mRNAs respond to a wide variety of *trans*-acting factors (e.g., intracellular concentration of metabolites (riboswitches), divalent ions, uncharged tRNAs (T-boxes), and RNA-binding proteins) and to environmental cues (temperature (thermosensors)). They act as direct sensors of the physical and metabolic states of the cell via conformational changes (Breaker, 2009; Ramesh and Winkler, 2010; Smith *et al.*, 2010; Narberhaus, 2010; Marzi *et al.*, 2008). Specific examples found in *S. aureus* are given below.

---

[1] sRNA is the generic name given for regulatory non-coding RNAs including *cis*-acting and *trans*-acting RNAs

2.1 RNA Thermosensors

These regulatory elements are often situated at the 5'-UTR of mRNAs encoding heat or cold-shock proteins, virulence factors, or proteins involved in bacteriophage development and lysogeny (Klinkert and Narberhaus, 2009). They all share a common mechanism and control translational initiation in response to temperature changes. At low temperatures, the RNA is in an inactive state due to the fact that the Shine-Dalgarno (SD) sequence is trapped in a hairpin structure that opens up gradually when the temperature increases (Chowdhury *et al.*, 2006). The most common RNA thermosensor is the ROSE element, which was discovered in many small heat-shock genes of *Rhizobiaceae* and in a wide variety of α- and γ-Proteobacteria ((Narberhaus *et al.*, 2006), this issue). A NMR study of the ROSE temperature-responsive element shows that the SD sequence forms several non-canonical base pairs with a conserved upstream sequence of the 5'UTR, which is the first to melt at 37 °C, making it accessible to the 30S ribosome subunit (Chowdhury *et al.*, 2006). A new type of RNA thermometer, named the FourU element, depends on a stretch of four uridines that pair with the SD sequence of the *Salmonella* heat shock *agsA* mRNA (Waldminghaus *et al.*, 2007). Based on these data, a database (Structures of Untranslated Regions In Bacteria) was established to search conserved RNA folds in the first 110 nucleotides upstream of each gene from 62 bacterial genomes. This program found the experimentally defined ROSE-like thermosensors in *Rhizobiaceae.* Several hits were predicted in the genome of *S. aureus* MW2, a community-acquired meticillin-resistant strain that causes fatal septicaemia and septic arthritis in the USA. Among these candidates, a typical "fourU" sequence was predicted to form base pairs with the SD region of the staphylococcal *groES* gene (Waldminghaus *et al.*, 2007). The functionality of these motifs remains to be demonstrated.

Interestingly, in the Gram-positive pathogen *Listeria monocytogenes*, virulence genes are controlled by PrfA, a transcriptional activator whose expression was shown to be thermo regulated by a mechanism similar to the ROSE thermosensor (Johansson *et al.*, 2002). More strikingly, two SAM-riboswitches acting in *trans* bind to the 5'UTR of *prfA* mRNA to repress its translation at the permissive temperature (Loh *et al.*, 2009). A similar regulatory mechanism has not yet been demonstrated for *S. aureus*, which is an opportunistic bacterium for which the temperature might not be so critical for virulence gene expression.

2.2 Riboswitches

The 5'UTR regions of certain mRNAs regulate gene expression at the levels of transcription attenuation or translation initiation in response to the intracellular concentration of metabolites (cofactors, vitamins, amino acids, nucleotides, second messenger cyclic di-GMP, Mg^{2+}). Riboswitches function as a feedback mechanism since the metabolite is often synthesized by the gene located downstream of the regulatory element and regulation occurs when the concentration of the metabolite exceeds a threshold. Recent structural advances allowed the description of the molecular mechanisms of riboswitch functions (reviewed in (Serganov, 2009)). Riboswitches consist of a highly structured sensor domain (or aptamer), which is recognized and stabilized by the ligand, and the expression platform that controls the expression of the downstream transcripts. Binding of the metabolite to the aptamer domain stabilizes the formation of a Rho-independent terminator to stop transcription or the formation of a structure that sequesters the ribosome-binding site (RBS) to inhibit translation. In *B. subtilis*, almost 5 % of the genes involved in metabolism are controlled by riboswitches (Barrick and Breaker, 2007). Based on sequence and structure conservation (Barrick and Breaker, 2007; Yao *et al.*, 2007), most of the riboswitches were found in the genome of *S. aureus* (Geissmann *et al.*, 2009a). Their genetic organization and schematic models of regulatory mechanisms are shown in Figure 1. They comprise S-adenosylmethionine (SAM) class 1 and class 2 riboswitches and riboswitches responding to thiamine pyrophosphate (TPP), flavin mononucleotide (FMN), lysine, glycine, purine, 7-aminomethyl-7-deazaguanine (preQ1), and glucosamine-6-phosphate (Glc-6P) (Figure 1A). Transcriptomic analysis showed that short transcripts corresponding to riboswitches were all detected, except for one of the TPP and lysine riboswitches (Beaume *et al.*, 2010; Bohn *et al.*, 2010). These transcripts most likely result from a premature transcription arrest since most of them (SAM, TPP, lysine, glycine, preQ, FMN) are characterized by the presence of anti-terminator/terminator structures (Figure 1A). However, most of the sRNAs stop just upstream the aptamer domain (Beaume *et al.*, 2010). It was recently shown that *B. subtilis* RNase Y initiates turnover of SAM riboswitches by inducing a cleavage just downstream the terminator structure (Shahbabian *et al.*, 2009). Interestingly, the homologous enzyme CvfA in *S. aureus* is required for virulence (Nagata *et al.*, 2008).

Two of the riboswitches (GlmS and FMN) regulate their downstream genes by a different mechanism (Figure 1). The regulatory region of *glmS* mRNA in various

---

next page:

▸ **Fig. 1. The distribution of riboswitches and of protein-mediated regulatory sites located upstream the target genes of the *S. aureus* N315 genome.** The aptamer domain recognized by the metabolite or by proteins (ribosomal proteins L20, L10, L19, L13, and the protein PyrR) are squared. (A) Binding of the metabolites or of regulatory proteins cause premature arrest of transcription. The terminators of transcription are schematized. For *glmS* riboswitch, binding of the glucosamine-6-phosphate confers to the sRNA a catalytic activity thus inducing degradation of the target mRNA. SAM, S-adenosyl-methionine; TPP, thiamine pyrophosphate; FMN, flavin mononucleotide; pre-Q, 7-aminomethyl-7-deazaguanine. (B) Binding of FMN or of the ribosomal protein L19 regulates translation of the downstream genes. (C) Schematic representation of the FMN riboswitch: FMN binds to the aptamer domain (Serganov *et al.*, 2009; Winkler *et al.*, 2002), which in turn stabilizes a stem-loop structure sequestrating the Shine-Dalgarno (SD) sequence. As a consequence, translation is repressed. (D) The regulatory leader regions of *S. aureus pyr*R and *pyr*P mRNAs. At high concentrations of UMP, PyrR binds to UMP and acquires the capacity to bind to a AGAGAG hairpin loop in the leaders of the *pyr* operon. This interaction favours the formation of a terminator to induce premature arrest of transcription (Lu *et al.*, 1996). Alternative pairings leading to the formation of a terminator or of an anti-terminator structure are schematized in red and blue. Black terminators represent the *cis*-acting leaders, which were detected by the deep sequencing approach (Beaume *et al.*, 2010)

Gram-positive bacteria possess unique characteristics that confer a catalytic activity to the 5'UTR in response to the binding of Glc-6P (Winkler *et al.*, 2004). The crystal structure of the *glmS* regulatory region from *Thermoanaerobacter tengcongensis* has revealed the pre-folded state of the active site and shown that its ribozyme activity is promoted by the binding of the ligand Glc-6P, which acts as a coenzyme (Klein and Ferre-D'Amare, 2006). As a consequence, this initial cleavage induces

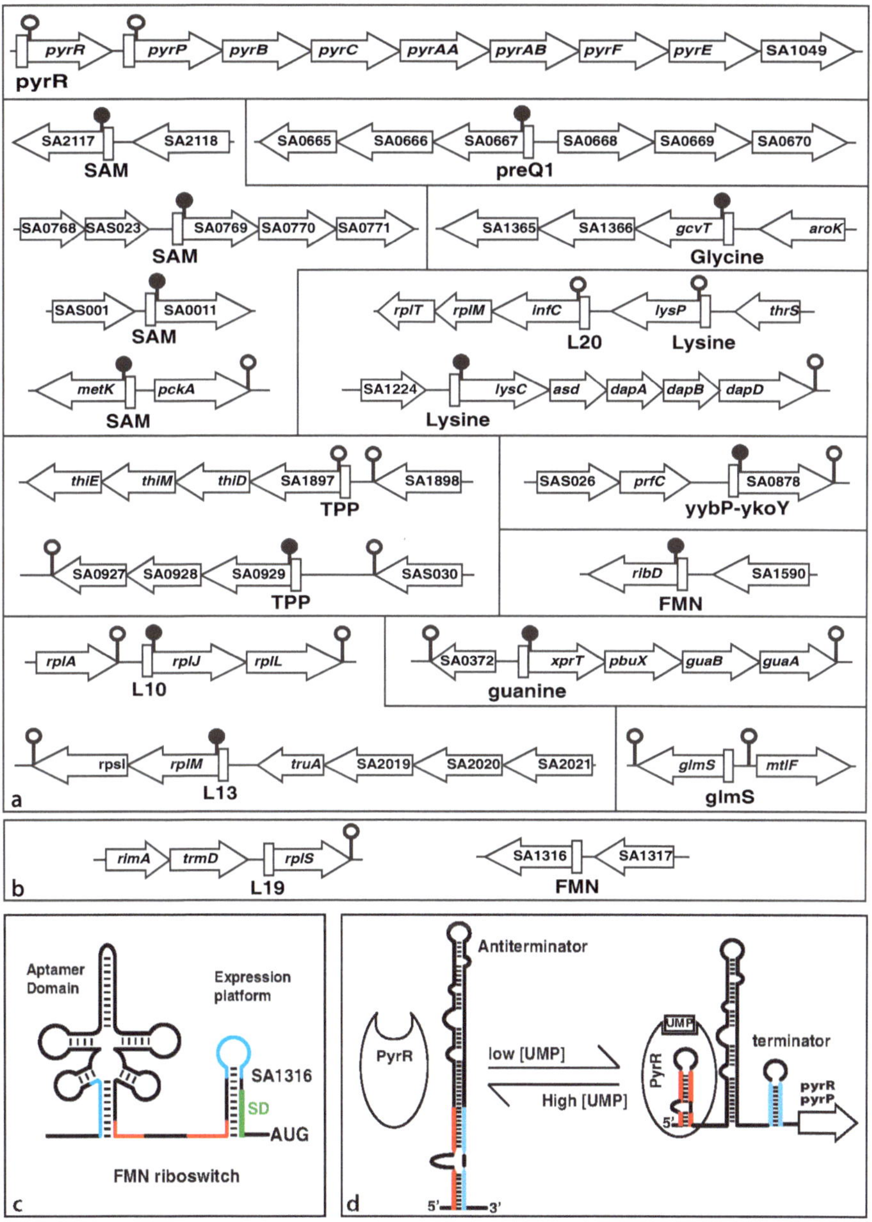

the rapid degradation of the mRNA (Collins *et al.*, 2007). The second example is illustrated by the FMN riboswitch, which controls the expression of a hypothetical protein (SA1316) at the translational level. Although the regulatory mechanism was not experimentally tested, it is proposed that the binding of the FMN would stabilize the formation of a hairpin structure sequestering the SD sequence (Figure 1C). Dynamic equilibrium could occur as riboswitches fold into alternative conformations, but some of the mRNAs would require *de novo* synthesis to adopt a different conformation. It was recently shown that alkali pH induces increased duration of RNA polymerase pauses on the nascent *E. coli alx* mRNA and promotes the folding of an active pH responsive element in the 5'UTR, enhancing translation initiation (Nechooshtan *et al.*, 2009). This pH responsive element is homologous to the yybP-ykoY riboswitch present in the 5'UTR of *S. aureus* SA0878 gene, which encodes a transporter protein. Whether the same pH-dependent effect is observable in *S. aureus* remains to be analyzed. In *S. aureus*, this *cis*-acting element is predicted to control transcription termination/anti-termination (Barrick *et al.*, 2004).

Many of these riboswitches regulate the synthesis of proteins involved in amino acid transport and biosynthesis, as well as co-factors and nucleotide biosynthesis. As riboswitches control essential genes and have evolved structures suitable for the recognition of small molecules, they appear to be tractable targets for anti-microbial compounds (Breaker, 2009). Rather recently, Mulhbacher *et al.* (2010) identified a pyrimidine derivative compound that binds to the guanine riboswitch to repress the expression of *guaA*. This compound shows bactericidal activity against *S. aureus* and significantly reduced mammary gland infection in the mouse.

2.3 Erythromycin-induced Translation Attenuation

The *erm* family of genes encoding rRNA methylases confers resistance to macrolide, lincosamide, and streptogramin B antibiotics by reducing the affinity between these antibiotics and ribosomes. Expression of the methylase is induced by erythromycin. *S. aureus ermA* and *ermC* genes are both controlled by a translational attenuation mechanism (Murphy, 1985). The *ermA* leader can adopt two alternative structures, one of which cannot be translated because the SD sequence is base-paired. In contrast to *ermC*, the leader of *ermA* encodes two small peptides (15 and 19 amino acids). The translation of these two peptides and the stalling of the erythromycin-bound ribosome in these two regions are both required, rendering the SD accessible for *ermA* mRNA translation (Murphy, 1985).

2.4 tRNA-mediated Riboswitches

The so-called T-box family controls the expression of downstream genes involved in tRNA aminoacylation and amino acid transport, and biosynthesis (Green *et al.*, 2010). The specificity of the regulation is based on pairings between the anticodon of the sensor uncharged tRNA and a codon-like sequence located in the specifier domain (Figure 2A). The NMR structure of the specifier region was recently shown to adopt a loop E motif formed by non-canonical base pairs, allowing an optimal presentation of the specifier triplet to pair with the anticodon of tRNA (Wang *et al.*, 2010). A second interaction common to all tRNAs involves the 3'NCCA end and a complementary sequence in the T-box region. In most cases, the binding of the tRNA sensor stabilizes an anti-terminator structure allowing continued transcription of the downstream genes (Figure 2). The *ileS* gene, encoding isoleucyl-tRNA synthetase, was the first *S. aureus* gene shown to be a member of the T-box family and is regulated by transcription anti-termination/termination (Grundy *et al.*, 1997). As expected, its expression was enhanced in response to amino acid limitation. As for

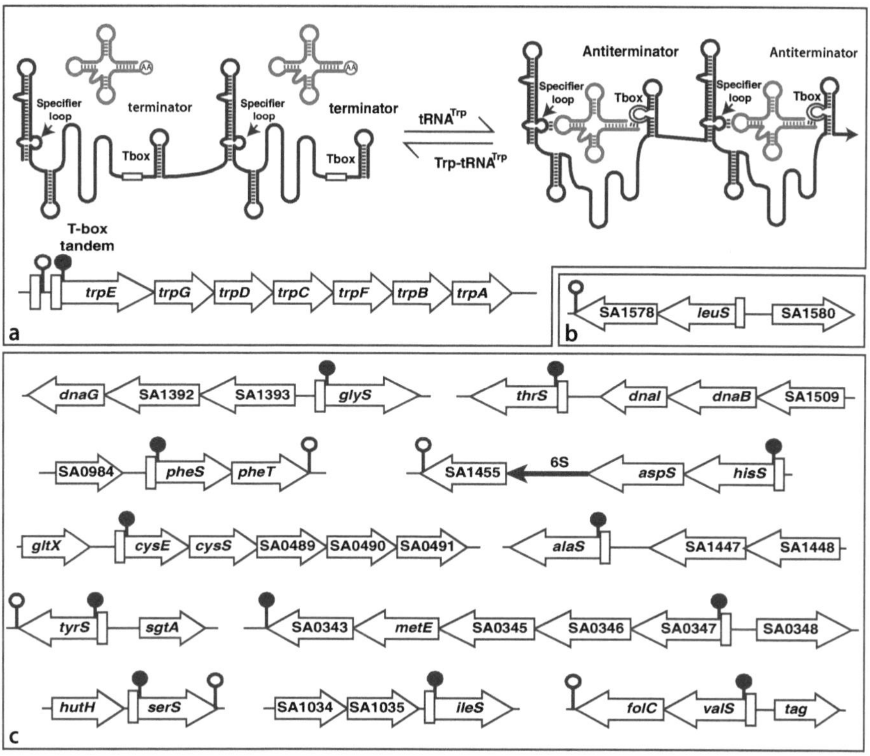

Fig. 2. tRNA-dependent riboswitches and operon arrangements of the *S. aureus* N315 genome. (A) Schematic representation of a tandem T-Box involved in the regulation of tryptophan synthesis. Non-aminoacylated $tRNA^{Trp}$ binds to the leader region of the *trpA-G* operon at two different sites (specifier loop, T-Box). This interaction stabilizes an anti-terminator structure allowing transcription of the entire *trp* operon. The picture is adapted from (Gutierrez-Preciado *et al.*, 2005). (B) The T-Box motif found upstream *leuS* gene encoding leucyl-tRNA synthetase. The prediction of the structure suggests that the $tRNA^{Leu}$ regulates the synthesis of the enzyme at the translational level. (C) tRNA-mediated anti-termination of transcription. The T-Box is boxed and the terminator of transcription is schematized. Black terminators represent the *cis*-acting leaders, which were detected by the deep sequencing approach (Beaume *et al.*, 2010)

riboswitches, most of the T-box elements were detected in the *S. aureus* N315 strain as the results of premature transcription events (Beaume *et al.*, 2010; Bohn *et al.*, 2010). A T-box in tandem was also found upstream of the operon encoding enzymes involved in tryptophan biosynthesis (Gutierrez-Preciado *et al.*, 2005) (Figure 2A). The structures of three of the *cis*-acting regions (HisS, LeuS, GlyS) were predicted to diverge from the consensus, and an operon "codon" specifier sequence could not be detected in an appropriate configuration (Beaume *et al.*, 2010). Only the T-box RNA leader, which regulates the synthesis of *S. aureus* leucyl-tRNA synthetase, is predicted to control translation initiation (Figure 2B).

2.5 Protein-mediated Transcription Termination/Anti-termination

RNA-binding anti-termination proteins control various genes and operons, including rRNA synthesis and the production of proteins involved in carbohydrate transport and utilization. A typical example is the regulation of the *ptsG* operon of *S. carnosus* by the transcriptional anti-terminator protein GlcT, which senses the concentration of glucose. Binding of the dimeric form of GlcT protein to the RAT motif (ribonucleic anti-terminator) of the 5'UTR of *glcA* mRNA prevents the formation of the terminator structure, thus allowing the transcription of the downstream gene (Knezevic *et al.*, 2000). Comparative sequence alignments also identified in *S. aureus* several large conserved leaders containing a terminator structure (Geissmann *et al.*, 2009a). This group includes several ribosomal proteins and the *pyr* operons (Figure 1). The 5'-UTRs of *pyr*R and *pyr*P mRNAs contain an AGAGAG hairpin loop motif that is recognized by the UMP-bound PyrR protein. In turn, PyrR-UMP induces premature arrest of transcription (Figure 1D). It was also shown that GMP can bind to PyrR but it antagonized the mRNA binding, suggesting a cross-regulation of the *pyr* operon by purines (Chander *et al.*, 2005). In Gram-positive bacteria, the synthesis of ribosomal protein operons (L10, L20, L13, L19) is regulated by a feedback mechanism similar to that of PyrR (Choonee *et al.*, 2007). As an example, the *S. aureus infC* leader carries a typical binding site for L20 ribosomal protein that mimics the 23S rRNA binding site. In *B. subtilis* and most likely in *S. aureus*, binding of L20 to the *infC* leader stabilizes the formation of a terminator structure to induce the premature arrest of transcription (Choonee *et al.*, 2007).

3 Small Non-coding RNAs Targeting mRNAs

3.1 Generalities

Efforts to identify ncRNAs in *S. aureus* and the aspects of ncRNA regulation have been uncovered only recently (Pichon and Felden, 2005; Geissmann *et al.*, 2009a; Abu-Qatouseh *et al.*, 2010; Chabelskaya *et al.*, 2010; Bohn *et al.*, 2010).

Even though the functions of most of the newly discovered *S. aureus* non-coding RNAs have not yet been determined, several of them mediate regulatory effects via base-pairing with their mRNA targets. As in many other bacteria (Waters and Storz, 2009), two antisense RNA systems have been identified in *S. aureus* (Pichon and Felden, 2005; Beaume *et al.*, 2010): they are either *cis*- or *trans*-encoded. In the first case, the same DNA sequence encodes both the antisense and the target RNA. These asRNAs are transcribed in an overlapping fashion from both strands and, hence, are fully complementary. This type of asRNA was shown for the first time to control replication of the rolling-circle-replication plasmid pT181 by transcriptional attenuation in *S. aureus* (Novick *et al.*, 1989; Brantl, 2007). Chromosome-encoded asRNAs were also detected in the *S. aureus* N315 genome (Beaume *et al.*, 2010). Interestingly, recent observations show that a long antisense RNA in *Salmonella*, which is required for virulence, was produced under the control of PhoPQ which is also essential for the transcription of the sense mRNA, but the efficiency of transcription of the asRNA is lower than for the sense (Lee and Groisman, 2010). This study suggests that the asRNA limits the levels of the protein produced by the sense mRNA as determined by the degree of the activation of PhoPQ. Since several asRNAs were found complementary of mRNA, which encode virulence factors in *S. aureus*, such a situation is expected to occur in *S. aureus* as well (Beaume *et al.*, 2010).

In the second case, the gene encoding the regulator is located at a locus different from that encoding the target(s), and thus the complementarity is not extensive. For this reason, these ncRNAs have the potential to regulate multiple targets. In all cases studied so far, base-pairing takes place within the 5'-UTR of the target mRNA (Boisset *et al.*, 2007; Geissmann *et al.*, 2009a; Chabelskaya *et al.*, 2010). Thus, the interaction results in the inhibition of ribosome binding and the arrest of translation initiation. In several cases, this process is coupled to RNA degradation. The expression of several ncRNAs was shown to be growth phase-dependent and stress-related (Geissmann *et al.*, 2009a; Bohn *et al.*, 2010). Some of these ncRNAs are expected to play a significant role in the establishment and survival of pathogenic bacteria within the host and eventually in the infectious process. The mechanism by which bacteria achieve this outcome is the turning on/off of regulatory ncRNAs, which leads to the adaptation of central metabolic pathways or altered expression of transport systems and cell wall-associated or excreted proteins. In contrast to *cis*-acting regulatory RNAs, the conservation of these ncRNAs is mainly restricted to *S. aureus* and to staphylococcal species.

Besides regulatory RNAs, several highly conserved bacterial ncRNAs were found to be expressed in *S. aureus* (Pichon and Felden, 2005; Beaume *et al.*, 2010). Briefly, i) 4.5S RNA, which constitutes the signal recognition particle when in complex with Ffh protein, is essential for membrane localization or secretion of proteins; ii) tmRNA composed of a tRNA and an mRNA, rescues ribosomes stalled on truncated mRNAs and tags incomplete peptides for degradation when in complex with SmpB protein; iii) RNase P is responsible for the maturation of the 5' end of tRNAs and has been reported to play a role in the maturation of riboswitches (Seif and Altman, 2008). (iv) Finally, *S. aureus* 6S RNA accumulates during the stationary phase of growth (Pichon and Felden, 2005; Beaume *et al.*, 2010). In *E. coli*, this

RNA interacts with the housekeeping form of RNA polymerase in complex with σ^{70} and inhibits expression from certain promoters in stationary phase (Wassarman, 2007). It has also been shown that 6S RNA is used as a template by the RNA polymerase to produce a short RNA, pRNA, which in turn releases the pRNA-6S RNA hybrid from the enzyme (Wassarman and Saecker, 2006). Since the pRNA has not yet been characterized in *S. aureus*, it remains to be shown whether the functions of *E. coli* and *S. aureus* 6S RNAs are conserved. Below we will provide examples of functions regulated by ncRNAs.

3.2 Pathogenicity Island-encoded sRNAs

An initial *in silico* prediction followed by experimental validation of a set of novel ncRNAs came from the work of Pichon and Felden (2005). They identified seven *bona fide* ncRNAs encoded from pathogenicity islands (SprA-G, for small pathogenicity island RNA). The expression pattern of these ncRNAs diverged considerably among four different *S. aureus* strains, demonstrating the capacity of the bacteria

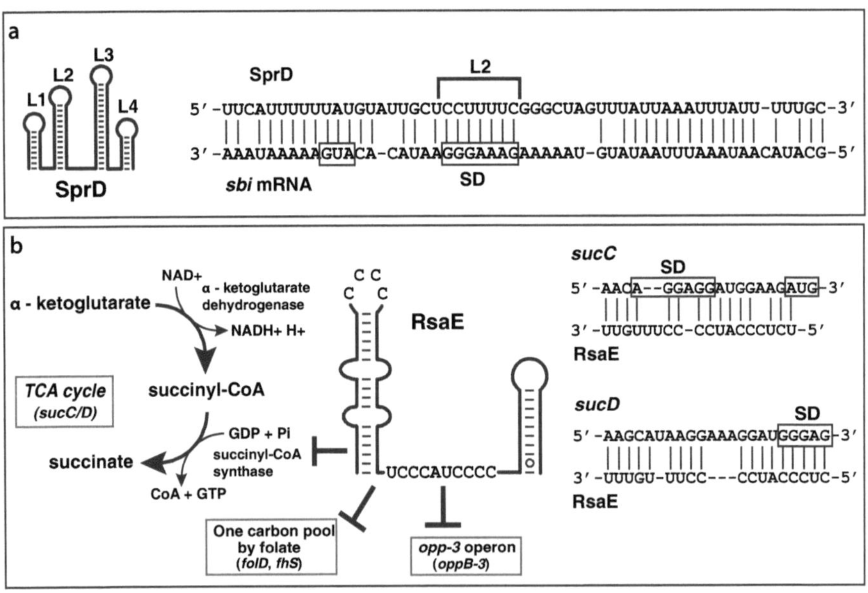

Fig. 3. Examples of two ncRNAs regulating translation of mRNA targets. (A) The pathogenicity island-encoded SprD sRNA binds to the leader sequence of *sbi* mRNA and blocks the Shine-Dalgarno (SD) sequence to repress translation. The formed duplex and the secondary structure of SprD are according to Chabelskaya *et al.* (2010). (B) *S. aureus* RsaE repress the synthesis of several enzymes involved in TCA cycle, in formatted pathway, and of the *opp3* operon involved in peptide and amino acid uptake. The RsaE uses the conserved C-rich motif to target the SD sequence of the mRNA targets, and examples of RsaE-mRNA duplexes are given. The secondary structure of RsaE is adapted from Geissmann *et al.* (2009a)

to evolve different means of response even though their genomes are equivalent. Notably, two of these novel ncRNAs are present in multiple copies in the genome and are usually located next to genes involved in virulence or antibiotic resistance. The mRNA targets of SprA were predicted, based on sequence homology, and one of them: SA2216, encodes an ABC transporter. Analysis of the transcriptome revealed that three ncRNAs are antisense of several Spr sRNAs (Beaume *et al.*, 2010). Two of these RNAs, antisense of SprA and SprA2, are predicted to be type I anti-toxin and SprA and SprA2 contain small open reading frames predicted to encode hydrophobic peptide type I toxins (Fozo *et al.*, 2010). More recently, it was shown that SprD contributes significantly to causing septic infections in a murine animal model (Chabelskaya *et al.*, 2010). One of its targets was identified as *sbi* mRNA, which encodes an immune evasion molecule that protects the bacteria against the host immune responses. SprD binds to the 5'UTR of *sbi* mRNA and inhibits the initiation of translation (Figure 3A). The two genes, sprD and *sbi*, are not encoded on the same pathogenicity island.

3.3 sRNA Stress Response and Metabolism

In an effort to characterize the heat shock, cold shock, stringent and SOS responses as well as their effect on RNA turnover in *S. aureus*, Dunman and colleagues identified a set of stable RNAs (Anderson *et al.*, 2006; Roberts *et al.*, 2006). These RNAs were found expressed in the log-phase of growth and under specific stress conditions, but some of these candidates may be co-transcribed as part of an operon. More recently, a combination of computational tools was used (based on conservation and sequence/structure characteristics) for the prediction of ncRNAs and expression was experimentally confirmed for 11 new ncRNAs (RsaA-K) among different *S. aureus* strains (Geissmann *et al.*, 2009a). These ncRNAs exhibited a tightly regulated pattern of expression, which was dependent on the growth phase and the stress applied. The expression of several of them was induced by σ^B (RsaA) or enhanced by the *agr* locus (RsaE). Their sequences and structures were either found to be specific to *S. aureus* strains or conserved amongst staphylococci species. Most of the sRNAs contained an UCCC motif in unpaired regions, which likely represents a characteristic signature for sRNAs that pair with the Shine-Dalgarno (SD) sequence of mRNAs, as has been shown for *S. aureus* RNAIII (Figure 3A, B). Some noteworthy points concern the functions regulated by ncRNAs and their mechanisms of action. RsaA constitutes a member of the σ^B regulon, which was shown to be an important modulator of metabolism and virulence gene expression (Bischoff *et al.*, 2004). RsaA can potentially form base pairs with mRNAs that are repressed by σ^B (the magnesium-citrate transporter, CitM, an APC amino-acid permease and the pyruvate carboxylase, PycA) (Geissmann *et al.*, 2009a). Thus, one can postulate that the repression of σ^B targets occurs indirectly, most likely through RsaA. Another remarkable point is the involvement of RsaE in several metabolic pathways (Figure 3B). The biological function of RsaE was demonstrated using combined transcriptome/proteome approaches (Geissmann *et al.*, 2009a; Bohn *et al.*, 2010). It was revealed that RsaE regulates several enzymes involved in amino acid and peptide transport (*opp-3* operon), cofactor synthesis, folate-dependent one-carbon metabolism, lipid and carbohydrate metabolism, and the TCA cycle. Direct repression was shown for at least three targets, *opp-3B* (amino acid and peptide transporter), *sucD* (succinyl-CoA synthetase of the TCA cycle) and SA0873 (unknown function) mRNAs (Figure 3B). In all three cases, the C-rich motif of RsaE pairs with the SD sequence of these mRNAs to prevent the formation of the ribosomal initiation complex (Geissmann *et al.*, 2009a). Moreover, the overproduction of RsaE induces a growth defect that is partially alleviated by acetate (Bohn *et al.*, 2010), suggesting that RsaE would coordinate down-regulation of the energy metabolism (TCA cycle), and purine biosynthesis when carbon sources become scarce. Interestingly, RsaE is conserved in *Staphylococcus*, *Macrococcus* and *Bacillus* (Geissmann *et al.*, 2009a), which all derive from a common Gram-positive ancestor. Thus it is not so surprising that this RNA, which is present in all three genera, regulates metabolic pathways. This notion is further supported by the fact that in *B. subtilis*, RsaE was predicted to regulate mRNAs encoding the carbon storage proteins CstA, SucC and the galactokinase GalK, which is involved in carbon metabolism (Geissmann *et al.*, 2009a; Rasmussen *et al.*, 2009; Irnov *et al.*, 2010). Due to the interplay between metabolism and virulence factor production (Somerville and Proctor, 2009), one wonders if RsaE plays an indirect role in virulence.

3.4 sRNA and Small Colony Variant

Abu-Qatouseh and colleagues (Abu-Qatouseh *et al.*, 2010) have recently identified a new set of ncRNAs that are differentially expressed between two isogenic variants of a clinical strain displaying the normal and the small colony variant (SCV) phenotypes. The SCV phenotype exhibits some particular characteristics of clinical importance such as persistent infection in the host cell, a reduced susceptibility to antibiotics, and the ability to cause latent or recurrent infections (Proctor *et al.*, 2006). SCVs have a slow growth rate, unusual colony morphology, and specific metabolic properties, including defects in pathways involved in electron transport (Proctor *et al.*, 2006). Therefore, it was of particular interest to evaluate the importance of ncRNA regulation in this phenotype. Following an approach in which cDNA libraries were made from samples collected at different growth stages for the two strains, researchers identified a total of 142 new ncRNAs and validated the expression of 18 of the candidates. The new ncRNAs originated from IGRs or were antisense or partially overlapping with ORFs. In addition, six candidate sRNAs encoded from pathogenicity islands or inserted phage sequences were identified although expression was confirmed for only one of them. The expression of the newly identified ncRNAs was growth phase regulated and phenotype specific for a subset of them comprising RNAIII. Certain ncRNAs were exclusively expressed in the WT strain like RNAIII or were up-regulated in the SCV like RsaA and the pH responsive element yybP-ykoY. The deregulation of sRNAs in the development of phenotypic/virulence variations in *S. aureus* was highlighted.

3.5 *S. aureus* Transcriptome

The most recent effort to reveal the totality of the ncRNAs encoded in the *S. aureus* genome comes from the work of Beaume *et al.* (2010). Here, high-throughput sequencing (HTS) was performed for the whole transcriptome at various stages of growth to uncover the full repertoire of ncRNAs in *S. aureus*. Moreover, expression of a subset of the newly identified sRNAs was examined under different stress conditions mimicking environments encountered in the host. Interestingly, this work showed that approximately 10% of the intergenic regions generate expressed transcripts corresponding to 160 sRNAs (including riboswitches and *cis*-acting elements) and 35 antisense RNAs (asRNAs). All of the previously known sRNAs were identified and 21 novel ncRNAs, encoded by autonomous IGR, were discovered. In addition, 32 sRNAs corresponding to long 5' or 3'UTRs of mRNAs were detected. Even though no particular sequence or structural features were described, this does not preclude these processed regions being endowed with regulatory functions. Most asRNAs are encoded in pathogenicity islands and are predicted to interact with mRNAs encoding cell wall and cell envelope proteins, as well as with proteins involved in replication, recombination and repair. Several asRNAs were also found within operons and most likely mediate discoordinate expression of genes within the operon. A noteworthy point made in this study is the presence of asRNAs at the ends of transposase genes and also in the vicinity of metabolic or virulence genes. Moreover, possible members of type I toxin-antitoxin families (Fozo *et al.*, 2010) were revealed, although this finding awaits further experimentation. In addition, this study revealed the existence of nine sRNAs carrying small open reading frames encoding predicted hydrophobic peptides. Finally, several overlapping 3' and 5' UTRs, which are known to be subject to transcriptional interference, were also detected. The newly identified ncRNAs were found conserved among *S. aureus* strains and even between staphylococcal species. Lowering the homology stringency cut-off showed conservation amongst *Bacillales* and Firmicutes, particularly for the *cis*-acting regulatory elements. All in all, the differential expression of ncRNAs emphasizes once again the importance of RNA regulation in the adaptation of the bacterium to new niches within the host.

4 RNAIII, a mRNA and a Regulatory RNA

4.1 Quorum Sensing and Virulence in *S. aureus*

The ability of bacteria to exchange information between cells has been studied in detail in *S. aureus* (Novick and Geisinger, 2008). The quorum-sensing system, which is encoded by the *agr* system, involves a small signalling peptide, the so-called auto-inducer peptide, which is sensed by the bacterial population to induce a substantial physiological rearrangement. The *agr* system has been recognized as a pivotal regulator of gene expression and as a potential target for therapeutic applications (Novick, 2003; Novick and Geisinger, 2008). The *agr* system consists of two divergent transcription units leading to RNAII and RNAIII. RNAII comprises a quorum-sensing cassette for a membrane protease AgrB, which matures and exports the extracellular auto-inducer peptide (AIP). It also encodes the two-component signal transduction system, the sensor kinase AgrC and its cognate response regulator AgrA. Upon a threshold concentration, the AIP binds to the kinase AgrC, which phosphorylates AgrA to auto-activate the transcription of the entire *agr* system. This feedback mechanism ensures that target gene expression is only modified at a certain threshold level of the bacterial cell density. Studies of the *agr* regulon in a meticillin-resistant *S. aureus* strain revealed that AgrA and RNAIII are the two intracellular effectors of the quorum-sensing system. While RNAIII coordinates the expression of the virulence factors (Novick *et al.*, 1993), AgrA regulates metabolic genes involved in carbohydrate and amino acid metabolism and phenol-soluble-modulin (PSM) cytolysin genes in a predominantly RNAIII-independent manner (Queck *et al.*, 2008). These regulations may be direct or indirect, i. e., even through the activation of other sRNAs, like RsaE (Geissmann *et al.*, 2009a). Queck and colleagues suggested that RNAIII-dependent control of virulence factors and quorum-sensing control of metabolism were originally separate regulatory circuits (Queck *et al.*, 2008). Since regulatory RNAs are appropriate for fast adaptive responses, the connection between RNAIII-dependent and quorum-sensing regulation likely evolved to allow for a faster cell density-dependent alteration of virulence factor expression during infection. RNAIII-independent regulation would be an ancient function of the *agr* quorum-sensing system in the non-pathogenic lifestyle of *S. aureus* in which *agr* enables cells to adapt to low nutrition conditions in a high cell density population. This type of model is in agreement with the fact that the evolution of virulent species of *S. aureus* is believed to be a phylogenetically more recent event. Hence, this system gives some clues regarding the stepwise occurrence of two different regulatory circuits that helped *S. aureus* in establishing its pathogeny. Other bacteria have also developed virulence regulatory circuits including regulatory RNAs from communication systems (Lenz *et al.*, 2004).

4.2 RNAIII Encodes a Small Toxin

RNAIII encodes for a PSM, haemolysin δ. (Novick *et al.*, 1993). Although both haemolysin δ and RNAIII play roles in virulence, they do not act synergistically. This finding was confirmed by the fact that *S. lugdunensis* RNAIII does not encode an sORF (Vandenesch *et al.*, 1993). Interestingly, the synthesis of haemolysin δ is delayed 1 h after the transcription of RNAIII (Novick, 2003). This result is not in line with the accessibility of the RBS sequence of *hld* since it was shown that the ribosome binds efficiently to RNAIII *in vitro* (Benito *et al.*, 2000). Thus, it is possible that a *trans*-acting factor could be involved either in repressing *hld* translation during the first hour, or in activating its translation.

4.3 RNAIII as the Regulator

S. aureus RNAIII promotes the switch between the expression of surface proteins and the synthesis of the excreted toxins in a time-dependent manner (Novick, 2003). The regulatory mechanisms exerted by RNAIII were largely documented in recent years (review in (Geissmann *et al.*, 2009b; Romby and Charpentier, 2010)). RNAIII is the longest regulatory RNA studied to date (514 nts long). It is comprised of 14 hairpin structures, which are organized into three functional domains. The 5' region of RNAIII is complementary to the leader region of *hla* mRNA, encoding haemolysin α(Morfeldt *et al.*, 1995). RNAIII most likely binds to the nascent mRNA, prevents the formation of an inhibitory structure that sequesters the RBS, and in turn facilitates the access of the 30S subunit to activate translation (Figure 4). Three conserved apical loops contain a redundant C-rich sequence with specific regulatory functions. These hairpins evolved to efficiently pair with the RBS of many mRNAs, thus promoting a variety of RNAIII-mRNA structures (Figure 4). Hairpin 13 binds in a very similar way to the SD of several mRNAs (*spa*, *sa1000*, *sa2353*) to form long duplexes that inhibit translation initiation and create binding sites for cleavage by the double-strand-specific Endoribonuclease III (Huntzinger *et al.*, 2005; Boisset *et al.*, 2007). RNAIII also uses two distant domains (hairpins 7 and 14) to bind to two hairpin loops of *rot* mRNA that encode the transcriptional repressor of toxin Rot, one of which contains the SD sequence (Geisinger *et al.*, 2006; Boisset *et al.*, 2007). The inhibition of *coa* mRNA translation is achieved by the formation of a long duplex between hairpin 13 and the SD sequence of *coa* mRNA, which is further stabilized by a loop-loop interaction involving hairpin 7 (Chevalier *et al.*, 2010). In all these examples, fast binding is initiated by a first contact, mediated by loop-loop interactions and specificity is attained either by propagating the initial contacts into extended duplexes or by the addition of a second loop-loop interaction (review, (Geissmann *et al.*, 2009a)). All of these duplexes are substrates for RNase III, which initiates the rapid degradation of the repressed mRNA. In contrast to sRNA-dependent regulation in *E. coli*, RNase III cleavage is required for efficient repression. Interestingly, RNase III is able to bind the RNAIII without inducing further cleavage (Huntzinger *et al.*, 2005). It is tempting to propose that RNAIII guides the RNase to couple repression of translation with degradation. Strikingly, *S. aureus* Hfq protein binds to RNAIII but has no detectable effect on RNAIII-target mRNA complex formation, *in vivo* RNAIII-dependent regulation, or RNAIII stability (Boisset *et al.*, 2007; Bohn *et al.*, 2007; Geissmann *et al.*, 2009a). Since RNAIII binds rapidly to its target mRNAs, it might be possible that the RNA structures compensate for the need for a helper protein. Furthermore, the *S. aureus hfq* gene is transcribed very poorly, and the deletion of *hfq* does not exhibit severe phenotypic defects as in *E. coli* (Bohn *et al.*, 2007). Thus, the elucidation of *S. aureus* Hfq function still remains a challenge, which also raises the question of whether *S. aureus* uses other helper proteins to facilitate gene expression regulation by sRNAs.

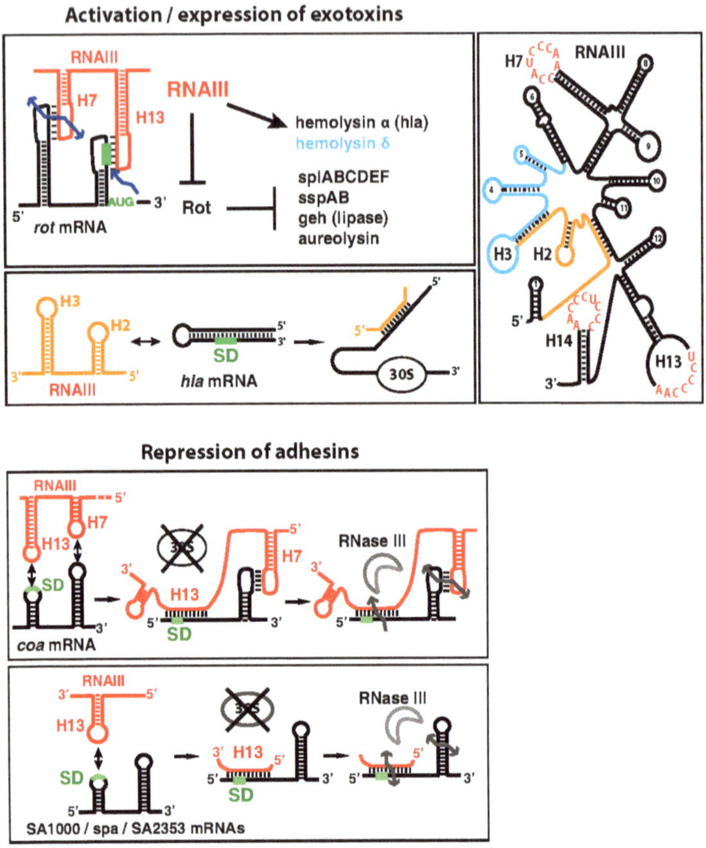

Fig. 4. Structure and mechanism of action of the quorum-sensing RNAIII from *S. aureus*. The secondary structure of RNAIII (middle) with the three hairpins 7, 13, and 14, adapted from Benito *et al.* (2000). The three functional domains of RNAIII are coloured in orange for the translational activator antisense RNA domain, in blue for the coding sequence expressing haemolysin δ, and in black the repressor antisense RNA domain, which contain the three hairpin loops rich in C-residues. (Left) Two hairpin 7 and 14 bind to the leader sequence of *rot* mRNA which encodes the repressor of exotoxins, and represses the translation of *rot* mRNA (Geisinger *et al.*, 2006; Boisset *et al.*, 2007). The 5' domain of RNAIII prevents the formation of an inhibitory structure of *hla* mRNA to activate the translation of haemolysin α (Morfeldt *et al.*, 1995). (Right) RNAIII uses different hairpin motifs to bind mRNAs encoding virulence factors (protein A, fibrinogen-binding protein SA1000, SA2353, coagulase). Formation of RNAIII-mRNA complexes blocks the access of the 30S ribosomal subunit and concomitantly recruits the Endoribonuclease III (RNase III) to cleave the repressed mRNAs (Huntzinger *et al.*, 2005; Boisset *et al.*, 2007; Chevalier *et al.*, 2010). SD is for Shine-Dalgarno sequence

4.4 RNAIII and its Regulatory Network

Studies done on RNAIII have provided some clues on the temporal expression of virulence factors in response to cell density. RNAIII acts directly at the post-transcriptional level by targeting mRNAs and indirectly at the transcriptional level by repressing the translation of Rot. Therefore RNAIII represses the synthesis of *spa* mRNA and activates the synthesis of haemolysin α at two different levels (Figure 4). Post-transcriptional regulation is important since many of the RNAIII-mRNA targets are quite stable (Boisset *et al.*, 2007; Huntzinger *et al.*, 2005). This type of feed-forward regulatory loop ensures a tight regulation of gene expression, a property that has been described for several sRNA-dependent regulatory effects (Shimoni *et al.*, 2009; Shimoni *et al.*, 2007). This complex network explains the delay between the repression of adhesin synthesis, which occurs directly through RNA-RNA pairing, and the induction of exoprotein synthesis, which first requires the inhibition of *rot* translation. Finally, it should be noted that RNAIII does not only modulate the expression of virulence factors, but also modulates the proteins involved in the metabolism of peptidoglycan, like the hydrolase LytM (Boisset *et al.*, 2007; Lioliou *et al.*, unpublished data) or teichoic acid (Dunman *et al.*, 2001). Therefore, RNAIII significantly contributes to the remodelling of the membrane and to cell wall integrity at high cell density.

A major problem in Staphylococcal infections is the extensive phenotypic variation of the clinical *S. aureus* isolates, which is not always due to genotypic diversity. Major strain-dependent differences in gene expression may be caused at the level of regulation, which is reflected by studies of the RNAIII regulon. Recently, it was shown that RNAIII expression may vary considerably among clinical isolates (Jelsbak *et al.*, 2010) and that the pattern of proteins regulated by RNAIII may not be absolutely conserved in all isolates (Blevins *et al.*, 2002; Cassat *et al.*, 2006; Queck *et al.*, 2008). The regulators that repressed RNAIII transcription or altered its stability are still not known. One of the future tasks for this area of study will be to understand how the different nodes of the networks are interconnected and how they influence each other. We may then be able to understand how modulating one node may affect the whole network and how *S. aureus* can take advantage of the system to develop variations in the pattern of virulence gene expression.

5 CRISPR in Defense Mechanism

Bacteria and archaea rely on different defense mechanisms to combat predation. Among them, the recently discovered, clustered regularly interspaced short palindromic repeats (CRISPR) defense system protects bacteria and archaea from foreign mobile genetic elements like viruses, plasmids and transposons (for reviews, (Horvath and Barrangou, 2010; Marraffini and Sontheimer, 2010a), see this issue). The CRISPRs are a novel class of repetitive DNA sequences that are present in almost 40% of bacteria and 90% of archaea (Grissa *et al.*, 2008). CRISPR sequences share certain similarities with RNA interference gene silencing used in plants to protect against viral infections. Briefly, the locus is comprised of an array of short direct repeats (24 to 47 nucleotides) disrupted by spacer elements (26 to 72 nucleotides), which are not conserved. The repeats are usually present in a given CRISPR array, occurring between 2 to 250 times. The leader sequence is directly adjacent to the repeats and carries the signals for the incorporation of new repeats and for the transcription of the precursor RNA. A cohort of proteins precedes the leader sequence, which comprises 2 to 6 core *cas* genes. In addition to the core set of Cas proteins, several species-specific CRISPR-associated genes can also be present. The CRISPR locus is constitutively expressed into a long pre-crRNA precursor that can span the entire locus. The pre-crRNA is endonucleolytically processed into fragments that comprise the spacer flanked by repeated sequences, thus producing the mature products. The CRISPR locus is absent in most sequenced *Staphylococci*, but has recently been found in a clinical isolate, RN62a from *S. epidermidis* (Marraffini and Sontheimer, 2008). The CRISPR locus of RN62a contains a spacer against both the coding and the non-coding sequence of the nickase gene. Analyses of this CRISPR locus in *S. epidermidis* showed that the defense system directly targets the DNA and not the mRNAs (Marraffini and Sontheimer, 2008). An elegant experiment showed that the insertion of an intron into the target nickase gene in a conjugative plasmid abolished immunity against plasmid conjugation even though the uninterrupted target sequence was regenerated in the spliced mRNA. The same authors further showed how the crRNA distinguishes the spacer DNA within the CRISPR locus from the foreign DNA, thus providing the basis for the self versus non-self DNA recognition mechanism (Marraffini and Sontheimer, 2010b). Therefore, the *S. epidermidis* CRISPR defense system prevents plasmid horizontal gene transfer, which is a major contributor of the spread of antibiotic resistance. These studies raised other questions: How do crRNAs interact with the DNA? What is the molecular mechanism of the CRISPR-mediated silencing of the DNA? How do the Cas proteins exert their action? How are new spacers selected? Are CRISPRs in staphylococcal strains restricted to just the defense against mobile genetic elements? It should also be noted that CRISPRs are also involved in other processes that are linked to virulence. For example, CRISPRs affect lysogeny of *Pseudomonas aeruginosa* by the DMS3 phage, which results in the loss of biofilm formation and swarming motility (Zegans *et al.*, 2009). Finally, Staphylococci harbour other repeated sequences, called STAR elements (Cramton *et al.*, 2000), which are GC-rich. Some of them carry features reminiscent of CRISPRs, although no *cas* gene was found (Geissmann *et al.*, 2009a). Two STAR repeats were found to be present in the leader region of mRNA as well as in the sprA2 ncRNA (Pichon and Felden, 2005; Geissmann *et al.*, 2009a). Pathogenicity island (SaPI) DNAs have been recently shown to be encapsidated in phage-like infectious particles when expressed, leading to intra and intergenic transfer (Tormo *et al.*, 2008). Recently, highly specific interactions between repressors encoding from the SaPI and helper-phage-encoded anti-repressors were shown to ensure that the SaPI would be transferred once induced (Tormo-Mas *et al.*, 2010). In addition to the CRISPR, this study shows another remarkable co-evolutionary adaptation of the bacteria and phage for controlling SaPI movement and/or mobilization. One cannot exclude that sRNAs might be involved in the regulation of these processes.

6 Perspectives

This review provided a first glimpse of sRNA functions in the major human pathogen, *S. aureus*. The constantly increasing database of sRNAs in *S. aureus* and the sustained discovery of their roles in the bacterium's physiology guarantee that a clearer understanding of mechanisms of pathogenesis is now possible. Analysis of phenotypes of *S. aureus* mutant strains for the sRNA genes, regulatory proteins, major processing enzymes and RNA-binding proteins, combined with methods to find and validate the targets should help to identify the sRNA functions that are specific to *S. aureus*, and those that are conserved among staphylococcal and Gram-positive bacteria. This information will help in designing the RNA-dependent regulatory circuits involved in bacterial adaptation and virulence.

The diversity of sRNAs described so far in *S. aureus* might predict that additional regulatory mechanisms are still to be discovered. Future research is, therefore, necessary to identify the exact mechanisms by which regulatory RNAs act on their targets. Up to now, all the sRNAs that target mRNAs bind to the RBS. Binding to the 3'-UTR or the coding sequence has not yet been observed as reported in *E. coli* (Opdyke *et al.*, 2004) and *S. typhimurium* (Pfeiffer *et al.*, 2009), respectively. Another point of interest is how sRNAs recognize their targets. Do they require specific RNA-binding proteins? Are there some specific recognition signatures on sRNAs that target mRNAs, like the C-rich motif of RNAIII, which is also found in several sRNAs (Geissmann *et al.*, 2009)? How widespread are multifunctional RNAs like *S. aureus* RNAIII? What is the contribution of the *cis*-acting sRNAs versus *trans*-acting sRNAs in stress response adaptation and virulence?

As for Gram-negative bacteria, *S. aureus* sRNAs modulate the stability of mRNAs that have direct consequences on virulence and stress responses. However, mechanisms of mRNA turnover have recently been well studied in *B. subtilis* (Condon, 2007), but little is still known for *S. aureus* (Anderson and Dunman, 2009), except for RNase III, which was found to be associated with the regulatory functions of RNAIII (Huntzinger *et al.*, 2005). Even less is known about *S. aureus* RNA-binding proteins associated with sRNAs. In contrast to *E. coli*, Hfq is not required for RNAIII functions and for sRNA stability (Boisset *et al.*, 2007; Geissmann *et al.*, 2009a), and the protein is not essential for virulence and stress adaptation (Geisinger *et al.*, 2006; Bohn *et al.*, 2007). It was shown that SarA, a pleiotropic regulatory protein involved in virulence, affected mRNA turnover (Roberts *et al.*, 2006). Whether SarA may indirectly affect mRNA degradation through the activation of sRNAs that bind to mRNAs is still an open question.

Most of the studies of the mechanism of action of the regulatory RNAs in *S. aureus* provided a static picture of gene regulation in a population of cells. However, identical genotype and growth environments are insufficient to ensure that individual cells within clonal microbial cultures will exhibit the same phenotype

(Gefen and Balaban, 2009). This heterogeneity and diversity is thought to be fundamental to fitness and development, and allows the microbial population to persist under external changes, and to exploit new niches. This fact is particularly true for the biofilm formation, which is a remarkable structure in which *S. aureus* organizes a multi-cellular way of life (Lasa, 2006). Another layer of complexity lies in the fact that individual cells use spatial regulation of gene products to regulate a variety of physiological events. Recent observations show that *Caulobacter crescentus* and *E. coli* mRNAs display limited dispersion from their site of transcription (Llopis *et al.*, 2010). This important finding implies that transcription and translation are spatially organized according to the chromosome organization. It is thus of importance to identify the mechanism by which the sRNAs will reach their mRNA targets and to determine at which point in the mRNA's lifetime they will exert their action.

Last but not least is the impact of *S. aureus* RNA-dependent regulation in virulence. Here we show that obvious links exist between sRNA, quorum sensing, metabolism and virulence. However, little is known about the role of *S. aureus* sRNAs in biofilm formation or during infection in model animals. We also do not know how the host microbial flora influences sRNA expression and their regulatory networks in *S. aureus*. Studies on the mechanisms of RNA-dependent regulation may also pave the way for the development of novel anti-microbial compounds and for biotechnological applications. It has already been proven feasible to find metabolite analogs that artificially modulate gene expression and alter bacterial growth (Sudarsan *et al.*, 2005; Blount *et al.*, 2007; Mulhbacher *et al.*, 2010). All in all, *S. aureus* has become an ideal model with which to deeply study mechanistics, physiological, virulence and evolutionary questions regarding regulatory RNAs.

Acknowledgments

We are thankful to Thomas Geissmann and François Vandenesch for stimulating discussions. This work was supported by the Centre National de la Recherche Scientifique (CNRS; PR), the Institut National de la Recherche Agronomique (INRA; CG), the Agence Nationale pour la Recherche (ANR09-BLAN-0024–01: PR), and the European Community (COST action BM-09–06; PR). EL is holder of a long-term fellowship from the FEBS, and CR received the support from the Région Alsace and from Délégation Générale pour l'Armement (DGA).

References

Abu-Qatouseh LF, Chinni SV, Seggewiss J, Proctor RA, Brosius J, Rozhdestvensky TS, Peters G, von Eiff C, Becker K (2010) Identification of differentially expressed small non-protein-coding RNAs in *Staphylococcus aureus* displaying both the normal and the small-colony variant phenotype. J Mol Med 88: 565–575.

Anderson KL, Dunman PM (2009) Messenger RNA Turnover Processes in Escherichia coli, Bacillus subtilis, and Emerging Studies in *Staphylococcus aureus*. Int J Microbiol ID 525491.

Anderson KL, Roberts C, Disz T, Vonstein V, Hwang K, Overbeek R, Olson PD, Projan SJ, Dunman PM (2006) Characterization of the *Staphylococcus aureus* heat shock, cold shock, stringent, and SOS responses and their effects on log-phase mRNA turnover. J Bacteriol 188: 6739–6756.

Barrick JE, Breaker RR (2007) The distributions, mechanisms, and structures of metabolite-binding riboswitches. Genome Biol 8: R239.

Barrick JE, Corbino KA, Winkler WC, Nahvi A, Mandal M, Collins J, Lee M, Roth A, Sudarsan N, Jona I, Wickiser JK, Breaker RR (2004) New RNA motifs suggest an expanded scope for riboswitches in bacterial genetic control. Proc Natl Acad Sci U S A 101: 6421–6426.

Beaume M, Hernandez D, Farinelli L, Deluen C, Linder P, Gaspin C, Romby P, Schrenzel J, Francois P (2010) Cartography of methicillin-resistant *S. aureus* transcripts: detection, orientation and temporal expression during growth phase and stress conditions. PLoS One 5: e10725.

Benito Y, Kolb FA, Romby P, Lina G, Etienne J, Vandenesch F (2000) Probing the structure of RNAIII, the *Staphylococcus aureus agr* regulatory RNA, and identification of the RNA domain involved in repression of protein A expression. RNA 6: 668–679.

Bischoff M, Dunman P, Kormanec J. , Macapagal, D., Murphy, E., Mounts, W., Berger-Bachi, B., and Projan, S. (2004) Microarray-based analysis of the *Staphylococcus aureus* sigmaB regulon. J Bacteriol 186: 4085–4099.

Blevins JS, Beenken KE, Elasri MO, Hurlburt BK, Smeltzer MS (2002) Strain-dependent differences in the regulatory roles of *sarA* and *agr* in *Staphylococcus aureus*. Infect Immun 70: 470–480.

Blount KF, Wang JX, Lim J, Sudarsan N, Breaker RR (2007) Antibacterial lysine analogs that target lysine riboswitches. Nat Chem Biol 3: 44–49.

Bohn C, Rigoulay C, Bouloc, P (2007) No detectable effect of RNA-binding protein Hfq absence in *Staphylococcus aureus*. BMC Microbiol 7: 10.

Bohn C, Rigoulay C, Chabelskaya S, Sharma CM, Marchais A, Skorski P, Borezee-Durant E, Barbet R, Jacquet E, Jacq A, Gautheret D, Felden B, Vogel J, Bouloc P (2010) Experimental discovery of small RNAs in Staphylococcus aureus reveals a riboregulator of central metabolism. Nucleic Acids Res

Boisset S, Geissmann T, Huntzinger E, Fechter P, Bendridi N, Possedko M, Chevalier C, Helfer AC, Benito Y, Jacquier A, Gaspin C, Vandenesch F, Romby P. (2007) Staphylococcus aureus RNAIII coordinately represses the synthesis of virulence factors and the transcription regulator Rot by an antisense mechanism. Genes Dev 21: 1353–1366.

Brantl S (2007) Regulatory mechanisms employed by cis-encoded antisense RNAs. Curr Opin Microbiol 10: 102–109.

Breaker RR (2009) Riboswitches: from ancient gene-control systems to modern drug targets. Future Microbiol 4: 771–773.

Cassat J, Dunman PM, Murphy E, Projan SJ, Beenken KE, Palm KJ, Yang SJ, Rice KC, Bayles KW, Smeltzer MS (2006) Transcriptional profiling of a *Staphylococcus aureus* clinical isolate and its isogenic agr and sarA mutants reveals global differences in comparison to the laboratory strain RN6390. Microbiology 152: 3075–3090.

Chabelskaya S, Gaillot O, Felden B (2010) A Staphylococcus aureus small RNA is required for bacterial virulence and regulates the expression of an immune-evasion molecule. PLoS Pathog 6: e1000927.

Chander P, Halbig KM, Miller JK, Fields CJ, Bonner HK, Grabner GK, Switzer RL, Smith JL (2005) Structure of the nucleotide complex of PyrR, the pyr attenuation protein from *Bacillus caldolyticus*, suggests dual regulation by pyrimidine and purine nucleotides. J Bacteriol 187: 1773–1782.

Cheung AL, Bayer AS, Zhang G, Gresham H, Xiong YQ (2004) Regulation of virulence determinants *in vitro* and *in vivo* in *Staphylococcus aureus*. FEMS Immunol Med Microbiol 40: 1–9.

Chevalier C, Boisset S, Romilly C, Masquida B, Fechter P, Geissmann T, Vandenesch F, Romby P (2010) *Staphylococcus aureus* RNAIII binds to two distant regions of *coa* mRNA to arrest translation and promote mRNA degradation. PLoS Pathog 6: e1000809.

Choonee N, Even S, Zig L, Putzer H (2007) Ribosomal protein L20 controls expression of the *Bacillus subtilis* infC operon via a transcription attenuation mechanism. Nucleic Acids Res 35: 1578–1588.

Chowdhury S, Maris C, Allain FH, Narberhaus F (2006) Molecular basis for temperature sensing by an RNA thermometer. EMBO J 25: 2487–2497.

Clatworthy AE, Pierson E, Hung DT (2007) Targeting virulence: a new paradigm for antimicrobial therapy. Nat Chem Biol 3: 541–548.

Collins JA, Irnov I, Baker S, Winkler WC (2007) Mechanism of mRNA destabilization by the *glmS* ribozyme. Genes Dev 21: 3356–3368.

Condon C (2007) Maturation and degradation of RNA in bacteria. Curr Opin Microbiol 10: 271–278.

Cramton SE, Schnell NF, Gotz F, Bruckner R (2000) Identification of a new repetitive element in *Staphylococcus aureus*. Infect Immun 68: 2344–2348.

Dunman PM, Murphy E, Haney S, Palacios D, Tucker-Kellogg G, Wu S, Brown EL, Zagursky RJ, Shlaes D, Projan SJ (2001) Transcription profiling-based identification of *Staphylococcus aureus* genes regulated by the *agr* and/or *sarA* loci. J Bacteriol 183: 7341–7353.

Fozo EM, Makarova KS, Shabalina SA, Yutin N, Koonin EV, Storz G (2010) Abundance of type I toxin-antitoxin systems in bacteria: searches for new candidates and discovery of novel families. Nucleic Acids Res 38: 3743–3759.

Gefen O, Balaban NQ (2009) The importance of being persistent: heterogeneity of bacterial populations under antibiotic stress. FEMS Microbiol Rev 33: 704–717.

Geisinger E, Adhikari RP, Jin R, Ross HF, Novick RP (2006) Inhibition of rot translation by RNAIII, a key feature of *agr* function. Mol Microbiol 61: 1038–1048.

Geissmann T, Chevalier C, Cros MJ, Boisset S, Fechter P, Noirot C, Schrenzel J, Francois P, Vandenesch F, Gaspin C, Romby P (2009a) A search for small noncoding RNAs in *Staphylococcus aureus* reveals a conserved sequence motif for regulation. Nucleic Acids Res 37: 7239–7257.

Geissmann T, Marzi S, Romby P (2009b) The role of mRNA structure in translational control in bacteria. RNA Biol 6: 153–160.

Green NJ, Grundy FJ, Henkin TM (2010) The T box mechanism: tRNA as a regulatory molecule. FEBS Lett 584: 318–324.

Grissa I, Vergnaud G, Pourcel C (2008) CRISPRcompar: a website to compare clustered regularly interspaced short palindromic repeats. Nucleic Acids Res 36: W145–8.

Grundy FJ, Haldeman MT, Hornblow GM, Ward JM, Chalker AF, Henkin TM (1997) The *Staphylococcus aureus ileS* gene, encoding isoleucyl-tRNA synthetase, is a member of the T-box family. J Bacteriol 179: 3767–3772.

Gutierrez-Preciado A, Jensen RA, Yanofsky C, Merino E (2005) New insights into regulation of the tryptophan biosynthetic operon in Gram-positive bacteria. Trends Genet 21: 432–436.

Horvath P, Barrangou R (2010) CRISPR/Cas, the immune system of bacteria and archaea. Science 327: 167–170.

Huntzinger E, Boisset S, Saveanu C. Benito Y, Geissmann T, Namane A, Lina G, Etienne J, Ehresmann B, Ehresmann C, Jacquier A, Vandenesch F, Romby P (2005) *Staphylococcus aureus* RNAIII and the endoribonuclease III coordinately regulate *spa* gene expression. EMBO J 24: 824–835.

Iost I, Dreyfus M (1995) The stability of *Escherichia coli lacZ* mRNA depends upon the simultaneity of its synthesis and translation. EMBO J 14: 3252–3261.

Irnov I, Sharma CM, Vogel J, Winkler WC (2010) Identification of regulatory RNAs in *Bacillus subtilis*. Nucleic Acids Res, in press.

Jelsbak L, Hemmingsen L, Donat S, Ohlsen K, Boye K, Westh H, Ingmer H, Frees D (2010) Growth phase-dependent regulation of the global virulence regulator Rot in clinical isolates of *Staphylococcus aureus*. Int J Med Microbiol 300: 229–236.

Johansson J, Mandin P, Renzoni A, Chiaruttini C, Springer M, Cossart P (2002) An RNA thermosensor controls expression of virulence genes in *Listeria monocytogenes*. Cell 110: 551–561.

Klein DJ, Ferre-D'Amare AR (2006) Structural basis of *glmS* ribozyme activation by glucosamine-6-phosphate. Science 313: 1752–1756.

Klinkert B, Narberhaus F (2009) Microbial thermosensors. Cell Mol Life Sci 66: 2661–2676.

Knezevic I, Bachem S, Sickmann A, Meyer HE, Stulke J, Hengstenberg W (2000) Regulation of the glucose-specific phosphotransferase system (PTS) of *Staphylococcus carnosus* by the antiterminator protein GlcT. Microbiology 146: 2333–2342.
Lasa I (2006) Towards the identification of the common features of bacterial biofilm development. Int Microbiol 9: 21–28.
Lee EJ, Groisman EA (2010) An antisense RNA that governs the expression kinetics of a multifunctional virulence gene. Mol Microbiol 76: 1020–1033.
Lenz DH, Mok KC, Lilley BN, Kulkarni RV, Wingreen NS, Bassler BL (2004) The small RNA chaperone Hfq and multiple small RNAs control quorum sensing in *Vibrio harveyi* and *Vibrio cholerae*. Cell 118: 69–82.
Livny J, Teonadi H, Livny M, Waldor MK (2008) High-throughput, kingdom-wide prediction and annotation of bacterial non-coding RNAs. PLoS One 3: e3197.
Llopis PM, Jackson AF, Sliusarenko O, Surovtsev I, Heinritz J, Emonet T, Jacobs-Wagner C (2010) Spatial organization of the flow of genetic information in bacteria. Nature 466: 77–81.
Loh E, Dussurget O, Gripenland J, Vaitkevicius K, Tiensuu T, Mandin P, Repoila F, Buchrieser C, Cossart P, Johansson J (2009) A trans-acting riboswitch controls expression of the virulence regulator PrfA in Listeria monocytogenes. Cell 139: 770–779.
Lu Y, Turner RJ, Switzer RL (1996) Function of RNA secondary structures in transcriptional attenuation of the *Bacillus subtilis pyr* operon. Proc Natl Acad Sci U S A 93: 14462–14467.
Marchais A, Naville M, Bohn C, Bouloc P, Gautheret D. (2009) Single-pass classification of all noncoding sequences in a bacterial genome using phylogenetic profiles. Genome Res 19: 1084–1092.
Marraffini LA, Sontheimer EJ (2008) CRISPR interference limits horizontal gene transfer in *staphylococci* by targeting DNA. Science 322: 1843–1845.
Marraffini LA, Sontheimer EJ (2010a) CRISPR interference: RNA-directed adaptive immunity in bacteria and archaea. Nat Rev Genet 11: 181–190.
Marraffini LA, Sontheimer EJ (2010b) Self versus non-self discrimination during CRISPR RNA-directed immunity. Nature 463: 568–571.
Marzi S, Fechter P, Chevalier C, Romby P, Geissmann T (2008) RNA switches regulate initiation of translation in bacteria. Biol Chem 389: 585–598.
Morfeldt E, Taylor D, von Gabain A, Arvidson S (1995) Activation of alpha-toxin translation in *Staphylococcus aureus* by the *trans*-encoded antisense RNA, RNAIII. EMBO J 14: 4569–4577.
Mulhbacher J, Brouillette E, Allard M, Fortier LC, Malouin F, Lafontaine DA (2010) Novel riboswitch ligand analogs as selective inhibitors of guanine-related metabolic pathways. PLoS Pathog 6: e1000865.
Murphy E (1985) Nucleotide sequence of *ermA*, a macrolide-lincosamide-streptogramin B determinant in *Staphylococcus aureus*. J Bacteriol 162: 633–640.
Nagata M, Kaito C, Sekimizu K (2008) Phosphodiesterase activity of CvfA is required for virulence in *Staphylococcus aureus*. J Biol Chem 283: 2176–2184.
Narberhaus F (2010) Translational control of bacterial heat shock and virulence genes by temperature-sensing mRNAs. RNA Biol 7:
Narberhaus F, Waldminghaus T, Chowdhury S (2006) RNA thermometers. FEMS Microbiol Rev 30: 3–16.
Nechooshtan G, Elgrably-Weiss M, Sheaffer A, Westhof E, Altuvia S (2009) A pH-responsive riboregulator. Genes Dev 23: 2650–2662.
Novick RP (2003) Autoinduction and signal transduction in the regulation of *staphylococcal* virulence. Mol Microbiol 48: 1429–1449.
Novick RP, Geisinger E (2008) Quorum sensing in *staphylococci*. Annu Rev Genet 42: 541–564.
Novick RP, Iordanescu S, Projan SJ, Kornblum J, Edelman I (1989) pT181 plasmid replication is regulated by a countertranscript-driven transcriptional attenuator. Cell 59: 395–404.
Novick RP, Ross HF, Projan SJ, Kornblum J, Kreiswirth B, Moghazeh, S (1993) Synthesis of staphylococcal virulence factors is controlled by a regulatory RNA molecule. EMBO J 12: 3967–3975.

Opdyke JA, Kang JG, Storz G (2004) GadY, a small-RNA regulator of acid response genes in *Escherichia coli*. J Bacteriol 186: 6698–6705.
Pfeiffer V, Papenfort K, Lucchini S, Hinton JC, Vogel J (2009) Coding sequence targeting by MicC RNA reveals bacterial mRNA silencing downstream of translational initiation. Nat Struct Mol Biol 16: 840–846.
Pichon C, Felden B (2005) Small RNA genes expressed from *Staphylococcus aureus* genomic and pathogenicity islands with specific expression among pathogenic strains. Proc Natl Acad Sci U S A 102: 14249–14254.
Proctor RA, von Eiff C, Kahl BC, Becker K, McNamara P, Herrmann M, Peters G (2006) Small colony variants: a pathogenic form of bacteria that facilitates persistent and recurrent infections. Nat Rev Microbiol 4: 295–305.
Proshkin S, Rahmouni AR, Mironov A, Nudler E (2010) Cooperation between translating ribosomes and RNA polymerase in transcription elongation. Science 328: 504–508.
Queck SY, Jameson-Lee M, Villaruz AE, Bach TH, Khan BA, Sturdevant DE, Ricklefs SM, Li M, Otto M (2008) RNAIII-independent target gene control by the *agr* quorum-sensing system: insight into the evolution of virulence regulation in *Staphylococcus aureus*. Mol Cell 32: 150–158.
Ramesh A, Winkler WC (2010) Magnesium-sensing riboswitches in bacteria. RNA Biol 7:
Rasmussen S, Nielsen HB, Jarmer H (2009) The transcriptionally active regions in the genome of *Bacillus subtilis*. Mol Microbiol 73: 1043–1057.
Roberts C, Anderson KL, Murphy E, Projan SJ, Mounts W, Hurlburt B, Smeltzer M, Overbeek R, Disz T, Dunman PM (2006) Characterizing the effect of the *Staphylococcus aureus* virulence factor regulator, SarA, on log-phase mRNA half-lives. J Bacteriol 188: 2593–2603.
Romby P, Charpentier E (2010) An overview of RNAs with regulatory functions in gram-positive bacteria. Cell Mol Life Sci 67: 217–237.
Seif E, Altman S (2008) RNase P cleaves the adenine riboswitch and stabilizes *pbuE* mRNA in *Bacillus subtilis*. RNA 14: 1237–1243.
Serganov A (2009) The long and the short of riboswitches. Curr Opin Struct Biol 19: 251–259.
Serganov A, Huang L, Patel DJ (2009) Coenzyme recognition and gene regulation by a flavin mononucleotide riboswitch. Nature 458: 233–237.
Shahbabian K, Jamalli A, Zig L, Putzer H (2009) RNase Y, a novel endoribonuclease, initiates riboswitch turnover in *Bacillus subtilis*. EMBO J 28: 3523–3533.
Shimoni Y, Altuvia S, Margalit H, Biham O (2009) Stochastic analysis of the SOS response in *Escherichia coli*. PLoS One 4: e5363.
Shimoni Y, Friedlander G, Hetzroni G, Niv G, Altuvia S, Biham O, Margalit H (2007) Regulation of gene expression by small non-coding RNAs: a quantitative view. Mol Syst Biol 3: 138.
Smith AM, Fuchs RT, Grundy FJ, Henkin TM (2010) Riboswitch RNAs: Regulation of gene expression by direct monitoring of a physiological signal. RNA Biol 7:
Somerville GA, Proctor RA (2009) At the crossroads of bacterial metabolism and virulence factor synthesis in *Staphylococci*. Microbiol Mol Biol Rev 73: 233–248.
Sudarsan N, Cohen-Chalamish S, Nakamura S, Emilsson GM, Breaker RR (2005) Thiamine pyrophosphate riboswitches are targets for the antimicrobial compound pyrithiamine. Chem Biol 12: 1325–1335.
Toledo-Arana A, Repoila F, Cossart P (2007) Small noncoding RNAs controlling pathogenesis. Curr Opin Microbiol 10: 182–188.
Tormo-Mas MA, Mir I, Shrestha A, Tallent SM, Campoy S, Lasa I, Barbe J, Novick RP, Christie GE, Penades JR (2010) Moonlighting bacteriophage proteins derepress staphylococcal pathogenicity islands. Nature 465: 779–782.
Tormo MA, Ferrer MD, Maiques E, Ubeda C, Selva L, Lasa I, Calvete JJ, Novick RP, Penades JR (2008) *Staphylococcus aureus* pathogenicity island DNA is packaged in particles composed of phage proteins. J Bacteriol 190: 2434–2440.
Vandenesch F, Projan SJ, Kreiswirth B, Etienne J, Novick RP (1993) *Agr*-related sequences in *Staphylococcus lugdunensis*. FEMS Microbiol Lett 111: 115–122.
Vogel J (2009) A rough guide to the non-coding RNA world of *Salmonella*. Mol Microbiol 71: 1–11.

Waldminghaus T, Gaubig LC, Narberhaus F (2007) Genome-wide bioinformatic prediction and experimental evaluation of potential RNA thermometers. Mol Genet Genomics 278: 555–564.

Wang J, Henkin TM, Nikonowicz EP (2010) NMR structure and dynamics of the Specifier Loop domain from the *Bacillus subtilis tyrS* T box leader RNA. Nucleic Acids Res 38: 3388–3398.

Wassarman KM (2007) 6S RNA: a small RNA regulator of transcription. Curr Opin Microbiol 10: 164–168.

Wassarman KM, Saecker RM (2006) Synthesis-mediated release of a small RNA inhibitor of RNA polymerase. Science 314: 1601–1603.

Waters LS, Storz G (2009) Regulatory RNAs in bacteria. Cell 136: 615–628.

Winkler WC, Cohen-Chalamish S, Breaker RR (2002) An mRNA structure that controls gene expression by binding FMN. Proc Natl Acad Sci U S A 99: 15908–15913.

Winkler WC, Nahvi A, Roth A, Collins AJ, Breaker RR (2004) Control of gene expression by a natural metabolite-responsive ribozyme. Nature 428: 281–286.

Wyatt MA, Wang W, Roux CM, Beasley FC, Heinrichs DE, Dunman PM, Magarvey NA (2010) *Staphylococcus aureus* Nonribosomal Peptide Secondary Metabolites Regulate Virulence. Science

Yao Z, Barrick J, Weinberg Z, Neph S, Breaker R, Tompa M, Ruzzo WL (2007) A computational pipeline for high- throughput discovery of *cis*-regulatory noncoding RNA in prokaryotes. PLoS Comput Biol 3: e126.

Zegans ME, Wagner JC, Cady KC, Murphy DM, Hammond JH, O'Toole GA (2009) Interaction between bacteriophage DMS3 and host CRISPR region inhibits group behaviors of *Pseudomonas aeruginosa*. J Bacteriol 191: 210–219.

Chapter 4
Pseudomonas Aeruginosa Small Regulatory RNAs

Elisabeth Sonnleitner, Theresa Sorger-Domenigg, and Udo Bläsi*

1 Introduction

Pseudomonas aeruginosa (family Pseudomonaceae) is a Gram-negative, rod shaped, aerobic, motile bacterium, which is known for its ability to survive in many habitats, primarily in water, soil, and vegetation. Being an opportunistic human pathogen, *P. aeruginosa* is the most common Gram-negative bacterium found in nosocomial infections. It causes urinary tract infections, respiratory system infections, dermatitis, soft tissue infections, bone and joint infections, gastrointestinal infections and a variety of systemic infections, predominantly in immunosuppressed patients. Particularly patients hospitalized with cancer, AIDS, cystic fibrosis and severe burns are prone to *P. aeruginosa* infections (van Delden and Iglewski 1998).

The adherence to host epithelial cells is mediated by type IV pili and several other non-pilus adhesins. In addition, flagella, which are primarily responsible for motility, can act as adhesins (Feldman *et al.* 1998). *P. aeruginosa* produces several extracellular products that after colonization can cause extensive tissue damage, bloodstream invasion, and dissemination. These include (i) exotoxin A (Wick *et al.* 1990), which is responsible for local tissue damage, bacterial invasion, and possibly immunosuppression, (ii) two haemolysins, phospholipase C and rhamnolipid, which contribute to tissue invasion by breaking down lipids and lecithin, and (iii) several proteases including LasB elastase, LasA elastase, as well as alkaline protease (Gambello *et al.* 1993, Toder *et al.* 1994). *P. aeruginosa* utilizes a type III secretion system as a powerful tool to directly inject virulence factors (ExoS, ExoT, ExoU and ExoY) into the host cell (Hauser 2009).

The bacterium is resistant to several structurally unrelated antibiotics. Resistance mechanisms include low outer membrane permeability, multidrug efflux pumps

* Max F. Perutz Laboratories, Department of Microbiology, Immunobiology and Genetics, University of Vienna, Dr. Bohrgasse 9, A-1030 Vienna, Austria

(tetracycline, imipenem, fluoroquinolones, aminoglycosides) and the production of antibiotic modifying enzymes (aminoglycosides, β-lactams) (Nakae 1995, Nikaido 1996, Piddock 2006). In addition, the production of exopolysaccharide and the ability of *P. aeruginosa* to form biofilms protects the bacteria from therapeutic intervention (Davies *et al.* 1998).

P. aeruginosa is a highly adaptable organism with a large dynamic genome (Mathee *et al.* 2008), approximately 10% of which is devoted to elements that regulate the production of virulence factors, secondary metabolites, expression of multidrug efflux pumps and biofilm formation. These regulatory functions include quorum sensing, the GacS-GacA two-component system, alternative sigma factors (RpoS, RpoN), and the post-transcriptional repressor protein RsmA (Juhas *et al.* 2005). *P. aeruginosa* possesses a complex quorum- sensing cascade enabling it to communicate, sense environmental changes, and modulate inflammatory and immune responses in mammals (Cooley *et al.* 2008). It consists of three interlinked systems (Las, Rhl, PQS) employing acyl-homoserinelactones and alkyl quinolones as signalling molecules (Williams and Camara 2009). The quorum-sensing systems are modulated by a large number of regulators, such as RpoS, RsmA, MvaT, PtxR, GacA, RpoN and Hfq (Williams and Camara 2009).

Small bacterial regulatory RNAs (sRNA), also known as riboregulators, are regulatory elements that modulate gene expression at the post-transcriptional level (Waters and Storz 2009). The majority of the identified sRNAs adjusts bacterial physiology in response to environmental changes or stress conditions. One class of sRNAs base-pairs with mRNAs and changes their translation efficiency and/or stability. The other class of sRNAs interacts with proteins and modulates their activity. In addition to binding to proteins and mRNA, sRNAs can target DNA, as exemplified by the clustered, regularly interspaced, short palindromic repeats (CRISPR) in *Streptococcus thermophilus* (van der Oost *et al.* 2009). sRNAs are either generated through processing or synthesized as primary transcripts. The size of "small" regulatory RNAs ranges from 50–500 nts. sRNAs are generally non-coding, but there are also sRNAs that encode polypeptides and concomitantly act as riboregulators. For instance, the *Staphylococcus aureus* RNAIII and *Escherichia coli* SgrS code for the δ-haemolysin and the SgrT protein, respectively (Janzon *et al.* 1989, Wadler and Vanderpool 2007). In pathogenic bacteria, such as *P. aeruginosa*, sRNAs are involved in regulation of virulence gene expression (Heurlier *et al.* 2004; Kay *et al.* 2006; Oglesby *et al.* 2008).

The RNA chaperone Hfq is often required for the stability/function of sRNAs. Hfq is a pleiotropic post-transcriptional regulator, which was first discovered as a host factor for bacteriophage Qβ (Franze de Fernandez *et al.* 1968). It belongs to the eukaryotic and archaeal family of Sm and Sm-like proteins, a group of RNA binding proteins (Møller *et al.* 2002, Sun *et al.* 2002, Zhang *et al.* 2002; see also chapter 8 of this book). Several studies have suggested that Hfq binds concomitantly to sRNAs and mRNAs, which in turn results in an increased local concentration, aiding the RNA-RNA interaction (Schumacher *et al.* 2002, Mikulecky *et al.* 2004, Vecerek *et al.* 2008). Furthermore, Hfq was also shown to protect the *Escherichia coli* sRNAs RyhB and DsrA RNAs from cleavage by RNase E (Massé *et al.* 2003, Moll *et al.* 2003), an endonuclease involved in mRNA decay as well as in processing of tRNA and rRNA precursors. This observation can probably be extended towards many riboregulators, as Hfq binding sites can coincide with recognition sites for RNase E (Massé *et al.* 2003, Moll *et al.* 2003, Sorger-Domenigg *et al.* 2007). For a detailed discussion of Hfq-associated regulatory small RNAs see also chapter 2 of this book.

This chapter provides an overview on current knowledge of different classes and functions of regulatory RNAs in *P. aeruginosa.*

2 Bacterial Regulatory RNAs and their Mode of Action

In bacteria three major classes of small regulatory RNAs are known: (i) sRNAs, which interact with target mRNAs by base-pairing, (ii) those, which bind to proteins and modulate their activity and (iii) clustered regulatory interspaced short palindromic repeats (CRISPRs), which may interact with DNA and cause degradation (Fig. 1).

Base-pairing sRNAs can be divided into *cis* and *trans*-acting sRNAs. The majority of *cis*-acting sRNAs, which are transcribed convergent to their target genes, are encoded by accessory elements, plasmids, transposons and phages. Only a few *cis*-acting RNAs have been found in the bacterial chromosome (Brantl 2007). These RNAs are usually small (50–300 nts) and regulate biological processes such as replication, conjugation, toxin-antitoxin modules, transposition, mRNA degradation and translation initiation. In addition, some control metabolic processes, such as the p3 RNA of *Clostridium acetobutylicum* (Fierro-Monti *et al.* 1992) or govern expression of genes encoding toxic proteins (e.g. RatA/TxpA of *B. subtilis* (Silvaggi *et al.* 2005)). The mechanisms of action include transcription attenuation, translation inhibition, inhibition of primer maturation, promotion or inhibition of mRNA degradation and inhibition of RNA pseudo-knot formation (Brantl 2007). A novel mechanism entails premature termination of transcription of the *cis*-encoded mRNA target *icsA* by RnaG in *Shigella flexneri* (Giangrossi *et al.* 2010). *Cis*-encoded sRNAs are fully complementary to their target mRNAs. Therefore, the RNA-RNA interaction is usually strong and does not require proteins that facilitate annealing, such as the RNA chaperone Hfq. Tiling array approaches (Toledo-Arana *et al.* 2009) as well as genomic SELEX (Lorenz *et al.* 2006, Lorenz *et al.* 2010) have suggested a number of chromosomally encoded *cis*-antisense RNAs (see also chapter 5 of this book). Long (>1000 nts) RNAs spanning more than one gene as well as *cis*-acting sRNAs that overlap with the 5' or 3' UTR of mRNAs have been detected (Toledo-Arana *et al.* 2009).

In contrast to the previously mentioned *cis*-encoded anti-sense RNAs, *trans*-acting sRNAs are generally encoded distantly from their target mRNA(s). The majority of the hitherto characterized sRNAs act as negative regulators by preventing ribosome loading onto the mRNA through base-pairing with, or in the vicinity of, the ribosome-binding site (rbs). As a result, the respective mRNA is prone to rapid decay (for reviews, Gottesman 2004, Kaberdin and Bläsi 2006, Papenfort and Vogel 2009). Variations of negative regulation by sRNAs exerted distantly from

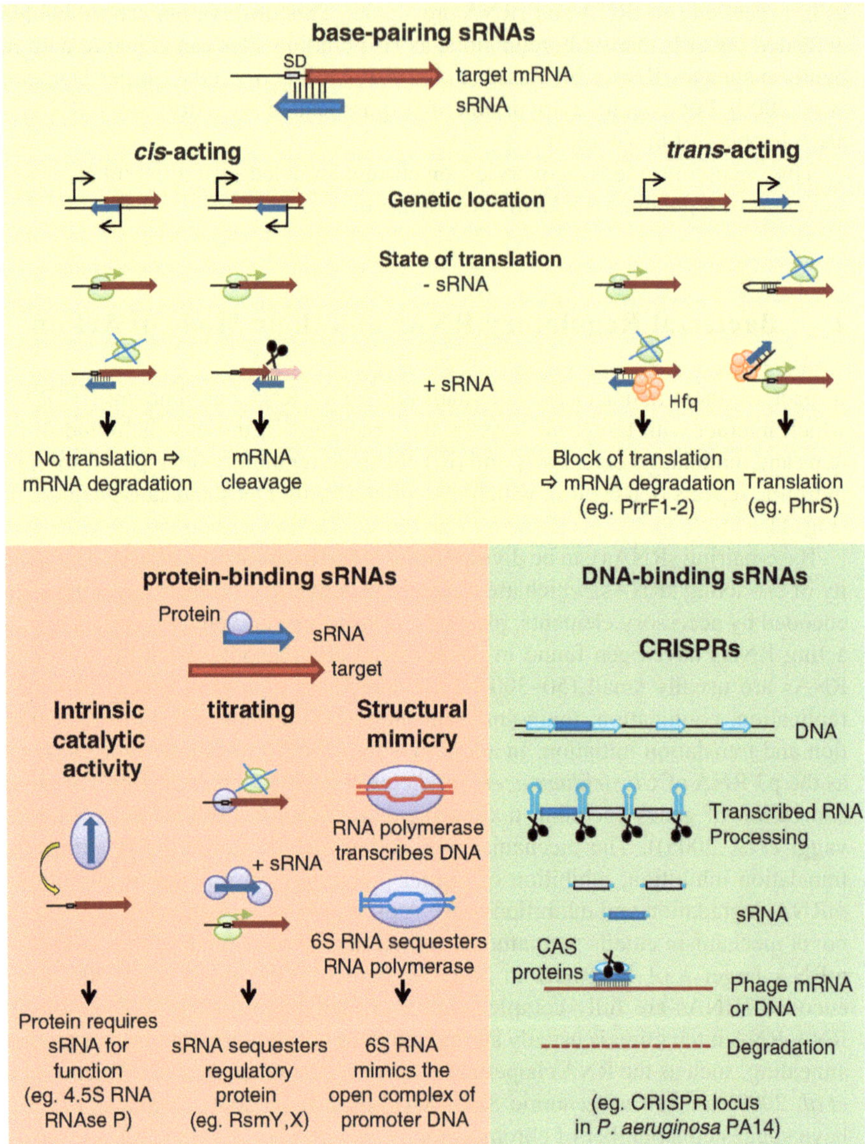

Fig. 1. Overview of the three classes of prokaryotic sRNAs and their mechanism. A detailed description is given in 2

the rbs can further include binding to, and obstructing of, an upstream ribosome loading site (Darfeuille *et al.* 2007), obstruction of a C/A rich element (Sharma *et al.* 2007) as well as binding to the immediate (Bouvier *et al.* 2008), or within the, mRNA coding region (Pfeiffer *et al.* 2009). Positive regulation by sRNAs appears to be less frequent. In this case the sRNAs act by means of an "anti-antisense" mechanism and compete with the formation of, or melt, intra-molecular secondary structures that block ribosome binding (Morfeldt *et al.* 1995, Repoila *et al.* 2003, Urban and Vogel 2008). In addition, indirect translation silencing by preventing translation of an upstream reading frame to which the target mRNA is positively translationally coupled has also been reported (Vecerek *et al.* 2007). In the case of *V. cholerae* Qrr sRNA, the same region can even be used to activate and repress mRNAs (Lenz *et al.* 2004). A hallmark of *trans*-acting sRNAs is their imperfect complementary with their mRNA targets. Therefore, some sRNAs can interact with several mRNA targets (Papenfort and Vogel 2009). The requirement of Hfq in riboregulation appears to be rooted in this limited complementarity. *Trans*-encoded sRNAs are generally synthesized under certain growth or stress conditions (Repoila and Darfeuille 2009). For instance *E. coli* sRNAs are synthesized during iron depletion, oxidative stress, outer membrane stress, elevated glycine levels, changes in glucose concentration, and elevated glucose-phosphate levels (De Lay and Gottesman 2009, Görke and Vogel 2008, Gottesman 2005, Johansen *et al.* 2006, Urbanowski *et al.* 2000).

Protein-binding RNAs can be either essential for the intrinsic catalytic activity of the RNA-Protein complex, like RNase P (Kazantsev and Pace 2006), or act by antagonizing the function of a regulatory protein through mimicking the structure of other nucleic acid targets. For example, the 6S RNA mimics the open complex of promoter DNA, and thus sequesters the σ^{70}-containing RNA polymerase (Wassarman 2007). The *E. coli* CsrB/C RNAs carry multiple GGA motifs, which are required to sequester the translational repressor CsrA, the central regulator of carbon storage. This interaction prevents binding of CsrA to translation initiation regions of target mRNAs (Babitzke and Romeo 2007).

Clustered regularly interspaced short palindromic repeats (CRISPRs) are a novel class of repetitive DNA elements, where the repeats are interspersed with spacers homologous to DNA from phages and plasmids (Sorek *et al.* 2008). Adjacent to those clusters CRISPR- associated (CAS) genes are encoded, which are needed for CRISPR activity. It has been shown that integration of a prophage into the CRISPR array confers resistance to subsequent phage infections (Barrangou *et al.* 2007). This interspaced DNA is transcribed into a long RNA, which is processed by the *cas* gene products into single spacers in a manner reminiscent of the eukaryotic RNAi pathway. They directly base-pair with cognate nucleic acid targets (Waters and Storz 2009). Although many details remain poorly understood, it has become clear that the CRISPR–Cas system is a specific, adaptable and inheritable protection system in Prokaryotes that provides them with immunity against accessory genetic elements (van der Oost *et al.* 2009). For a detailed presentation of the CRISPR/Cas and CRISPR/Cmr systems of archaea see chapter 9 of this book.

3 *P. aeruginosa* Housekeeping RNAs

P. aeruginosa shares a number of "housekeeping RNAs" with *E. coli* and other bacteria, including 6S RNA (Sonnleitner *et al.* 2008, Vogel *et al.* 1987), which inhibits transcription from certain σ^{70} promoters in stationary phase (Wassarman 2007).

6S RNA accumulates during exponential growth and reaches elevated levels when cells enter stationary phase. As mentioned above, it sequesters the σ^{70}-containing RNA polymerase by mimicking the open complex structure of promoter DNA, and thus provokes a reprogramming of the transcriptional activity. At outgrowth from stationary phase the NTP levels increase to a threshold and 6S RNA is used as a template for transcription of *de novo* RNA (pRNAs), which leads to dissociation of the 6S-pRNA complex. As a consequence, the σ^{70}-dependent promoters can be re-used (Wassarman 2007). 6S RNA was experimentally verified in *P. aeruginosa* (Vogel *et al.* 1987) and appeared in a bioinformatic screen (Sonnleitner *et al.* 2008). The primary and secondary structure of *P.aeruginosa* 6S RNA is similar to that of *E. coli* (Vogel *et al.* 1987), and therefore a similar mechanism can be hypothesized.

4.5S RNA is part of the signal recognition particle (SRP) and is essential in order for the SRP to target proteins to the cytoplasmatic membrane. SRP is a complex of Ffh protein and 4.5S RNA, which binds to the hydrophobic signal sequence of nascent membrane proteins. The ribosome-SRP complex moves to the cytoplasmic membrane, where the nascent membrane protein is co-translationally inserted into the membrane. This mechanism prevents misfolding or unfolding of hydrophobic proteins in the cytoplasm (Keenan *et al.* 2001). *P. aeruginosa* 4.5S RNA was identified in two bioinformatic screens (Gonzalez *et al.* 2008, Sonnleitner *et al.* 2008). *P. aeruginosa* 4.5S RNA has 75% sequence identity with *E. coli* 4.5S RNA, and thus is likely to have the same function in both organisms. *E. coli* 4.5S RNA is processed from a longer precursor by RNase P, an enzyme that contains a catalytic RNA. *Pseudomonas* RNase P RNA was first isolated from *P. fluorescens* (James *et al.* 1988) but was also detected in three bioinformatic screens (Gonzalez *et al.* 2008, Livny *et al.* 2006, Sonnleitner *et al.* 2008) in *P. aeruginosa.*

Transfer-messenger RNA (tmRNA) has features of tRNA and mRNA (Keiler 2007). In case of ribosome stalling, it is charged with alanine and enters the ribosomal A-site. The nascent polypeptide is then transferred to the alanyl-tmRNA and a small ORF, encoded by the mRNA part of tmRNA, is translated. The tmRNA RNA codons specify an amino acid tag, which subjects the truncated polypeptide to decay. In case of premature termination or mRNA truncation this mechanism ensures the release of stalled ribosomes and degradation of the truncated and tagged polypeptide. The tag sequence of *P. aeruginosa* tmRNA is similar to that of *E. coli* (Williams and Bartel 1996).

4 Protein Sequestering RNAs

4.1 RsmY/Z

Many γ-proteobacteria regulate the synthesis of secondary metabolites, extracellular enzymes and carbon storage compounds *via* a conserved system, termed the Gac/Rsm system in *P. aeruginosa.* The RNAs of the Rsm system, RsmY/Z (orthologues of *E. coli* CsrB/C) and RsmX (*P. fluorescens*), modulate the activity of the translational repressor protein RsmA (Lapouge *et al.* 2008). In Pseudomonads, the homodimers RsmA and RsmE (*P. fluorescens*) negatively control expression of virulence factors by binding to GGA motifs in the leader of their target mRNAs (Lapouge *et al.* 2007, Schubert *et al.* 2007). In addition, the stationary and stress sigma factor *rpoS* was shown in *P. fluorescens* to be either directly or indirectly under control of RsmA, which largely expands the number of genes and processes under control of the Rsm system (Heeb *et al.* 2005).

Expression of the sRNAs RsmY/Z is induced by the GacA/GacS two-component system (Fig. 2), which is modulated by the two hybrid sensor kinases RetS and LadS (Ventre *et al.* 2006). RetS has been shown to be required for the transcription of genes associated with cytotoxicity and acute infections (Goodman *et al.* 2004, Laskowski *et al.* 2004, Ventre *et al.* 2006). On the other hand, RetS negatively controls the transcription of genes that are associated with the synthesis of biofilm oligosaccharides (Goodman *et al.* 2004, Ventre *et al.* 2006). It directly interacts with GacS (Goodman *et al.* 2009) and acts as an antagonist of GacA/S by blocking *rsmZ* transcription (Ventre *et al.* 2006). RetS seems to be temperature-sensitive in *P. fluorescens* (Humair *et al.* 2009). The kinases LadS and GacS, on the other hand, directly counteract RetS, as they stimulate the expression of the *psl* and *pel* operons by up-regulating the expression of the RsmA-antagonist RsmZ (Kay *et al.* 2006, Ventre *et al.* 2006). This signal transduction network in which the activities of the signal-receiving sensor kinases LadS, RetS, and GacS regulate expression of virulence at the transcriptional and posttranscriptional level provides a "switch" between acute and chronic infection phenotypes (Ventre *et al.* 2006). In addition, it has been shown that the histidine phosphotransfer protein B (HptB), which operates as a phosphorylation relay, acts only on *rsmY via* GacA, most probably by regulating the putative anti-anti σ-factor PA3347 (Fig. 2) (Bordi *et al.* 2010, Hsu *et al.* 2008).

In addition, the sRNAs RsmY/Z can inhibit their own transcription through a feedback mechanism by interfering with the GacA/S system, most probably by allosterically inhibiting GacA phosphorylation or by binding to the putative GacA box (Lapouge *et al.* 2008). It was shown that the A+T-rich DNA segment upstream of *rsmZ* is bound and silenced by MvaT and MvaU (Fig. 2), the global gene regulators of the H-NS family (Brencic *et al.* 2009). In addition, in *P. aeruginosa* the sigma factor RpoN affects GacA expression in a negative manner (Heurlier *et al.* 2003). The GacA-regulated type of sRNAs are present in several bacterial species and vary in size between 100 and 479 nts, the largest known being RsmB of *Erwinia carotovora* (Babitzke and Romeo 2007). All of them possess multiple GGA motifs that are located in single-stranded RNA regions, mostly in loops or between stem-loop structures. These motifs were shown to be required for interaction with RsmA (Valverde *et al.* 2004). While *P. aeruginosa* contains two RsmA-binding sRNAs, RsmY and RsmZ (Kay *et al.* 2006), *P. fluorescens* contains three RsmX, RsmY and RsmZ (Kay *et al.* 2005), all of which function as antagonists of RsmA. The synthesis of the sRNAs as well as of RsmA and/or RsmE increases throughout growth, whereas *rsmZ* expression is observed later in the growth cycle (Kay *et al.* 2005). The *P. fluorescens* PsrA protein was recently shown to activate transcription of both the *rpoS* as well as the *rsmZ* gene (Humair *et al.* 2010). Both

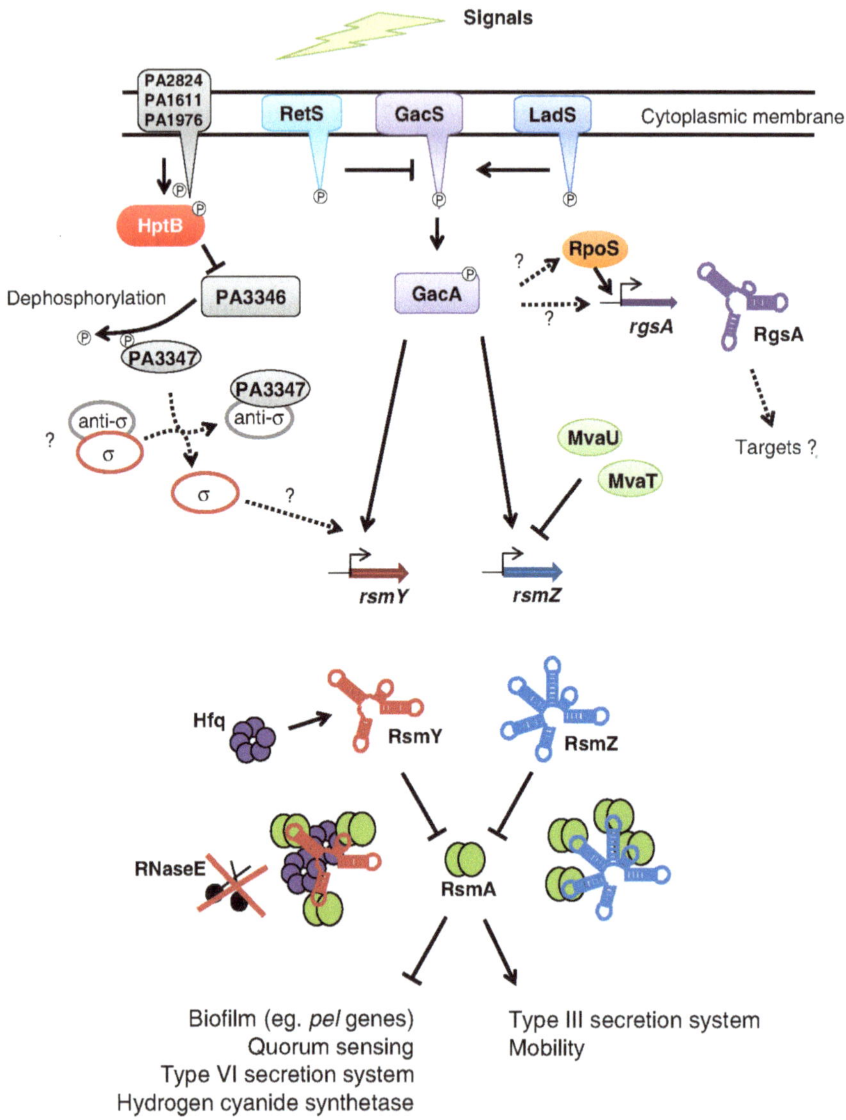

Fig. 2. Gac/Rsm regulatory network in *P. aeruginosa*. RsmA is a translational regulator, which can act as a translational repressor (Brencic and Lory 2009). The two sRNAs RsmY and RsmZ antagonize RsmA function by sequestering the protein. Hfq stabilizes RsmY by blocking RNase E cleavage (Sonnleitner *et al.* 2006; Sorger-Domenigg *et al.* 2007). The expression of the two sRNAs is under control of the two-component system GacA/S, which is regulated by RetS and LadS. Moreover, RsmY/Z transcription is regulated by other factors, which repress specifically one of the sRNA genes: MvaU and MvaT act on *rsmZ* and the HptB system acts on *rsmY*, respectively. HptB is activated by three sensor kinases PA2842, PA1611 and PA1976 (Hsu *et al.* 2008). In the absence of HptB, PA3346 dephosphorylates PA3347, which shows homology to anti-anti sigma factors. However, neither the corresponding sigma nor the anti-sigma factor has been identified (Bordi *et al.* 2010; Hsu *et al.* 2008). The Gac-system is also required for expression of the sRNA RgsA, which does not sequester RsmA. Nevertheless, the regulation may be indirect through RpoS (Gonzales *et al.* 2008)

RsmA and RsmE stabilize the sRNAs, possibly by protecting them from nucleases (Kay *et al.* 2006, Reimmann *et al.* 2005). Moreover, it was shown that binding of Hfq stabilizes RsmY in *P. aeruginosa* (Sonnleitner *et al.* 2006) and that RsmA and Hfq can concomitantly bind to RsmY (Sorger-Domenigg *et al.* 2007) (Fig. 2). Binding of Hfq to RsmY was shown to protect the sRNA from RNase E cleavage *in vitro* (Sorger-Domenigg *et al.* 2007). This effect of Hfq on the sRNA could only be observed for *P. aeruginosa* RsmY and not for RsmZ (Sonnleitner *et al.* 2006). This complex regulatory network allows *P. aeruginosa* to adapt quickly to changing environmental cues and to set the stage for the type of infection that will develop.

4.2 CrcZ

Catabolite repression (CR) prevents expression of genes required for degradation of the less-preferred carbon source in the presence of the preferred substrate (Görke and Stülke 2008, Ullmann 1996). In many bacteria, CR affects transcription of target genes, but in Pseudomonads CR seems to operate post-transcriptionally (Moreno *et al.* 2007; Sonnleitner *et al.* 2009). In *P. aeruginosa* the RNA binding protein Crc is responsible for regulation of functions involved in carbon degradation pathways, e.g. *amiE* encoding an aliphatic amidase (Sonnleitner *et al.* 2009). Crc acts as a translational repressor by binding to CA-rich sequences (CA-motifs) in the vicinity of the rbs of target mRNAs (Sonnleitner *et al.* 2009). The apparent analogy to RsmA in terms of translational repression suggests that the function of Crc might be antagonized by RNAs. A search of intergenic regions for CA motifs led to the discovery of the RNA CrcZ (Sonnleitner *et al.* 2009), which is transcribed in inverse orientation to the previously published P30 sRNA (Livny *et al.* 2006). It contains five CA-motifs (Sonnleitner *et al.* 2009). The sRNA CrcZ was shown to antagonize Crc in catabolite repression and CrcZ expression was shown to be governed by the CbrA/B two-component system in a RpoN-dependent manner (Sonnleitner *et al.* 2009) (Fig. 3). The level of induction depends on the carbon source, whereas CrcZ levels are low in the presence of a preferred carbon (C) source like succinate, elevated in the presence of an intermediate C-source like glucose and high in the presence of a poor C-source such as mannitol (Sonnleitner *et al.* 2009). CrcZ binds to Crc and thereby alleviates the negative control exerted by Crc on target mRNAs (Fig. 3). Interestingly, CrcZ does not have a rho-independent terminator stem-loop, and thus might depend on Rho protein (Sonnleitner *et al.* 2009).

5 *Verified and Candidate P. Aeruginosa* Base-pairing sRNAs

5.1 Prrf 1/2

Iron serves as a cofactor of many enzymes, which are involved in metabolic pathways such as the tricarboxylic acid (TCA) cycle, respiration, DNA synthesis and

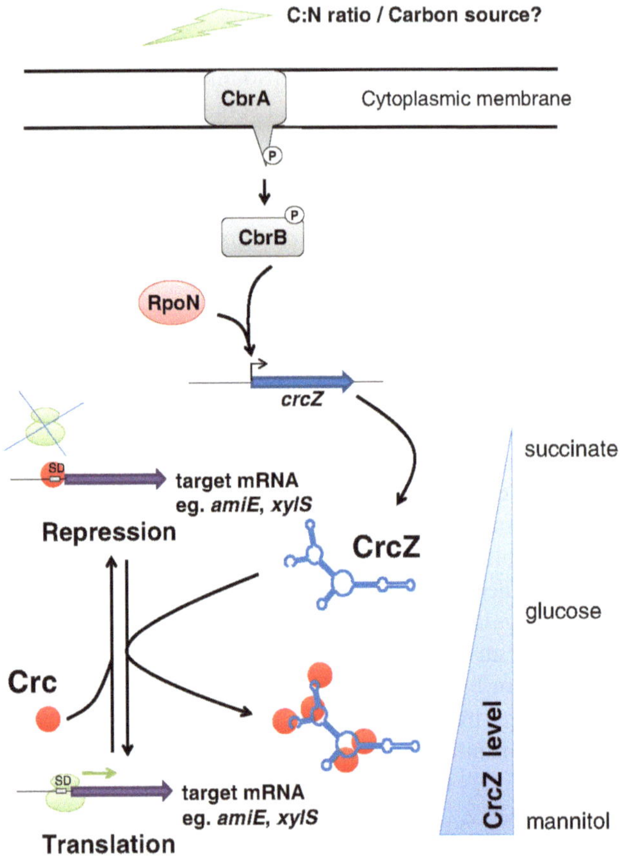

Fig. 3. Model for CR in *P. aeruginosa* (modified from Sonnleitner *et al.* 2009). The *crcZ* gene is controlled by the two-component system CbrAB and RpoN, which permits a differential synthesis of CrcZ according to different C-sources. When a preferred C-source is present, CrcZ levels are low and Crc binds to target mRNAs, most of which are involved in degradation of non-preferred C-sources. In contrast, when *P. aeruginosa* grows on a non-preferred C-source, CrcZ is up-regulated and binds to, and sequesters, Crc, which results in translation of target genes. When cells grow in the presence of glucose, intermediate levels of CrcZ are present, which leads to both repression and activation of target genes, respectively (Sonnleitner *et al.* 2009)

the synthesis of metabolites. Moreover, iron acquisition is a central prerequisite for successful establishment of pathogens in mammalian hosts. In most Gram-negative bacteria the ferric uptake regulator Fur negatively controls the expression of genes involved in iron acquisition and of iron-regulated genes involved in virulence (Hantke 2001). Fur can also regulate iron-metabolism in an indirect manner *via* extra-cytoplasmatic sigma factors or by the sRNA RyhB (Massé and Gottesman 2002, Massé *et al.* 2007). During iron starvation *E. coli* Fur becomes inactive and the sRNA RyhB, which is under negative control of Fur, is synthesized. In the presence of Hfq, RyhB binds to its target mRNAs, and thereby blocks synthesis of iron-containing enzymes (Massé *et al.* 2003). In this way RyhB leads to iron sparing in the cell (Jacques *et al.* 2006). The *P. aeruginosa* intergenic regions were queried for Fur boxes followed by an up to 200 nt long spacer and a rho-independent terminator. This search revealed two RyhB orthologs, PrrF1 and PrrF2 (Wilderman *et al.* 2004). They are located in tandem, share 95 % identity and each of them is preceded by a Fur-box (Wilderman *et al.* 2004). Moreover, two putative *prrF* homologues were found in *Pseudomonas putida*, *Pseudomonas fluorescens*, and *Pseudomonas syringae* (Wilderman *et al.* 2004). A microarray analysis revealed a large number of candidate PrrF-regulated genes. Among them were several genes involved in iron storage and oxidative stress protection, but the majority of PrrF-repressed genes appear to encode enzymes that participate in aerobic and anaerobic metabolism

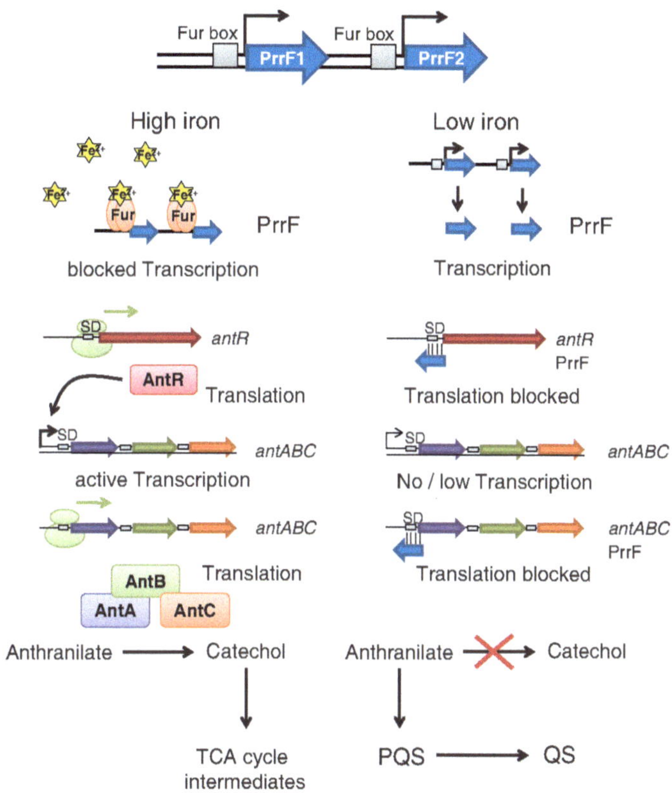

Fig. 4. Model of PrrF regulation, linking iron response to quorum sensing (modified from Oglesby *et al.* 2008). Under iron-replete conditions, transcription of *prrF*1 and 2 is repressed by Fur, which leads to translation of target genes like *antR*. AntR is a transcriptional activator for the *antABC* operon, which is involved in anthranilate degradation to catechol and tricaboxylic acid cycle (TCA) intermediates. When iron is scarce, Fur becomes inactive and *prrF*1/2 are transcribed. Oglesby *et al.* (2008) suggested that the interaction of PrrF1/2 with *antR* and *antA* inhibits translation. In the absence of AntR, the *antABC* genes are poorly transcribed. Thus, anthranilate accumulates and can be used as a precursor for PQS synthesis. In this way, PrrF seems to activate indirectly the PQS quorum sensing system

(Oglesby *et al.* 2008, Vasil 2007). Several of these functions are involved in the TCA cycle (aconitase A, aconitase B, succinate dehydrogenase), which are likewise repressed by RyhB in *E. coli* (Masse and Gottesman 2002, Masse *et al.* 2005). Moreover, Oglesby *et al.* (2008) showed that PrrF represses genes required for degradation of anthranilate (*antABC* and *catBCA*), a precursor of the Pseudomonas quinolone quorum-sensing signal (PQS). Their studies have suggested that repression of *antABC* occurs *via* direct interaction of PrrF with the mRNA, encoding both AntA and its activator, AntR (Fig. 4; Oglesby *et al.* 2008). In this way, iron seems to affect quorum sensing through regulation of PQS production.

5.2 RgsA

The *P. aeruginosa* sRNA RgsA (regulation by GacA and stress; also named 1698 or P16), was identified in two bioinformatic screens (Gonzalez *et al.* 2008, Livny *et al.* 2006). RgsA is expressed during stationary phase of growth in a manner dependent on the stationary sigma factor RpoS. It is probably indirectly activated by GacA (Fig. 2) (Gonzalez *et al.* 2008). Further experiments revealed a decreased survival of RgsA mutants under oxidative stress (Gonzalez *et al.* 2008). RgsA is highly conserved in other Pseudomonads and contains a GGA motif, which is typical for RsmA-binding RNAs. However, it appears to be unable to sequester RsmA in *P. aeruginosa* or RsmE in *P. fluorescens* (Gonzalez *et al.* 2008). It remains to be shown whether RgsA does indeed act by base-pairing and RgsA targets remain to be identified.

5.3 PhrS and PhrD

The sRNAs PhrD and PhrS (PhrS corresponds to P20 in Livny *et al.* (2006) and to 1887 in Gonzales *et al.* (2008)) were detected using a shotgun-cloning approach (RNomics) in combination with Hfq co-immunopreciptation (Sonnleitner *et al.* 2008) and by biocomputation (Livny *et al.*, 2006, Gonzales *et al.* 2008). They encode 72 and 212 nt long RNAs, respectively (Sonnleitner *et al.* 2008). Both of them display reduced steady state levels in an *hfq* mutant, but their stability was unaffected, which suggested that Hfq is somehow required for their transcription. Several efforts to over-express PhrD failed, which could suggest that PhrD represses a function detrimental to cell viability (Sonnleitner *et al.* 2008).

Plasmid-mediated over-expression of *phrS*, followed by a proteome analysis, revealed GroEL, OprD and PA5153 (putative periplasmic binding protein) as potential targets of PhrS (Sonnleitner *et al.* 2008). However, further studies are required to establish whether this regulation is direct or indirect. Moreover, PhrS activates PqsR synthesis, one of the key quorum sensing regulators in *P. aeruginosa*. PhrS acts by stimulating translation of a short open reading frame to which the *pqsR* gene is translationally coupled. Expression of *phrS* is induced by the oxygen-responsive regulator ANR when the oxygen supply decreases. PhrS is the first bacterial sRNA that provides a regulatory link between oxygen availability and quorum sensing, which may impact on oxygen-limited growth in *P. aeruginosa* biofilms (Sonnleitner *et al.*, in press). Inspection of the PhrS sequence disclosed an ORF encoding a 37 aa peptide, which was shown to be translated, but the function remains to be elucidated (Sonnleitner *et al.* 2008).

6 CRISPR

CRISPR elements were located in several sequenced Pseudomonads (*P. aeruginosa* UCBPP-PA14/ LESB58/ PA7, *P. entomophila* L48, *P. fluorescens* Pf-5, *P. mendocina* ymp, P. *putida* F1/ KT2440/ W619, *P stutzeri* A1501, *P. syringae pv. phaseolicola* 1448A/ *pv. syringae* B728a) and harbour sequences with similarity to CRISPR elements. However, no CRISPR regions were detected in *P. aeruginosa* PAO1, *P fluorescens* PfO-1 and SPW25, *P. putida* GB-1 and in *P. synringae pv. tomato* str. DC3000. It was shown in *P. aeruginosa* PA14 that lysogenic infection with the phage DMS3 inhibits biofilm formation as well as swarming motility, and that this inhibition requires the CRISPR locus in the host (Zegans *et al.* 2009). The authors hypothesized that the bacterium might "quarantine" itself from biofilms to reduce the risk of infecting the larger community. Alternatively, inhibition of biofilm formation and swarming could provide a survival advantage to phage DMS3 (Zegans *et al.* 2009). The CRISPR locus in *P. aeruginosa* PA14 is flanked by *cas* genes. Structural analysis of Cas1 revealed that it is a metal-dependent DNA-specific endonuclease and possibly involved in processing of foreign DNA (Wiedenheft *et al.* 2009).

7 Uncharacterized P. Aeruginosa sRNAs

The *in silico* analyses by Livny *et al.* (2006), which were based on the search for rho-independent terminators in intergenic regions, yielded 38 sRNA candidates, 17 of which could be detected by Northern blot analysis.

Gonzalez *et al.* (2008) analyzed intergenic regions >50 nts for sequence conservation among pseudomonads and subjected them to QRNA analysis (Rivas and Eddy 2001). Among the 162 intergenic regions, 32 contained tRNA and rRNA. Out of the remaining 130, rho-independent terminators and potential promoter elements were discerned for 49 loci. 14 sRNAs were reproducibly detected by Northern blot analysis. Among these were the previously described RNase P RNA, 4.5S RNA, PrrF1 and PrrF2 and three sRNAs, which were detected before by Livny *et al.* (2006). In summary, this approach resulted in 8 novel sRNAs.

In addition, Sonnleitner *et al.* (2008) used the algorithm RNAz (Washietl *et al.* 2005), which is based on the notion that structured RNAs fold into more stable secondary structures than the genomic background sequence of the same composition.

This approach revealed seven sRNA loci of which two were shown to be transcribed. In addition, a shotgun-cloning approach in combination with Hfq co-immunoprecipitation revealed the sRNA candidates PhrX and PhrY (Sonnleitner *et al.* 2008).

8 Conclusion

All different classes of bacterial regulatory RNAs are present in *P. aeruginosa*. However, the function of only a small number of *P. aeruginosa* base-pairing sRNAs has thus far been studied. All *P. aeruginosa* sRNAs described above were either identified *in silico* and/or by virtue of their binding to Hfq. Given the large genome of *P. aeruginosa,* it can be anticipated that the number of sRNAs by far exceeds those that have been described and that several riboregulators are involved in pathogenicity. The identification of such sRNAs will require an unbiased approach under growth conditions that mimic acute or chronic infections.

References

Babitzke P, Romeo T (2007) CsrB sRNA family: sequestration of RNA-binding regulatory proteins. Curr Opin Microbiol 10: 156–163.

Barrangou R, Fremaux C, Deveau H, Richards M, Boyaval P, Moineau S, Romero DA, Horvath P (2007) CRISPR provides acquired resistance against viruses in prokaryotes. Science 315: 1709–1712.

Bordi C, Lamy MC, Ventre I, Termine E, Hachani A, Fillet S, Roche B, Bleves S, Méjean V, Lazdunski A, Filloux A (2010) Regulatory RNAs and the HptB/RetS signalling pathways finetune *Pseudomonas aeruginosa* pathogenesis. Mol Microbiol 76: 1427–1443.

Bouvier M, Sharma CM, Mika F, Nierhaus KH, Vogel J (2008) Small RNA binding to 5' mRNA coding region inhibits translational initiation. Mol Cell 32: 827–837.

Brantl S (2007) Regulatory mechanisms employed by cis-encoded antisense RNAs. Curr Opin Microbiol 10: 102–109.

Brencic A, Lory S (2009) Determination of the regulon and identification of novel mRNA targets of *Pseudomonas aeruginosa* RsmA. Mol Microbiol 72: 612–632.

Brencic A, McFarland KA, McManus HR, Castang S, Mogno I, Dove SL, Lory S (2009) The GacS/GacA signal transduction system of *Pseudomonas aeruginosa* acts exclusively through its control over the transcription of the RsmY and RsmZ regulatory small RNAs. Mol Microbiol 73: 434–445.

Cooley M, Chhabra SR, Williams P (2008) N-Acylhomoserine lactone-mediated quorum sensing: a twist in the tail and a blow for host immunity. Chem Biol 15: 1141–1147.

Darfeuille F, Unoson C, Vogel J, Wagner EG (2007) An antisense RNA inhibits translation by competing with standby ribosomes. Mol Cell 26: 381–392.

Davies DG, Parsek MR, Pearson JP, Iglewski BH, Costerton JW, Greenberg EP (1998) The involvement of cell-to-cell signals in the development of a bacterial biofilm. Science 280: 295–298.

De Lay N, Gottesman S (2009) The Crp-activated small noncoding regulatory RNA CyaR (RyeE) links nutritional status to group behavior. J Bacteriol 191: 461–476.

Feldman M, Bryan R, Rajan S, Scheffler L, Brunnert S, Tang H, Prince A (1998) Role of flagella in pathogenesis of *Pseudomonas aeruginosa* pulmonary infection. Infect Immun 66: 43–51.

Fierro-Monti IP, Reid SJ, Woods DR (1992) Differential expression of a Clostridium acetobutylicum antisense RNA: implications for regulation of glutamine synthetase. J Bacteriol 174: 7642–7647.

Franze de Fernandez MT, Eoyang L, August JT (1968) Factor fraction required for the synthesis of bacteriophage Qbeta-RNA. Nature 219: 588–590.
Gambello MJ, Kaye S, Iglewski BH (1993) LasR of *Pseudomonas aeruginosa* is a transcriptional activator of the alkaline protease gene (*apr*) and an enhancer of exotoxin A expression. Infect Immun 61: 1180–1184.
Giangrossi M, Prosseda G, Tran CN, Brandi A, Colonna B, Falconi M (2010) A novel antisense RNA regulates at transcriptional level the virulence gene *icsA* of *Shigella flexneri*. Nucleic Acids Res 38: 3362–3375.
Gonzalez N, Heeb S, Valverde C, Kay E, Reimmann C, Junier T, Haas D (2008) Genome-wide search reveals a novel GacA-regulated small RNA in *Pseudomonas* species. BMC Genomics 9: 167.
Goodman AL, Kulasekara B, Rietsch A, Boyd D, Smith RS, Lory S (2004) A signaling network reciprocally regulates genes associated with acute infection and chronic persistence in *Pseudomonas aeruginosa*. Dev Cell 7: 745–754.
Goodman AL, Merighi M, Hyodo M, Ventre I, Filloux A, Lory S (2009) Direct interaction between sensor kinase proteins mediates acute and chronic disease phenotypes in a bacterial pathogen. Genes Dev 23: 249–259.
Görke B, Stülke J (2008) Carbon catabolite repression in bacteria: many ways to make the most out of nutrients. Nat Rev Microbiol 6: 613–624.
Görke B, Vogel J (2008) Noncoding RNA control of the making and breaking of sugars. Genes Dev 22: 2914–2925.
Gottesman S (2004) The small RNA regulators of *Escherichia coli*: roles and mechanisms. Annu Rev Microbiol 58: 303–328.
Gottesman S (2005) Micros for microbes: non-coding regulatory RNAs in bacteria. Trends Genet 21: 399–404.
Hantke K (2001) Iron and metal regulation in bacteria. Curr Opin Microbiol 4: 172–177.
Hauser AR (2009) The type III secretion system of *Pseudomonas aeruginosa*: infection by injection. Nat Rev Microbiol 7: 654–665.
Heeb S, Valverde C, Gigot-Bonnefoy C, Haas D (2005) Role of the stress sigma factor RpoS in GacA/RsmA-controlled secondary metabolism and resistance to oxidative stress in *Pseudomonas fluorescens* CHA0. FEMS Microbiol Lett 243: 251–258.
Heurlier K, Denervaud V, Pessi G, Reimmann C, Haas D (2003) Negative control of quorum sensing by RpoN (sigma54) in *Pseudomonas aeruginosa* PAO1. J Bacteriol 185: 2227–2235.
Heurlier K, Williams F, Heeb S, Dormond C, Pessi G, Singer D, Cámara M, Williams P, Haas D (2004) Positive control of swarming, rhamnolipid synthesis, and lipase production by the posttranscriptional RsmA/RsmZ system in *Pseudomonas aeruginosa* PAO1. J Bacteriol 186: 2936–2945.
Hsu JL, Chen HC, Peng HL, Chang HY (2008) Characterization of the histidine-containing phosphotransfer protein B-mediated multistep phosphorelay system in *Pseudomonas aeruginosa* PAO1. J Biol Chem 283: 9933–9944.
Humair B, Gonzalez N, Mossialos D, Reimmann C, Haas D (2009) Temperature-responsive sensing regulates biocontrol factor expression in *Pseudomonas fluorescens* CHA0. ISME J 3: 955–965.
Humair B, Wackwitz B, Haas D (2010) GacA-controlled activation of promoters for small RNA genes in *Pseudomonas fluorescens*. Appl Environ Microbiol 76: 1497–1506.
Jacques JF, Jang S, Prevost K, Desnoyers G, Desmarais M, Imlay J, Massé E (2006) RyhB small RNA modulates the free intracellular iron pool and is essential for normal growth during iron limitation in *Escherichia coli*. Mol Microbiol 62: 1181–1190.
James BD, Olsen GJ, Liu JS, Pace NR (1988) The secondary structure of ribonuclease P RNA, the catalytic element of a ribonucleoprotein enzyme. Cell 52: 19–26.
Janzon L, Lofdahl S, Arvidson S (1989) Identification and nucleotide sequence of the delta-lysin gene, *hld*, adjacent to the accessory gene regulator (*agr*) of *Staphylococcus aureus*. Mol Gen Genet 219: 480–485.
Johansen J, Rasmussen AA, Overgaard M, Valentin-Hansen P (2006) Conserved small non-coding RNAs that belong to the sigmaE regulon: role in down-regulation of outer membrane proteins. J Mol Biol 364: 1–8.

Juhas M, Eberl L, Tümmler B (2005) Quorum sensing: the power of cooperation in the world of *Pseudomonas*. Environ Microbiol 7: 459–471.

Kaberdin VR, Bläsi U (2006) Translation initiation and the fate of bacterial mRNAs. FEMS Microbiol Rev 30: 967–979.

Kay E, Dubuis C, Haas D (2005) Three small RNAs jointly ensure secondary metabolism and biocontrol in *Pseudomonas fluorescens* CHA0. Proc Natl Acad Sci U S A 102: 17136–17141.

Kay E, Humair B, Denervaud V, Riedel K, Spahr S, Eberl L, Valverde C, Haas D (2006) Two GacA-dependent small RNAs modulate the quorum-sensing response in *Pseudomonas aeruginosa*. J Bacteriol 188: 6026–6033.

Kazantsev AV, Pace NR (2006) Bacterial RNase P: a new view of an ancient enzyme. Nat Rev Microbiol 4: 729–740.

Keenan RJ, Freymann DM, Stroud RM, Walter P (2001) The signal recognition particle. Annu Rev Biochem 70: 755–775.

Keiler KC (2007) Physiology of tmRNA: what gets tagged and why? Curr Opin Microbiol 10: 169–175.

Lapouge K, Schubert M, Allain FH, Haas D (2008) Gac/Rsm signal transduction pathway of gamma-proteobacteria: from RNA recognition to regulation of social behaviour. Mol Microbiol 67: 241–253.

Lapouge K, Sineva E, Lindell M, Starke K, Baker CS, Babitzke P, Haas D (2007) Mechanism of *hcnA* mRNA recognition in the Gac/Rsm signal transduction pathway of *Pseudomonas fluorescens*. Mol Microbiol 66: 341–356.

Laskowski MA, Osborn E, Kazmierczak BI (2004) A novel sensor kinase-response regulator hybrid regulates type III secretion and is required for virulence in *Pseudomonas aeruginosa*. Mol Microbiol 54: 1090–1103.

Lenz DH, Mok KC, Lilley BN, Kulkarni RV, Wingreen NS, Bassler BL (2004) The small RNA chaperone Hfq and multiple small RNAs control quorum sensing in *Vibrio harveyi* and *Vibrio cholerae*. Cell 118: 69–82.

Livny J, Brencic A, Lory S, Waldor MK (2006) Identification of 17 *Pseudomonas aeruginosa* sRNAs and prediction of sRNA-encoding genes in 10 diverse pathogens using the bioinformatic tool sRNAPredict2. Nucleic Acids Res 34: 3484–3493.

Lorenz C, von Pelchrzim F, Schroeder R (2006) Genomic systematic evolution of ligands by exponential enrichment (Genomic SELEX) for the identification of protein-binding RNAs independent of their expression levels. Nat Protoc 1: 2204–2212.

Lorenz C, Gesell T, Zimmermann B, Schoeberl U, Bilusic I, Rajkowitsch L, Waldsich C, von Haeseler A, Schroeder R (2010) Genomic SELEX for Hfq-binding RNAs identifies genomic aptamers predominantly in antisense transcripts. Nucleic Acids Res. doi:10.1093/nar/gkq032

Massé E, Gottesman S (2002) A small RNA regulates the expression of genes involved in iron metabolism in Escherichia coli. Proc Natl Acad Sci U S A 99: 4620–4625.

Massé E, Vanderpool CK, Gottesman S (2005) Effect of RyhB small RNA on global iron use in *Escherichia coli*. J Bacteriol 187: 6962–6971.

Massé E, Salvail H, Desnoyers G, Arguin M (2007) Small RNAs controlling iron metabolism. Curr Opin Microbiol 10: 140–145.

Massé E, Escorcia FE, Gottesman S (2003) Coupled degradation of a small regulatory RNA and its mRNA targets in *Escherichia coli*. Genes Dev 17: 2374–2383.

Mathee K, Narasimhan G, Valdes C, Qiu X, Matewish JM, Koehrsen M, Rokas A, Yandava CN, Engels R, Zeng E, Olavarietta R, Doud M, Smith RS, Montgomery P, White JR, Godfrey PA, Kodira C, Birren B, Galagan JE, Lory S (2008) Dynamics of *Pseudomonas aeruginosa* genome evolution. Proc Natl Acad Sci U S A 105: 3100–3105.

Mikulecky PJ, Kaw MK, Brescia CC, Takach JC, Sledjeski DD, Feig AL (2004) *Escherichia coli* Hfq has distinct interaction surfaces for DsrA, *rpoS* and poly(A) RNAs. Nat Struct Mol Biol 11: 1206–1214.

Moll I, Afonyushkin T, Vytvytska O, Kaberdin VR, Bläsi U (2003) Coincident Hfq binding and RNase E cleavage sites on mRNA and small regulatory RNAs. RNA 9: 1308–1314.

Møller T, Franch T, Hojrup P, Keene DR, Bachinger HP, Brennan RG, Valentin-Hansen P (2002) Hfq: a bacterial Sm-like protein that mediates RNA-RNA interaction. Mol Cell 9: 23–30.

Moreno R, Ruiz-Manzano A, Yuste L, Rojo F (2007) The *Pseudomonas putida* Crc global regulator is an RNA binding protein that inhibits translation of the AlkS transcriptional regulator. Mol Microbiol. 64:665–675.

Morfeldt E, Taylor D, von Gabain A, Arvidson S (1995) Activation of alpha-toxin translation in *Staphylococcus aureus* by the trans-encoded antisense RNA, RNAIII. EMBO J 14: 4569–4577.

Nakae T (1995) Role of membrane permeability in determining antibiotic resistance in *Pseudomonas aeruginosa*. Microbiol Immunol 39: 221–229.

Nikaido H (1996) Multidrug efflux pumps of gram-negative bacteria. J Bacteriol 178: 5853–5859.

Oglesby AG, Farrow JM, 3rd, Lee JH, Tomaras AP, Greenberg EP, Pesci EC, Vasil ML (2008) The influence of iron on *Pseudomonas aeruginosa* physiology: a regulatory link between iron and quorum sensing. J Biol Chem 283: 15558–15567.

Papenfort K, Vogel J (2009) Multiple target regulation by small noncoding RNAs rewires gene expression at the post-transcriptional level. Res Microbiol 160: 278–287.

Pfeiffer V, Papenfort K, Lucchini S, Hinton JC, Vogel J (2009) Coding sequence targeting by MicC RNA reveals bacterial mRNA silencing downstream of translational initiation. Nat Struct Mol Biol 16: 840–846.

Piddock LJ (2006) Multidrug-resistance efflux pumps – not just for resistance. Nat Rev Microbiol 4: 629–636.

Reimmann C, Valverde C, Kay E, Haas D (2005) Posttranscriptional repression of GacS/GacA-controlled genes by the RNA-binding protein RsmE acting together with RsmA in the biocontrol strain *Pseudomonas fluorescens* CHA0. J Bacteriol 187: 276–285.

Repoila F, Darfeuille F (2009) Small regulatory non-coding RNAs in bacteria: physiology and mechanistic aspects. Biol Cell 101: 117–131.

Repoila F, Majdalani N, Gottesman S (2003) Small non-coding RNAs, co-ordinators of adaptation processes in *Escherichia coli*: the RpoS paradigm. Mol Microbiol 48: 855–861.

Rivas E, Eddy SR (2001) Noncoding RNA gene detection using comparative sequence analysis. BMC Bioinformatics 2: 8.

Schubert M, Lapouge K, Duss O, Oberstrass FC, Jelesarov I, Haas D, Allain FH (2007) Molecular basis of messenger RNA recognition by the specific bacterial repressing clamp RsmA/CsrA. Nat Struct Mol Biol 14: 807–813.

Schumacher MA, Pearson RF, Møller T, Valentin-Hansen P, Brennan RG (2002) Structures of the pleiotropic translational regulator Hfq and an Hfq-RNA complex: a bacterial Sm-like protein. EMBO J. 21: 3546–3556.

Sharma CM, Darfeuille F, Plantinga TH, Vogel J (2007) A small RNA regulates multiple ABC transporter mRNAs by targeting C/A-rich elements inside and upstream of ribosome-binding sites. Genes Dev 21: 2804–2817.

Silvaggi JM, Perkins JB, Losick R (2005) Small untranslated RNA antitoxin in *Bacillus subtilis*. J Bacteriol 187: 6641–6650.

Sonnleitner E, Abdou L, Haas D (2009) Small RNA as global regulator of carbon catabolite repression in *Pseudomonas aeruginosa*. Proc Natl Acad Sci U S A 106: 21866–21871.

Sonnleitner E, Schuster M, Sorger-Domenigg T, Greenberg EP, Bläsi U (2006) Hfq-dependent alterations of the transcriptome profile and effects on quorum sensing in *Pseudomonas aeruginosa*. Mol Microbiol 59: 1542–1558.

Sonnleitner E, Sorger-Domenigg T, Madej MJ, Findeiss S, Hackermüller J, Hüttenhofer A, Stadler PF, Bläsi U, Moll I (2008) Detection of small RNAs in *Pseudomonas aeruginosa* by RNomics and structure-based bioinformatic tools. Microbiology 154: 3175–3187.

Sorek R, Kunin V, Hugenholtz P (2008) CRISPR--a widespread system that provides acquired resistance against phages in bacteria and archaea. Nat Rev Microbiol 6: 181–186.

Sorger-Domenigg T, Sonnleitner E, Kaberdin VR, Bläsi U (2007) Distinct and overlapping binding sites of *Pseudomonas aeruginosa* Hfq and RsmA proteins on the non-coding RNA RsmY. Biochem Biophys Res Commun 352: 769–773.

Sun X, Zhulin I, Wartell RM (2002) Predicted structure and phyletic distribution of the RNA-binding protein Hfq. Nucleic Acids Res 30: 3662–3671.

Toder DS, Ferrell SJ, Nezezon JL, Rust L, Iglewski BH (1994) *lasA* and *lasB* genes of *Pseudomonas aeruginosa*: analysis of transcription and gene product activity. Infect Immun 62: 1320–1327.

Toledo-Arana A, Dussurget O, Nikitas G, Sesto N, Guet-Revillet H, Balestrino D, Loh E, Gripenland J, Tiensuu T, Vaitkevicius K, Barthelemy M, Vergassola M, Nahori MA, Soubigou G, Régnault B, Coppée JY, Lecuit M, Johansson J, Cossart P (2009) The *Listeria* transcriptional landscape from saprophytism to virulence. Nature 459: 950–956.

Ullmann A (1996) Catabolite repression: a story without end. Res Microbiol 147: 455–458.

Urban JH, Vogel J (2008) Two seemingly homologous noncoding RNAs act hierarchically to activate *glmS* mRNA translation. PLoS Biol 6: e64.

Urbanowski ML, Stauffer LT, Stauffer GV (2000) The *gcvB* gene encodes a small untranslated RNA involved in expression of the dipeptide and oligopeptide transport systems in *Escherichia coli*. Mol Microbiol 37: 856–868.

Valverde C, Lindell M, Wagner EG, Haas D (2004) A repeated GGA motif is critical for the activity and stability of the riboregulator RsmY of *Pseudomonas fluorescens*. J Biol Chem 279: 25066–25074.

van Delden C, Iglewski BH (1998) Cell-to-cell signaling and *Pseudomonas aeruginosa* infections. Emerg Infect Dis 4: 551–560.

van der Oost J, Jore MM, Westra ER, Lundgren M, Brouns SJ (2009). CRISPR-based adaptive and heritable immunity in prokaryotes. Trends Biochem Sci 34: 401–407.

Vasil ML (2007) How we learnt about iron acquisition in *Pseudomonas aeruginosa*: a series of very fortunate events. Biometals 20: 587–601.

Vecerek B, Moll I, Bläsi U (2007) Control of Fur synthesis by the non-coding RNA RyhB and iron-responsive decoding. EMBO J 26: 965–975.

Vecerek B, Rajkowitsch L, Sonnleitner E, Schroeder R, Bläsi U (2008). The C-terminal domain of *Escherichia coli* Hfq is required for regulation. Nucleic Acids Res 36: 133–143.

Ventre I, Goodman AL, Vallet-Gely I, Vasseur P, Soscia C, Molin S, Bleves S, Lazdunski A, Lory S, Filloux A. (2006) Multiple sensors control reciprocal expression of *Pseudomonas aeruginosa* regulatory RNA and virulence genes. Proc Natl Acad Sci U S A 103: 171–176.

Vogel DW, Hartmann RK, Struck JC, Ulbrich N, Erdmann VA (1987) The sequence of the 6S RNA gene of *Pseudomonas aeruginosa*. Nucleic Acids Res 15: 4583–4591.

Wadler CS, Vanderpool CK (2007) A dual function for a bacterial small RNA: SgrS performs base pairing-dependent regulation and encodes a functional polypeptide. Proc Natl Acad Sci U S A 104: 20454–20459.

Washietl S, Hofacker IL, Stadler PF (2005) Fast and reliable prediction of noncoding RNAs. Proc Natl Acad Sci U S A 102: 2454–2459.

Wassarman KM (2007) 6S RNA: a small RNA regulator of transcription. Curr Opin Microbiol 10: 164–168.

Waters LS, Storz G (2009) Regulatory RNAs in bacteria. Cell 136: 615–628.

Wick MJ, Hamood AN, Iglewski BH (1990) Analysis of the structure-function relationship of *Pseudomonas aeruginosa* exotoxin A. Mol Microbiol 4: 527–535.

Wiedenheft B, Zhou K, Jinek M, Coyle SM, Ma W, Doudna JA (2009) Structural basis for DNase activity of a conserved protein implicated in CRISPR-mediated genome defense. Structure 17: 904–912.

Wilderman PJ, Sowa NA, FitzGerald DJ, FitzGerald PC, Gottesman S, Ochsner UA, Vasil ML (2004) Identification of tandem duplicate regulatory small RNAs in *Pseudomonas aeruginosa* involved in iron homeostasis. Proc Natl Acad Sci U S A 101: 9792–9797.

Williams KP, Bartel DP (1996) Phylogenetic analysis of tmRNA secondary structure. RNA 2: 1306–1310.

Williams P, Camara M (2009) Quorum sensing and environmental adaptation in *Pseudomonas aeruginosa*: a tale of regulatory networks and multifunctional signal molecules. Curr Opin Microbiol 12: 182–191.

Zegans ME, Wagner JC, Cady KC, Murphy DM, Hammond JH, O'Toole GA (2009) Interaction between bacteriophage DMS3 and host CRISPR region inhibits group behaviors of *Pseudomonas aeruginosa*. J Bacteriol 191: 210–219.

Zhang A, Wassarman KM, Ortega J, Steven AC, Storz G (2002) The Sm-like Hfq protein increases OxyS RNA interaction with target mRNAs. Mol Cell 9: 11–22.

Chapter 5
Natural Antisense Transcripts in Bacteria

Jens Georg and Wolfgang R. Hess*

1 Defining Features of an Antisense Transcript

This chapter focuses on special types of transcripts, those which originate from the complementary strand of an mRNA or functional RNA. As a consequence, these RNA molecules are, partially or to the extent of their full length, complementary to their target. Antisense transcripts are also called *cis*-antisense RNAs to differentiate them from *trans*-acting non-coding RNAs, which frequently act on their targets through short regions of complementarity and therefore are sometimes also referred to as antisense RNAs. The promoter and the first transcribed nucleotides of a non-coding RNA may also be located within an intergenic spacer and only later extend into the region of complementarity to another transcript. Antisense transcription sometimes results from read-through of protein-coding genes over a long distance, as observed for the antisense RNA to the ferric uptake regulator *furA* in *Anabaena* PCC7120. This gene is overlapped along its full length by the extremely long 3' UTR of gene *alr1690* (Hernandez *et al.* 2005). Last but not least, a perfectly complementary antisense RNA may have additional targets in *trans*. All these cases are considered here as antisense RNAs, as long as there is a sense transcript expressed from the opposite strand of DNA.

In consequence, bacterial antisense transcripts can be characterized by the following features:

[1] Antisense RNAs are fully or to a large extent complementary to their target.

[2] Resulting from the extensive complementarity, antisense transcripts can form energetically much more stable duplexes than *trans*-acting non-coding RNAs. This should have implications for their binding kinetics; moreover, the target interactions are more likely to be independent of proteins such as Hfq, which is frequently

* Institute of Biology 3, Genetics & Experimental Bioinformatics,
Albert Ludwig University of Freiburg, Schänzlestr. 1, D-79104 Freiburg, Germany

involved in the function of *trans*-acting regulatory RNAs (see also Chapters 2 and 8 of this book and note the Hfq effect on the RNA-OUT antisense RNA (Ross *et al.* 2010)). On the other hand, other protein factors, such as RNA helicases and RNA binding proteins, may be involved in actively manipulating the association between sense and antisense RNA.

[3] Compared to the majority of *trans*-acting non-coding RNAs, antisense RNAs seem, on average, to be longer. Many of the former are about 80 to 150 nt in length, whereas antisense RNAs are frequently several hundred nt and some examples are even several kb long. The antisense RNA AmgR in *Salmonella enterica* is 1.2 kb (Lee and Groisman 2010) and an antisense RNA reported from *Listeria monocytogenesis* is more than 2 kb long, overlapping as many as three genes (Toledo-Arana *et al.* 2009). Antisense transcripts as long as 3.5 kb and 7 kb were found in *Prochlorococcus* MED4 (Stazic *et al.* 2011).

As a result of these characteristics, certain assumptions can be made about the distinct effects that are uniquely linked to antisense RNAs:

[1] Inheritance: An antisense RNA originates from the identical genomic region as its target. Thus, during conjugation or horizontal gene transfer events both are automatically transferred together.

[2] Genome capacity: Because no additional space is required, an organism can theoretically afford one or several antisense RNA regulators for every single gene, without the need to increase genome size. The recent transcriptome studies with high-throughput methods show that this situation is not too far from reality in some bacteria (Sharma *et al.* 2010).

[3] Evolution: Compared to other regulators, new antisense RNAs should originate easily during evolution because a few point mutations already yield an active promoter, whilst the sequence complementarity is automatic.

[4] Steric effects: Due to the *cis* position, antisense and target RNAs are transcribed in close proximity, resulting in a high local concentration of both types of molecules and facilitating efficient interaction. Indeed, recent observations hint towards limited diffusion of transcripts from their site of transcription, suggesting spatial effects play a biological role, at least in *E. coli* and *Caulobacter crescentus* (Montero Llopis *et al.* 2010).

[5] Effects from transcriptional interference: Regulation through transcriptional interference (Shearwin *et al.* 2005; Sneppen *et al.* 2005) relies on transcription from closely spaced promoters, either on the same strand of DNA or facing each other on complementary strands. The latter case inevitably results in the generation of antisense RNAs. It has been shown that transcriptional interference can act additively to enhance the regulatory effect of antisense RNA on its target (Giangrossi *et al.* 2010). Moreover, mathematical modeling suggests that in the transcriptional interference collision mechanism the distance between promoters facing each other is relevant (Sneppen *et al.* 2005), which would be consistent with the quite remarkable length of some identified antisense transcripts.

2 Antisense RNAs were Discovered in Bacteria

Although a much higher number of antisense transcripts has now been described in Eukaryotes, natural antisense transcripts were originally discovered in bacteria and several fundamental mechanisms through which antisense RNAs exert their functions were elucidated early on. It is worthwhile briefly reviewing these mechanisms because many of the aspects involved are now being discovered all over again when studying the molecular mechanisms of recently found antisense RNAs from the bacterial chromosome. For details of phage and plasmid-derived antisense RNA based regulatory mechanisms discovered early on, the excellent reviews by Eguchi *et al.* (1991), Simons and Kleckner (1988) and Wagner and Simons (1994) should be consulted.

2.1 Known Facts About Antisense RNAs from Bacteriophages, Plasmids and Transposons

Antisense transcription was postulated first in the bidirectional transcription and regulation of the bacteriophage λ gene *cro* (Spiegelman *et al.* 1972). It took more than ten years before the codegradation mechanism involved was fully elucidated. The 77 nt OOP antisense RNA is complementary to the 3' end of the λ cII-repressor mRNA. Overexpression of the OOP transcript from a plasmid results in an RNase III-dependent cleavage of the cII mRNA, initially at two sites, one in the 3' end of the coding region, and one in the cII and O gene intergenic region (Krinke and Wulff 1987; Krinke and Wulff 1990; Krinke *et al.* 1991).

In plasmid ColE1, a 108 nt antisense RNA called RNA I regulates maturation of the ColE1 primer for DNA replication. This primer, RNA II, is required for initiation of replication. Upon annealing to its DNA target sequence, it is cleaved by RNase H at a position located 550 nt from its 5'end. DNA synthesis of the leading strand begins at the newly formed RNA II 3' end resulting from this cleavage (Itoh and Tomizawa 1980; Tomizawa *et al.* 1977). The RNA I regulatory antisense RNA overlaps the 5' end of RNA II (Tomizawa *et al.* 1981) and upon binding to its target, a conformational change in RNA II is triggered (Masukata and Tomizawa 1986). As a consequence of this change, RNA II is prevented from hybridisation to DNA and maturation by RNase H, consequently DNA replication is inhibited (Tomizawa *et al.* 1977; Tomizawa *et al.* 1981).

But antisense RNA can also confer stochastic effects as can be seen in the control of plasmid incompatibility of ColE1-type plasmids (Tomizawa and Itoh 1981). In this case, the regulatory RNA I molecules from each plasmid also act in *trans* on the RNA II from the other plasmid. Slightly different copy numbers of each plasmid are found in different cells, resulting from the randomness of replication and segregation. Thus, the two plasmids exist in different cells in different ratios and as a consequence, the ratios between the respective RNA II/RNA I molecules also differ. These ratios effectively lead to the loss of one or the other plasmid, observed as incompatibility.

There is one more aspect of the ColE1 systems from which general insight may be gained, i.e., the fact that the interaction between an antisense RNA and its target is a dynamic process that competes with intramolecular folding of either RNA. In the ColE1 case, it is important that RNA II forms a distinct structure, otherwise it will not mature and not hybridise correctly to the origin of replication (Masukata and Tomizawa 1984, 1986). It is highly interesting that the inhibitory binding of RNA I must occur co-transcriptionally, after synthesis of the first 100 nt of RNA II but before the first ~150 nt are completed. RNA I can still bind at later times, but then RNAII is already committed to form a stable hybrid with DNA. Thus, a very important aspect of antisense RNA function is that it must be present not only at the right site but also at the right time.

Inhibition of the initiation of translation is another fundamental mechanism through which antisense RNAs exert their function. This mechanism was first documented for RNA-OUT, which controls expression of the transposase gene *tnp* (also called RNA-IN) of the IS10 element. RNA-OUT originates from a promoter close to the 5' end of the *tnp* reading frame and overlaps the 5' end of the *tnp* mRNA by 35 nucleotides. This overlap includes the *tnp* ribosome-binding site. Therefore, pairing of RNA-OUT with the *tnp* mRNA blocks ribosome binding, accounting for the inhibitory effect mediated by RNA-OUT (Case *et al.* 1990; Ma and Simons 1990; Simons and Kleckner 1983). Interestingly, it was later shown that the pairing rates between RNA-OUT and the *tnp* mRNA *in vivo* are very close to being optimal (Jain 1995) and that pairing between the *tnp* mRNA and RNA-OUT may be facilitated by Hfq (Ross *et al.* 2010).

From these early studies fundamental insight has been obtained into some functional principles governing the activity of bacterial antisense RNAs:

[1] Antisense RNA can target other non-coding RNAs with a physiological function (example control of plasmid replication).

[2] Interplay between an antisense RNA and the processing machinery (example RNase H cleavage of RNA II in plasmid replication).

[3] Control of gene expression through codegradation (bacteriophage λ gene *cro*).

[4] Antisense RNAs mediate stochastic effects with genetic and physiological relevance (plasmid incompatibility).

[5] Time does matter in the interaction of two RNA molecules (dynamics of RNA I/RNA II interaction during transcription of RNA II).

[6] Antisense RNA overlapping the 5' UTR of an mRNA may prevent initiation of translation (example RNA-OUT in control of IS10 activity).

3 Antisense Transcripts Come in High Numbers and Occur Throughout the Bacterial Kingdom

A closer look at the bacterial kingdom shows that antisense RNAs are present in a wide range of individual species and families. If bacteria in which at least a single antisense RNA was detected are plotted on a 16S rRNA tree (Fig. 1), the ubiquitous presence of antisense RNAs throughout the bacterial kingdom becomes obvious.

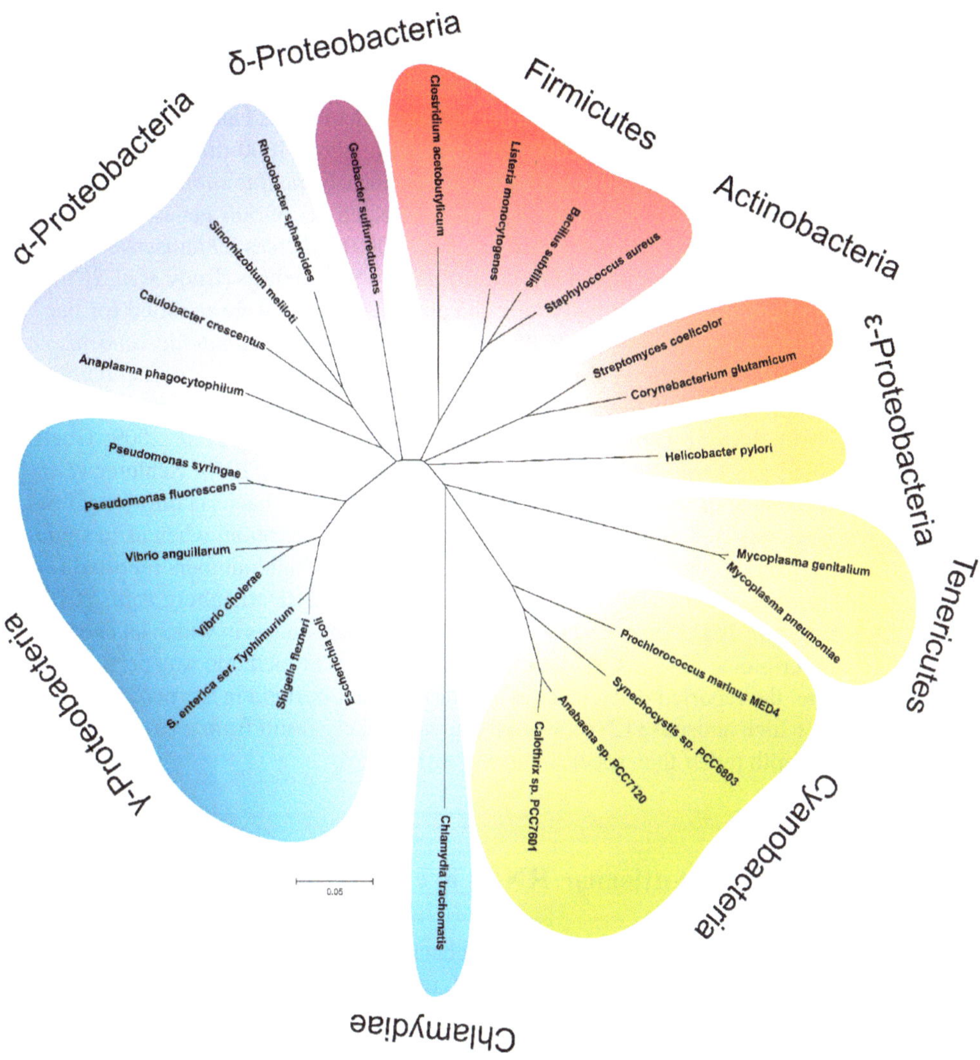

Fig. 1. Antisense transcription occurs throughout the bacterial kingdom. Bacteria for which at least a single antisense RNA was reported are plotted along a phylogenetic tree based on 16S rRNA sequences (Albrecht *et al.* 2010; Andre *et al.* 2008; Beaume *et al.* 2010; Berghoff *et al.* 2009; Csiszar *et al.* 1987; D'Alia *et al.* 2010; Dühring *et al.* 2006; Eiamphungporn and Helmann 2009; Filiatrault *et al.* 2010; Fozo *et al.* 2008; Georg *et al.* 2009; Giangrossi *et al.* 2010; Guell *et al.* 2009; Hernandez *et al.* 2005; Kawano *et al.* 2005; Landt *et al.* 2008; Lee *et al.* 2001; Lee and Groisman 2010; Liu *et al.* 2009; Lluch-Senar *et al.* 2007; Mitschke *et al.* 2011; Nelson *et al.* 2008; Opdyke *et al.* 2004; Qiu *et al.* 2010; Rasmussen *et al.* 2009; Schluter *et al.* 2010; Selinger *et al.* 2000; Sharma *et al.* 2010; Silby and Levy 2008; Silvaggi *et al.* 2005, 2006; Sittka *et al.* 2008; Steglich *et al.* 2008; Stork *et al.* 2007; Swiercz *et al.* 2008; Toledo-Arana *et al.* 2009; Vogel *et al.* 2003; Xiao *et al.* 2009; Yachie *et al.* 2006; Zemanova *et al.* 2008).

With the advent of high throughput RNomics methods such as tiling microarrays, direct RNA labeling methods and, in particular, RNA deep sequencing, the numbers of identified antisense transcripts have started to increase dramatically. A recent transcriptome analysis confirmed earlier reports of widespread antisense transcription in *E. coli* (Selinger *et al.* 2000) by identifying about 1000 different antisense RNAs (Dornenburg *et al.* 2010). The transcriptome of *Bacillus subtilis* was found to encompass antisense RNAs for 18% of a selected set of 506 genes (Lee *et al.* 2001) and 2.9% of all genes (Rasmussen *et al.* 2009), whereas 29 antisense RNAs were identified independently in *Bacillus* by dRNA-Seq analysis (Irnov *et al.* 2010). High numbers of chromosomally encoded antisense RNAs were reported for bacteria as different as *Synechocystis* PCC6803, *Mycoplasma pneumoniae*, *Sinorhizobium meliloti*, *Geobacter sulfurreducens*, *Vibrio cholerae*, *Chlamydia trachomatis*, *Pseudomonas syringae*, *Staphylococcus aureus* and *Helicobacter pylori* (Albrecht *et al.* 2010; Beaume *et al.* 2010; Filiatrault *et al.* 2010; Georg *et al.* 2009; Guell *et al.* 2009; Liu *et al.* 2009; Mitschke *et al.* 2011; Qiu *et al.* 2010; Schluter *et al.* 2010; Sharma *et al.* 2010). It should be noted that for some bacteria only very few genes were reported to be associated with antisense transcription, whereas in some these numbers go into the hundreds and thousands. The high numbers reported for *E. coli*, *Helicobacter pylori* and *Synechocystis* PCC6803 (Dornenburg *et al.* 2010; Mitschke *et al.* 2011; Sharma *et al.* 2010) all result from studies in which an exhaustive transcriptome analysis or a combination of different methods was performed. Therefore, the reported discrepancies among different bacteria in the percentage of genes in which antisense RNAs were found very likely result from a strong bias in analysis depth rather than species-specific differences.

4 Bacterial Antisense RNAs are Functionally Important

Functional relevance for antisense RNA in bacteria was first demonstrated for extrachromosomal elements such as the ColE1 replicon and bacteriophage λ. A more recent example is provided by the RnaG antisense RNA, which controls expression of the plasmid pVIN-located virulence gene *icsA* of *Shigella flexneri* by a combination of transcriptional interference and transcription attenuation (Giangrossi *et al.* 2010). Another example are the RNAα and RNAβ antisense RNAs of the *Vibrio anguillarum* virulence plasmid pJM1, which control expression of the *fatDCBAangRT* iron transport-biosynthesis operon (Chen and Crosa 1996; Salinas *et al.* 1993; Stork *et al.* 2007; Tolmasky and Crosa 1995; Waldbeser *et al.* 1995; Waldbeser *et al.* 1993); in the case of RNAβ, the regulatory effect is mediated by termination of transcription.

However, many antisense RNAs originating from chromosomal loci are now known to be functionally relevant too. The cyanobacterial regulator IsrR is another example of the involvement of antisense RNA in the control of iron-dependent processes (Dühring *et al.* 2006). The *E. coli* GadY antisense RNA controls the *gadX* gene by guiding the cleavage of the bicistronic *gadXW* transcript, resulting in two monocistronic mRNAs with an enhanced stability compared to their precursor (Opdyke *et al.* 2010; Tramonti *et al.* 2008).

A whole group of genes whose expression has to be tightly controlled, code for the toxic component of toxin-antitoxin modules. Although bacterial toxin-antitoxin systems were initially thought to control the stability of extrachromosomal elements such as plasmids by postsegregational killing, it is now evident that bacterial chromosomes may encode dozens of such modules (Fozo *et al.* 2008, 2010; Makarova *et al.* 2009; Pandey and Gerdes 2005). Mechanistically, two major groups of toxin-antitoxin systems can be differentiated. Interestingly, toxin-antitoxin systems classified as type I frequently employ antisense RNA (rather than a small protein which is typical for type II systems) as the antidote molecule. These type I antisense RNA antidotes frequently repress the expression of the toxic proteins by base pairing across the ribosome binding site, leading to a block in translation, frequently followed by rapid mRNA decay. Functionally characterized examples for this group of genes in enterobacteria include the antisense RNA SymR which controls the SOS response-induced protein SymE (Kawano *et al.* 2007) and the five closely related antisense RNAs SibA to SibE (Fozo *et al.* 2008), each of which (except SibA) controls in a highly sequence-specific manner its respective cognate target, IbsA to IbsE (Han *et al.* 2010).

4.1 How Bacterial Antisense RNAs Exert their Function

In addition to the mechanisms unraveled early on, new functional principles of antisense RNA have been recognized. Sometimes, earlier established mechanisms have become extended as a result of more recent findings. Control of an mRNA by codegradation was first shown for the RNase III dependent cleavage of the bacteriophage λ cII mRNA (Krinke and Wulff 1987, 1990; Krinke *et al.* 1991). In recent years, several candidates for chromosome encoded antisense/mRNA pairs were found in which codegradation may play a role and it is becoming evident that this regulation provides specific features and has unique advantages.

This has been exemplified in the *isiA*/IsrR sense/antisense pair of *Synechocystis* PCC6803 (Dühring *et al.* 2006). IsrR controls the accumulation of the *isiA* mRNA, and thereby the amount of the iron stress-induced protein A (IsiA). This protein binds chlorophyll and forms a 18mer ring-like structure around the photosystem I (PSI) trimer under prolonged iron depletion and under several other stress conditions. IsiA absorbs light, contributing to either light harvesting or energy dissipation (protection) of the photosynthetic machinery (Yeremenko *et al.* 2004). The expression of IsiA is a major strategy by which cyanobacteria respond to Fe deficiency and avoid oxidative stress (Chauhan *et al.* 2011; see also chapter 1 of this book for a discussion of small RNAs in the context of oxidative stress response). But these changes imply a major re-organization of the photosynthesis machinery; consequently, the expression of *isiA* is tightly regulated. The 177 nt IsrR is an important component in this regulation. It is complementary to the central region of the *isiA* mRNA. Under standard (no stress) conditions, IsrR is one of the most strongly accumulating transcripts in the cell (Georg *et al.* 2009). The accumulation of the *isiA* mRNA and IsrR occurs inversely to each other. If expressed inde-

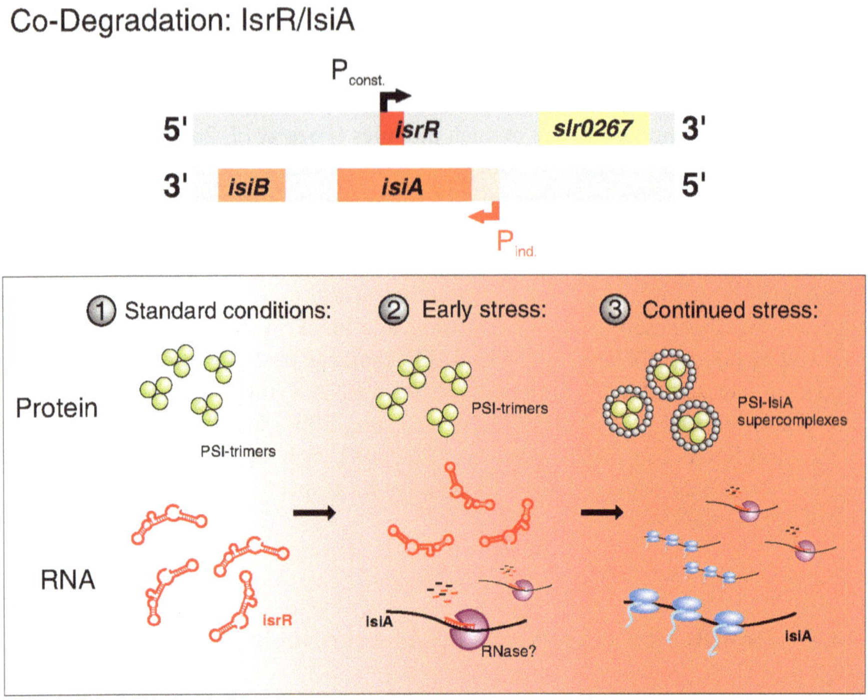

Fig. 2. Sense-antisense RNA interaction in the IsrR-*isiA* system of *Synechocystis* PCC 6803. The antisense RNA IsrR originates from the central part of the *isiA* gene from a constitutive promoter (Pconst.). The *isiA* gene is transcribed from the inducible promoter (Pind.) and codes for the iron stress induced protein A, which forms ring-like structures around photosystem I (PSI) under continuous stress. (1) Under standard growth conditions, the steady state concentration of IsrR [IsrR] is high, resulting from a low-to-moderate constitutive transcription rate and a very high half-life. Under these conditions, *isiA* is not transcribed, or is transcribed only at low levels. (2) In early stress conditions, *isiA* transcription becomes activated. However, as long as [IsrR] greatly exceeds the amount of the *isiA* mRNA [isiA], the message is codegraded and no protein is made. The trimeric PSI complexes remain free of IsiA protein. (3) When the stress conditions persist, IsrR is still transcribed but it is outnumbered by isiA. Consequently, its turnover is very high and it becomes titrated out. The mRNA accumulates, translation occurs and PSI-IsiA supercomplexes are formed. The structure of IsrR is drawn to resemble the secondary structure predicted by RNAfold.

pendently from each other, both transcripts appear to be very stable, with a half-life of more than 45 minutes for IsrR (Legewie *et al.* 2008). When both species are expressed simultaneously, they form RNA duplexes which are immediately degraded by a mechanism not yet identified (Dühring *et al.* 2006). Consequently, the message cannot accumulate until the number of *isiA* mRNA molecules titrates out the number of IsrR molecules (Fig. 2). This mode of regulation is known as “threshold linear response” (Levine *et al.* 2007; Levine and Hwa 2008).

In addition to the delay of *isiA* expression during early stress, an additional effect caused by the codegradation of both transcripts is the faster depletion of *isiA* mRNA during the recovery phase, when its transcription ceases.

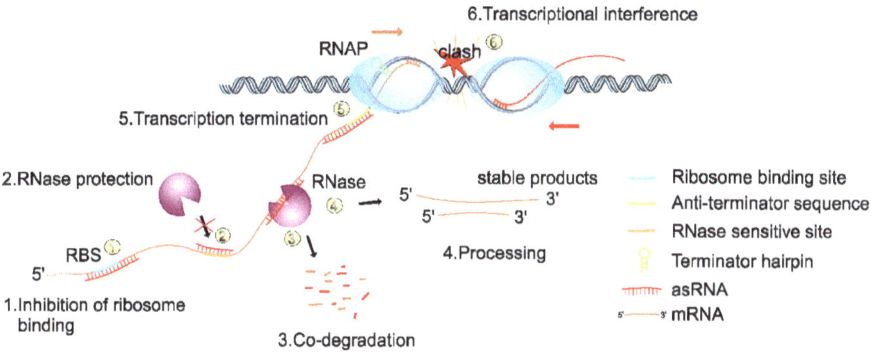

Fig. 3. Established gene regulatory mechanisms in bacteria that involve antisense RNAs. Transcription of a gene by the RNA polymerase (RNAP) is shown. Regulatory mechanisms include antisense RNA-dependent post-transcriptional processes (spheres 1–4) and co-transcriptional regulation (spheres 5–6). Functional regions in the growing mRNA are marked with colored lines (blue: Ribosome binding site, yellow: RNase sensitive region, green: Anti-terminator sequence).

Other more recently established functional mechanisms of how bacterial antisense RNAs exert their activity are summarized in Fig. 3. Antisense RNA can impact transcription directly as well as act post-transcriptionally. Antisense RNA can selectively stabilize its complementary RNA as well as guide distinct and specific processing events.

These mechanisms are:
1. Inhibition of translation. The antisense RNA binds the target mRNA at the ribosome-binding site (RBS) or at other regions, in this way the initiation of translation and possibly also elongation is inhibited sterically.
2. RNase protection: Binding of an antisense RNA to RNase sensitive sites prevents RNase binding and specific degradation. This results in an enhanced half-life of the mRNA.
3, 4. Co-degradation and specific processing: Binding of an antisense RNA creates new RNase recognition sites. This could result in complete degradation e.g. by the dsRNA-specific RNase III. Alternatively, the binding of the antisense RNA triggers specific processing with stable products of maturation.
5. Transcription termination (proposed mechanism): Without the antisense RNA the mRNA forms an anti-terminator secondary structure, which prevents the formation of a Rho-independent terminator hairpin. The antisense RNA interacts with the anti-terminator sequence and the terminator can be formed.
6. Transcriptional interference: In the case of transcriptional interference different RNAP complexes influence each other. RNAPs transcribing divergent overlapping genes can clash into each other (collision mechanism), which results in transcription abortion and yields antisense RNA as a by-product. Other mechanisms of transcriptional interference are promoter occlusion and "sitting duck" interference (not shown).

5 Outlook

In this chapter we have traced a line from the discovery of antisense transcription in bacteria and the initial establishment of functional principles, such as codegradation and inhibition of initiation of translation, to the high-throughput transcriptomics-based discovery of large numbers of antisense transcripts in bacteria, which has taken place more recently. Several recently published reviews provide additional details on the discovery and characterization of bacterial *cis*-antisense transcripts (Georg and Hess 2011; Sorek and Cossart 2010; Thomason and Storz 2010).

The major challenge now is the functional characterization of all these antisense RNAs. The questions are: Is there a biological function for all of them and what are the involved molecular mechanisms? It is evident that techniques such as dRNA-seq (Sharma *et al.* 2010), tiling microarrays and direct sequencing of RNA (Ozsolak *et al.* 2009) will in the near future establish the architecture of the transcriptome for model organisms representing all major groups of prokaryotes. Consequently, the results from these methods are already dramatically changing our view of how common and widespread antisense transcription really is. The emerging functional data as reviewed in this chapter suggests that antisense RNAs in bacteria mediate a plethora of effects and are involved in far more processes than previously anticipated. Thus, the functional impact of antisense RNAs should be considered in developing new strategies against pathogenic bacteria or when engineering bacteria for biotechnology. An ultimate question will be how prominent antisense RNAs are as a part of the regulatory network.

References

Albrecht M, Sharma CM, Reinhardt R, Vogel J, Rudel T (2010) Deep sequencing-based discovery of the *Chlamydia trachomatis* transcriptome. Nucleic Acids Res 38: 868–877

Andre G, Even S, Putzer H, Burguiere P, Croux C, Danchin A *et al.* (2008) S-box and T-box riboswitches and antisense RNA control a sulfur metabolic operon of *Clostridium acetobutylicum*. Nucleic Acids Res 36: 5955–5969

Beaume M, Hernandez D, Farinelli L, Deluen C, Linder P, Gaspin C *et al.* (2010) Cartography of methicillin-resistant *S. aureus* transcripts: detection, orientation and temporal expression during growth phase and stress conditions. PLoS One 5: e10725

Berghoff BA, Glaeser J, Sharma CM, Vogel J, Klug G (2009) Photooxidative stress-induced and abundant small RNAs in *Rhodobacter sphaeroides*. Mol Microbiol 74: 1497–1512

Case CC, Simons EL, Simons RW (1990) The IS10 transposase mRNA is destabilized during antisense RNA control. EMBO J 9: 1259–1266

Chauhan D, Folea IM, Jolley CC, Kouril R, Lubner CE, Lin S *et al.* (2011) A novel photosynthetic strategy for adaptation to low-iron aquatic environments. Biochemistry 50: 686–692

Chen Q, Crosa JH (1996) Antisense RNA, *fur*, iron, and the regulation of iron transport genes in *Vibrio anguillarum*. J Biol Chem 271: 18885–18891

Csiszar K, Houmard J, Damerval T, Tandeau de Marsac N (1987) Transcriptional analysis of the cyanobacterial *gvpABC* operon in differentiated cells: occurrence of an antisense RNA complementary to three overlapping transcripts. Gene 60: 29–37

D'Alia D, Nieselt K, Steigele S, Muller J, Verburg I, Takano E (2010) Noncoding RNA of glutamine synthetase I modulates antibiotic production in *Streptomyces coelicolor* A3(2). J Bacteriol 192: 1160–1164

Dornenburg JE, DeVita AM, Palumbo MJ, Wade JT (2010) Widespread antisense transcription in *Escherichia coli*. mBio 1: e00024–00010

Dühring U, Axmann IM, Hess WR, Wilde A (2006) An internal antisense RNA regulates expression of the photosynthesis gene *isiA*. Proc Nat Acad Sci USA 103: 7054–7058

Eguchi Y, Itoh T, Tomizawa J (1991) Antisense RNA. Annu Rev Biochem 60: 631–652

Eiamphungporn W, Helmann JD (2009) Extracytoplasmic function sigma factors regulate expression of the *Bacillus subtilis yabE* gene via a *cis*-acting antisense RNA. J Bacteriol 191: 1101–1105

Filiatrault MJ, Stodghill PV, Bronstein PA, Moll S, Lindeberg M, Grills G *et al.* (2010) Transcriptome analysis of *Pseudomonas syringae* identifies new genes, noncoding RNAs, and antisense activity. J Bacteriol 192: 2359–2372

Fozo EM, Kawano M, Fontaine F, Kaya Y, Mendieta KS, Jones KL *et al.* (2008) Repression of small toxic protein synthesis by the Sib and OhsC small RNAs. Mol Microbiol 70: 1076–1093

Fozo EM, Makarova KS, Shabalina SA, Yutin N, Koonin EV, Storz G (2010) Abundance of type I toxin-antitoxin systems in bacteria: searches for new candidates and discovery of novel families. Nucleic Acids Res 38: 3743–3759

Georg J, Hess WR (2011) Cis-antisense RNA: Another level of gene regulation in bacteria. Mol Microbiol Rev 75: in press

Georg J, Voss B, Scholz I, Mitschke J, Wilde A, Hess WR (2009) Evidence for a major role of antisense RNAs in cyanobacterial gene regulation. Mol Syst Biol 5: 305

Giangrossi M, Prosseda G, Tran CN, Brandi A, Colonna B, Falconi M (2010) A novel antisense RNA regulates at transcriptional level the virulence gene *icsA* of *Shigella flexneri*. Nucleic Acids Res 38: 3362–3375

Guell M, van Noort V, Yus E, Chen W-H, Leigh-Bell J, Michalodimitrakis K *et al.* (2009) Transcriptome complexity in a genome-reduced bacterium. Science 326:1268–1271

Han K, Kim K-s, Bak G, Park H, Lee Y (2010) Recognition and discrimination of target mRNAs by Sib RNAs, a cis-encoded sRNA family. Nucleic Acids Research 38: 5851–5866

Hernandez JA, Muro-Pastor AM, Flores E, Bes MT, Peleato ML, Fillat MF (2005) Identification of a *furA cis* antisense RNA in the cyanobacterium *Anabaena* sp. PCC 7120. J Mol Biol 355: 325–334

Irnov I, Sharma CM, Vogel Jr, Winkler WC (2010) Identification of regulatory RNAs in *Bacillus subtilis*. Nucleic Acids Res 38: 6637–6651

Itoh T, Tomizawa J (1980) Formation of an RNA primer for initiation of replication of ColE1 DNA by ribonuclease H. Proc Natl Acad Sci USA 77: 2450–2454

Jain C (1995) IS10 antisense control in vivo is affected by mutations throughout the region of complementarity between the interacting RNAs. J Mol Biol 246: 585–594

Kawano M, Aravind L, Storz G (2007) An antisense RNA controls synthesis of an SOS-induced toxin evolved from an antitoxin. Mol Microbiol 64: 738–754

Kawano M, Reynolds AA, Miranda-Rios J, Storz G (2005) Detection of 5'- and 3'-UTR-derived small RNAs and *cis*-encoded antisense RNAs in *Escherichia coli*. Nucleic Acids Res 33: 1040–1050

Krinke L, Mahoney M, Wulff DL (1991) The role of the OOP antisense RNA in coliphage λ development. Mol Microbiol 5: 1265–1272

Krinke L, Wulff DL (1987) OOP RNA, produced from multicopy plasmids, inhibits lambda cII gene expression through an RNase III-dependent mechanism. Genes Dev 1: 1005–1013

Krinke L, Wulff DL (1990) RNase III-dependent hydrolysis of lambda cII-O gene mRNA mediated by lambda OOP antisense RNA. Genes Dev 4: 2223–2233

Landt SG, Abeliuk E, McGrath PT, Lesley JA, McAdams HH, Shapiro L (2008) Small non-coding RNAs in *Caulobacter crescentus*. Mol Microbiol 68: 600–614

Lee EJ, Groisman EA (2010) An antisense RNA that governs the expression kinetics of a multifunctional virulence gene. Mol Microbiol 76: 1020–1033

Lee JM, Zhang S, Saha S, Santa Anna S, Jiang C, Perkins J (2001) RNA expression analysis using an antisense *Bacillus subtilis* genome array. J Bacteriol 183: 7371–7380
Legewie S, Dienst D, Wilde A, Herzel H, Axmann IM (2008) Small RNAs establish delays and temporal thresholds in gene expression. Biophys J 95: 3232–3238
Levine E, Zhang Z, Kuhlman T, Hwa T (2007) Quantitative characteristics of gene regulation by small RNA. PLOS Biol 5: e229
Levine E, Hwa T (2008) Small RNAs establish gene expression thresholds. Curr Opin Microbiol 11: 574–579
Liu JM, Livny J, Lawrence MS, Kimball MD, Waldor MK, Camilli A (2009) Experimental discovery of sRNAs in *Vibrio cholerae* by direct cloning, 5S/tRNA depletion and parallel sequencing. Nucleic Acids Res 37:e46
Lluch-Senar M, Vallmitjana M, Querol E, Pinol J (2007) A new promoterless reporter vector reveals antisense transcription in *Mycoplasma genitalium*. Microbiology 153: 2743–2752
Ma C, Simons RW (1990) The IS10 antisense RNA blocks ribosome binding at the transposase translation initiation site. EMBO J 9: 1267–1274
Makarova KS, Wolf YI, Koonin EV (2009) Comprehensive comparative-genomic analysis of type 2 toxin-antitoxin systems and related mobile stress response systems in prokaryotes. Biol Direct 4: 19
Masukata H, Tomizawa J (1984) Effects of point mutations on formation and structure of the RNA primer for ColE1 DNA replication. Cell 36: 513–522
Masukata H, Tomizawa J (1986) Control of primer formation for ColE1 plasmid replication: conformational change of the primer transcript. Cell 44: 125–136
Mitschke J, Georg J, Scholz I, Sharma C, Dienst D, Bantscheff J *et al.* (2011) An experimentally anchored map of transcriptional start sites in the model cyanobacterium *Synechocystis* sp. PCC 6803. Proc Natl Acad Sci USA 108: 2124–2129
Montero Llopis P, Jackson AF, Sliusarenko O, Surovtsev I, Heinritz J, Emonet T *et al.* (2010) Spatial organization of the flow of genetic information in bacteria. Nature 466: 77–81
Nelson CM, Herron MJ, Felsheim RF, Schloeder BR, Grindle SM, Chavez AO *et al.* (2008) Whole genome transcription profiling of *Anaplasma phagocytophilum* in human and tick host cells by tiling array analysis. BMC Genomics 9: 364
Opdyke JA, Fozo EM, Hemm MR, Storz G (2010) RNase III participates in GadY-dependent cleavage of the *gadX-gadW* mRNA. J Mol Biol 406: 29–43
Opdyke JA, Kang JG, Storz G (2004) GadY, a small-RNA regulator of acid response genes in *Escherichia coli*. J Bacteriol 186: 6698–6705
Ozsolak F, Platt AR, Jones DR, Reifenberger JG, Sass LE, McInerney P *et al.* (2009) Direct RNA sequencing. Nature 461: 814–818
Pandey DP, Gerdes K (2005) Toxin-antitoxin loci are highly abundant in free-living but lost from host-associated prokaryotes. Nucleic Acids Res 33: 966–976
Qiu Y, Cho BK, Park YS, Lovley D, Palsson BO, Zengler K (2010) Structural and operational complexity of the *Geobacter sulfurreducens* genome. Genome Res 9: 1304–1311
Rasmussen S, Nielsen HB, Jarmer H (2009) The transcriptionally active regions in the genome of *Bacillus subtilis*. Mol Microbiol 73: 1043–1057
Ross JA, Wardle SJ, Haniford DB (2010) Tn10/IS10 transposition is downregulated at the level of transposase expression by the RNA-binding protein Hfq. Mol Microbiol 78 :607–621
Salinas PC, Waldbeser LS, Crosa JH (1993) Regulation of the expression of bacterial iron transport genes: possible role of an antisense RNA as a repressor. Gene 123:33–38
Schluter JP, Reinkensmeier J, Daschkey S, Evguenieva-Hackenberg E, Janssen S, Janicke S *et al.* (2010) A genome-wide survey of sRNAs in the symbiotic nitrogen-fixing alpha-proteobacterium *Sinorhizobium meliloti*. BMC Genomics 11: 245
Selinger DW, Cheung KJ, Mei R, Johansson EM, Richmond CS, Blattner FR *et al.* (2000) RNA expression analysis using a 30 base pair resolution *Escherichia coli* genome array. Nat Biotechnol 18: 1262–1268
Sharma CM, Hoffmann S, Darfeuille F, Reignier J, Findeiss S, Sittka A *et al.* (2010) The primary transcriptome of the major human pathogen *Helicobacter pylori*. Nature 464: 250–255

Shearwin KE, Callen BP, Egan JB (2005) Transcriptional interference--a crash course. Trends Genet 21:339–345

Silby MW, Levy SB (2008) Overlapping protein-encoding genes in *Pseudomonas fluorescens* Pf0–1. PLoS Genet 4: e1000094

Silvaggi JM, Perkins JB, Losick R (2005) Small untranslated RNA antitoxin in *Bacillus subtilis*. J Bacteriol 187: 6641–6650

Silvaggi JM, Perkins JB, Losick R (2006) Genes for small, noncoding RNAs under sporulation control in *Bacillus subtilis*. J Bacteriol 188: 532–541

Simons RW, Kleckner N (1983) Translational control of IS10 transposition. Cell 34: 683–691

Simons RW, Kleckner N (1988) Biological regulation by antisense RNA in prokaryotes. Annu Rev Genet 22: 567–600

Sittka A, Lucchini S, Papenfort K, Sharma CM, Rolle K, Binnewies TT *et al.* (2008) Deep sequencing analysis of small noncoding RNA and mRNA targets of the global post-transcriptional regulator, Hfq. PLoS Genet 4: e1000163

Sneppen K, Dodd IB, Shearwin KE, Palmer AC, Schubert RA, Callen BP *et al.* (2005) A mathematical model for transcriptional interference by RNA polymerase traffic in *Escherichia coli*. J Mol Biol 346: 399–409

Sorek R, Cossart P (2010) Prokaryotic transcriptomics: a new view on regulation, physiology and pathogenicity. Nat Rev Genet 11: 9–16

Spiegelman WG, Reichardt LF, Yaniv M, Heinemann SF, Kaiser AD, Eisen H (1972) Bidirectional transcription and the regulation of Phage lambda repressor synthesis. Proc Natl Acad Sci USA 69: 3156–3160

Stazic D, Lindell D, Steglich C (2011) Antisense RNA protects from RNase E degradation by RNA-RNA duplex formation during phage infection. Nucleic Acids Res Epub ahead of print: 10.1093/nar/gkr037

Steglich C, Futschik ME, Lindell D, Voss B, Chisholm SW, Hess WR (2008) The challenge of regulation in a minimal phototroph: Non-coding RNAs in *Prochlorococcus*. PLoS Genetics 4: e1000173

Stork M, Di Lorenzo M, Welch TJ, Crosa JH (2007) Transcription termination within the iron transport-biosynthesis operon of *Vibrio anguillarum* requires an antisense RNA. J Bacteriol 189:3479–3488

Swiercz JP, Hindra, Bobek J, Haiser HJ, Di Berardo C, Tjaden B *et al.* (2008) Small non-coding RNAs in *Streptomyces coelicolor*. Nucleic Acids Res 36: 7240–7251

Thomason MK, Storz G (2010) Bacterial antisense RNAs: How many are there and what are they doing? Annu Rev Genet 44: 167–188

Toledo-Arana A, Dussurget O, Nikitas G, Sesto N, Guet-Revillet H, Balestrino D *et al.* (2009) The *Listeria* transcriptional landscape from saprophytism to virulence. Nature 459: 950–956

Tolmasky ME, Crosa JH (1995) Iron transport genes of the pJM1-mediated iron uptake system of *Vibrio anguillarum* are included in a transposon like structure. Plasmid 33: 180–190

Tomizawa J, Itoh T (1981) Plasmid incompatibility determined by interaction of RNAI with primer transcript. Proc Natl Acad Sci USA 78: 6096–6100

Tomizawa J, Itoh T, Seizer G, Som T (1981) Inhibition of ColEI RNA primer formation by a plasmid-specified small RNA. Proc Natl Acad Sci USA 78: 1421–1425

Tomizawa J, Ohmori H, Bird RE (1977) Origin of replication of colicin El plasmid DNA. Proc Natl Acad Sci USA 74: 1865–1869

Tramonti A, De Canio M, De Biase D (2008) GadX/GadW-dependent regulation of the *Escherichia coli* acid fitness island: transcriptional control at the *gadY-gadW* divergent promoters and identification of four novel 42 bp GadX/GadW-specific binding sites. Mol Microbiol 70: 965–982

Vogel J, Bartels V, Tang TH, Churakov G, Slagter-Jager JG, Hüttenhofer A *et al.* (2003) RNomics in Escherichia coli detects new sRNA species and indicates parallel transcriptional output in bacteria. Nucleic Acids Res 31: 6435–6443

Wagner EG, Simons RW (1994) Antisense RNA control in bacteria, phages, and plasmids. Ann Rev Microbiol 48: 713–742

Waldbeser LS, Chen Q, Crosa JH (1995) Antisense RNA regulation of the *fatB* iron transport protein gene in *Vibrio anguillarum*. Mol Microbiol 17: 747–756

Waldbeser LS, Tolmasky ME, Actis LA, Crosa JH (1993) Mechanisms for negative regulation by iron of the *fatA* outer membrane protein gene expression in *Vibrio anguillarum* 775. J Biol Chem 268: 10433–10439

Xiao B, Li W, Guo G, Li BS, Liu Z, Tang B *et al.* (2009) Screening and identification of natural antisense transcripts in *Helicobacter pylori* by a novel approach based on RNase I protection assay. Mol Biol Rep 36: 1853–1858

Yachie N, Numata K, Saito R, Kanai A, Tomita M (2006) Prediction of non-coding and antisense RNA genes in *Escherichia coli* with Gapped Markov Model. Gene 372: 171–181

Yeremenko N, Kouril R, Ihalainen JA, D'Haene S, van Oosterwijk N, Andrizhiyevskaya EG *et al.* (2004) Supramolecular organization and dual function of the IsiA chlorophyll-binding protein in cyanobacteria. Biochemistry 43: 10308–10313

Zemanova M, Kaderabkova P, Patek M, Knoppova M, Silar R, Nesvera J (2008) Chromosomally encoded small antisense RNA in *Corynebacterium glutamicum*. FEMS Microbiol Lett 279: 195–201

Chapter 6
6S RNA: A Regulator of Transcription

Karen M. Wassarman*

1 6S RNA – The Early Years

6S RNA was first identified in total RNA from *E. coli* due to its high abundance [Hindley, 1967]. It was one of the first small RNAs to be sequenced [Brownlee, 1971] and it was shown to be quite stable [Lee *et al.*, 1978]. In contrast to many other stable RNAs, its accumulation is not subject to stringent response [Ikemura and Dahlberg, 1973]. Limited sequence and structural homology to 7SL RNA led to an early hypothesis that 6S RNA was a bacterial signal recognition particle RNA (7SL or SRP RNA) involved in protein transport in eukaryotes [Walter and Blobel, 1983]. This hypothesis turned out to be incorrect as cells lacking 6S RNA did not exhibit protein secretion defects [Lee *et al.*, 1985] and 4.5S RNA was later identified as the bacterial SRP RNA [Poritz *et al.*, 1990; Ribes *et al.*, 1990]. Nevertheless, these studies demonstrated that 6S RNA, encoded by the *ssrS* gene, is dispensable for growth [Lee *et al.*, 1985]. In fact, cells lacking 6S RNA were found to grow indistinguishably from wild type under tested conditions, which unfortunately did not yield insight into 6S RNA function. Overexpression studies likewise did not uncover growth defects [Hsu *et al.*, 1985]. Additional characterization of 6S RNA revealed that it existed in an RNA-protein complex, although the nature of the protein components were not readily identifiable at that time [Lee *et al.*, 1978].

* University of Wisconsin-Madison, Department of Bacteriology

2 6S RNA Interactions with RNA Polymerase

2.1 6S RNA-RNA Polymerase: *In Vivo* Analysis

6S RNA functional studies pick up again with this RNA:protein complex many years later when the availability of the full *E. coli* genome sequence [Blattner *et al.*, 1997] and sensitive mass spectrometry techniques [Küster and Mann, 1998] allowed identification of individual proteins from polyacrylamide gels. The α subunit of RNA polymerase was found to co-peak with 6S RNA in glycerol gradients, suggesting they might exist in a complex [Wassarman and Storz, 2000]. The presence of a 6S RNA-RNA polymerase complex was confirmed by efficient and specific co-immunoprecipitation of 6S RNA with RNA polymerase from cell lysate. Likewise, an antisense oligonucleotide specific for 6S RNA efficiently co-selected RNA polymerase, including the σ^{70} subunit.

E. coli RNA polymerase consists of a core enzyme ($\alpha_2\beta\beta'\omega$), which requires an additional specificity subunit (σ) to generate the holoenzyme form of the enzyme (σ-RNA polymerase) required for transcription initiation. *E. coli* has seven σ factors [reviewed in Gruber and Gross, 2003]; σ^{70} is the highly abundant, housekeeping σ factor required at all times of growth. Co-immunoprecipitation and co-selection experiments demonstrated that 6S RNA interacts with the σ^{70}-RNA polymerase, but not with RNA polymerases containing other σ factors [Wassarman and Storz, 2000]. Crosslinking studies in cell extract suggested a direct interaction between 6S RNA and the σ^{70} subunit of RNA polymerase, as well as with core subunits [Wassarman and Storz, 2000].

6S RNA accumulates steadily through growth, varying from ~1,000 copies per cell during early exponential phase to ~10,000 copies per cell in late stationary phase [Lee *et al.*, 1978; Wassarman and Storz, 2000]. Gycerol gradient fractionation, immunoprecipitation and anti-sense oligo selection experiments revealed that essentially all of 6S RNA present from early exponential to late stationary phase is bound to RNA polymerase [Wassarman and Storz, 2000]. In addition, the vast majority (>75%) of σ^{70} is bound by 6S RNA in late stationary phase when 6S RNA levels are maximal. Thus, the σ^{70}-RNA polymerase:6S RNA complex is the predominant form of σ^{70}-RNA polymerase in late stationary phase. In addition, the first functional analysis demonstrated that 6S RNA binding to RNA polymerase results in down-regulation of transcription, at least at the σ^{70}-dependent promoter of the endogenous gene examined (*rsd*) [Wassarman and Storz, 2000]. See Section 3 for more details about 6S RNA-dependent regulation of transcription.

2.2 6S RNA-RNA Polymerase: *In Vitro* Analysis

Next, *in vitro* approaches were used to study details of the 6S RNA and RNA polymerase interaction. Co-immunoprecipitation and native gel electrophoresis demonstrated that *in vitro* generated 6S RNA and purified σ^{70}-RNA polymerase form complexes efficiently in the absence of other factors [Trotochaud and Wassarman, 2005; Gildehaus *et al.*, 2007]. Next, the specificity of RNA binding was tested by addition of competitors (tRNA, total yeast RNA, or heparin) or by examining the binding of RNA polymerase to 6S RNA compared to other RNAs, including an inactive mutant 6S RNA [6S(M5) RNA, see Fig. 1]. 6S RNA bound tightly to σ^{70}-RNA polymerase in the presence of competitors, suggesting the binding is specific [Trotochaud and Wassarman, 2005; Gildehaus *et al.*, 2007]. The 6S(M5) RNA mutant did not bind σ^{70}-RNA polymerase in the presence of competitors, consistent with *in vivo* results showing that this RNA does not co-immunoprecipitate with RNA polymerase [Trotochaud and Wassarman, 2005]. 6S(M5) RNA did show a low level of binding to RNA polymerase in the absence of competitors; this binding was interpreted as nonspecific as it was observed with several unrelated RNAs including 5S rRNA.

The question remained whether 6S RNA also could interact with other forms of RNA polymerase. Any potential complexes with 6S RNA and core RNA polymerase

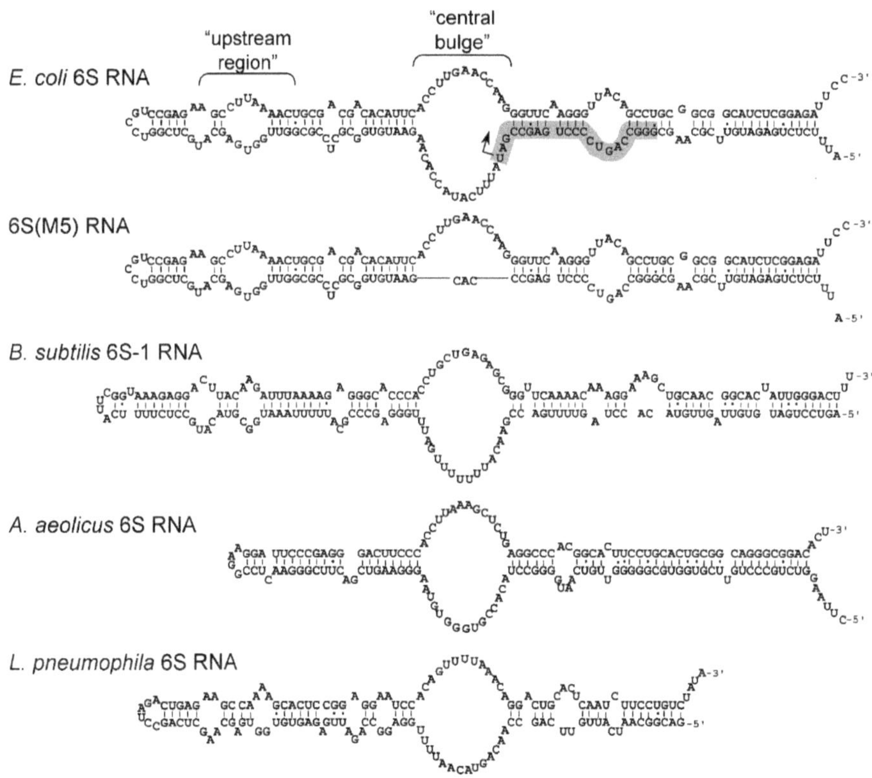

Fig. 1. 6S RNAs from *E. coli*, *B. subtilis*, *A. aeolicus*, and *L. pneumophila* in predicted secondary structures [Barrick *et al.*, 2005; Trotochaud and Wassarman, 2005; Willkhomm *et al.*, 2005; Faucher *et al.*, 2010]. The mutant 6S(M5) RNA also is shown [Trotochaud and Wassarman, 2005]. An arrow on the *E. coli* 6S RNA indicates where pRNA synthesis initiates; shaded in gray is the region that templates the longest observed pRNA. Regions referred to as "upstream region" and "central bulge" are indicated

lacking σ^{70} would not have been separated from σ^{70}-RNA polymerase complexes by co-immunoprecipitation or fractionation experiments. *In vitro*, only very weak binding of 6S RNA to core RNA polymerase was detected. This weak binding was sensitive to competition and was similar to binding of 6S(M5) RNA, suggesting it is nonspecific [Trotochaud and Wassarman, 2005; Gildehaus *et al.*, 2007]. In addition, no detectable 6S RNA binding was observed to free σ^{70} *in vitro* [Trotochaud and Wassarman, 2005; Gildehaus *et al.*, 2007], consistent with *in vivo* observations [Wassarman and Storz, 2000].

6S RNA complexes containing alternative σ factors were not observed *in vivo*, but the σ^{70}-RNA polymerase is very abundant and essentially saturates the available 6S RNA. Therefore, *in vitro* binding further investigated the possibility that 6S RNA might interact with other forms of RNA polymerase. A low level of 6S RNA binding to σ^{S}-RNA polymerase was observed in the absence of competitors [Trotochaud and Wassarman, 2005; Gildehaus *et al.*, 2007]. This complex was more resistant to competitors than core RNA polymerase, but less than the σ^{70}-RNA polymerase complex, and therefore was interpreted to be weak, but 6S RNA-specific binding in one report [Gildehaus *et al.*, 2007]. A second report found a similar extent of binding of 6S RNA and 6S(M5) RNA to σ^{S}-RNA polymerase (as well as σ^{32}-RNA polymerase), and therefore concluded that any interaction is nonspecific [Trotochaud and Wassarman, 2005]. In both cases, it is clear that 6S RNA has a strong preference for binding to σ^{70}-RNA polymerase. Given that *in vivo* complexes containing σ^{S}-RNA polymerase and 6S RNA have not been detected, we will use the interpretation that 6S RNA does not specifically interact with σ^{S}-RNA polymerase directly *in vivo*.

2.3 6S RNA: A Mimic of Promoter DNA Near the Active Site

Further experiments examined the details of the 6S RNA:σ^{70}-RNA polymerase interaction, in part to gain an understanding of how 6S RNA regulates transcription. Regions of 6S RNA important for binding to σ^{70}-RNA polymerase were first suggested by conserved features in 6S RNA [Trotochaud and Wassarman, 2005; Barrick *et al.*, 2005]. The overall secondary structure of 6S RNAs is well conserved among different species, although the primary sequence is not. The predicted structure is mostly double-stranded with a large single-stranded central region, which was largely confirmed by direct structure probing experiments on the *E. coli* 6S RNA and *B. subtilis* 6S RNAs (Fig. 1). This secondary structure is reminiscent of the conformation of DNA in the open complex (when the DNA surrounding the start site of transcription is melted), and therefore directly led to a model in which 6S RNA interactions with σ^{70}-RNA polymerase might mimic promoter DNA interactions. Consistent with this model, decreasing the size of the central bulge [e.g. 6S(M5) RNA] eliminates binding to σ^{70}-RNA polymerase *in vitro* and *in vivo* [Trotochaud and Wassarman, 2005; Wassarman and Storz, 2000]. Interestingly, the sequence of nucleotides in the central region are less important, and binding appears to be predominantly driven by structure [Trotochaud and Wassarman, 2005].

In agreement with this model, it was shown that 6S RNA is near the active site of RNA polymerase. The location of the active site on 6S RNA was within the central bubble at a location analogous to where the active site resides on promoter DNA in an open complex [Wassarman and Saecker, 2006]. Crosslinks between 6S RNA and σ^{70} were mapped near the central bulge of 6S RNA, consistent with σ^{70} interactions in this region of promoter DNA [Gildehaus *et al.*, 2007]. Native gel electrophoresis monitoring complex formation demonstrated that 6S RNA binding to RNA polymerase blocks promoter DNA from binding, and conversely that promoter DNA binding to σ^{70}-RNA polymerase blocks 6S RNA binding [Wassarman and Saecker, 2006; Gildehaus *et al.*, 2007], consistent with the proposal that both bind similarly to σ^{70}-RNA polymerase.

2.4 6S RNA: A Template for RNA Synthesis

Definitive evidence supporting the model that 6S RNA binds to σ^{70}-RNA polymerase similarly to promoter DNA came with the discovery that 6S RNA can be used as a template by σ^{70}-RNA polymerase to produce a product RNA (pRNA, also called dnRNA)[Wassarman and Saecker, 2006; Gildehaus *et al.*, 2007]. Initial steps in pRNA synthesis, as monitored *in vitro* using purified components, are analogous to early steps in transcription: short abortive products are generated, there is a transition to make longer RNAs, followed by the release of σ^{70}. 6S RNA is released from RNA polymerase after 14 -22 nt of pRNA synthesis as a hybrid with the pRNA [Wassarman and Saecker, 2006]. Interestingly, the pRNA length does not represent a "run-off" transcript, marking a departure from DNA-templated RNA synthesis. Given the release of 6S RNA:pRNA hybrids, we suggest the transition to a stable elongation complex is prevented, at least in part, by the inability of the central bulge in 6S RNA to reanneal, resulting in a persistent extended hybrid between template and nascent RNA. Similar destabilization of RNA polymerase complexes has been observed during transcription under conditions in which an extended hybrid is formed between template DNA and nascent RNA [Naryshkina *et al.*, 2006; Toulokhonov and Landick, 2006; Zenkin *et al.*, 2006].

It was speculated that pRNA synthesis *in vivo* could be a mechanism to relieve σ^{70}-RNA polymerase from 6S RNA control during outgrowth from stationary phase. Indeed, pRNA synthesis mediated release of 6S RNA from RNA polymerase was observed during outgrowth *in vivo* [Wassarman and Saecker, 2005; Wurm *et al.*, 2010] (see Fig. 2). Outgrowth was the only time when a significant proportion of free 6S RNA (i.e. not bound to σ^{70}-RNA polymerase) was detected by glycerol gradient fractionation of cell lysate, and 6S RNA levels are rapidly decreased upon outgrowth [Wassarman and Saecker, 2006]. In addition to solving a long-standing question regarding how σ^{70}-RNA polymerase is released from 6S RNA regulation during outgrowth, pRNA synthesis is compelling proof that the 6S RNA interacts with σ^{70}-RNA polymerase similarly enough to promoter DNA to be used as a template for RNA synthesis.

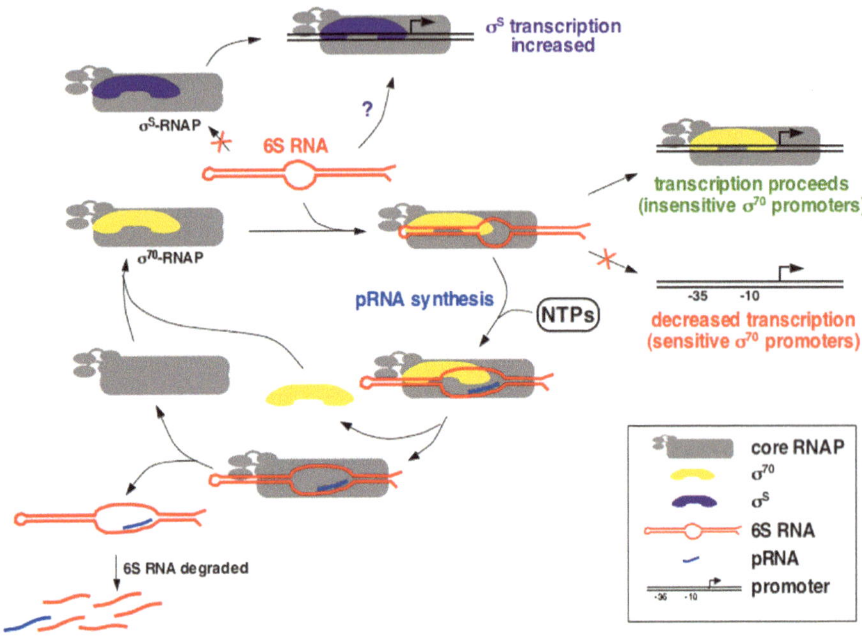

Fig. 2. A model illustrating 6S RNA interactions with RNA polymerase, the effects on transcription and the removal of 6S RNA from σ^{70}-RNA polymerase through pRNA synthesis. RNAP = RNA polymerase; NTPs = nucleotides; other symbols are described in key

A second type of RNA polymerase activity on small RNAs, including 6S RNA, has been reported [Windbichler *et al.*, 2008]. In this case, RNAs bind to RNA polymerase and RNA polymerase adds nucleotides to RNA ends. We point out that this activity is not templated nucleotide addition but instead uses the small RNA as a substrate, and it is not specific for 6S RNA. Therefore, this activity does not require the type of interaction between RNA and RNA polymerase we describe here for 6S RNA.

2.5 The 6S RNA Upstream Region and σ^{70} Region 4.2 Does *Not* Mimic Promoter DNA Interactions

To identify regions of σ^{70}-RNA polymerase that directly mediate 6S RNA binding, studies turned to σ^{70}. σ^{70} and related σ factors bind to DNA directly, within the context of σ^{70}-RNA polymerase, and provide much of the specificity for promoter recognition and binding. σ^{70}-dependent promoters are recognized primarily through two core promoter sequences: the -10 element and the -35 element, named for their approximate position upstream of the start site of transcription. Each promoter element is recognized by a region of σ^{70}; the -10 element by σ^{70} region 2.4 and the -35 element by σ^{70} region 4.2 [Dombroski *et al.*, 1992; Campbell *et al.*, 2002; Murakami *et al.*, 2002]. Deletion of region 4 from σ^{70} in the context of σ^{70}-RNA polymerase abolishes binding to 6S RNA, suggesting it is important for binding 6S RNA as well as the -35 element in promoters [Cavanagh *et al.*, 2008].

The location of the RNA polymerase active site on 6S RNA, as well as the region templating pRNA synthesis, provides the orientation of 6S RNA within σ^{70}-RNA polymerase and suggests the "upstream" stem of 6S RNA is directed towards σ^{70}-region 4.2 (see Fig. 1). Although it is difficult to model a precise location on 6S RNA for interaction with region 4.2 due to bulged nucleotides in the RNA, it is clear that predicted structures do not resemble the double-stranded -35 element. However, prediction of RNA structures can be misleading, leaving the possibility that the upstream region of 6S RNA could mimic B-form DNA such as observed for the NF-κB binding RNA aptamer [Huang *et al.*, 2003; Reiter *et al.*, 2008].

The contribution of residues within region 4.2 of σ^{70} to binding to 6S RNA were tested by examining σ^{70}-RNA polymerase variants for their ability to bind 6S RNA [Klocko and Wassarman, 2009]. Interestingly, residues important for RNA binding overlap with but are not coincident with residues important for DNA binding. The region important for 6S RNA binding is expanded compared to the DNA binding region, suggesting σ^{70} region 4.2 recognizes these nucleic acids differently. Intriguingly, many of the residues that are important for 6S RNA binding also are important for σ^{70}-RNA polymerase to interact with trans-acting transcription factors [Dove *et al.*, 2003] suggesting the presence or absence of transcription factors may alter the accessibility of this region of σ^{70} for binding to 6S RNA.

As it appears unlikely that the upstream region of 6S RNA is mimicking the double-stranded -35 element, questions remain regarding what structure and/or sequence elements are required in this region for binding to σ^{70}-RNA polymerase. Two similar secondary structure models have been proposed for the *E. coli* 6S RNA [Barrick *et al.*, 2005; Trotochaud and Wassarman, 2005], but they differ in details at the upstream region of 6S RNA. Re-examining the RNA structure probing data in both papers focusing on the upstream region did not strongly support or contradict either model (KMW, unpublished). One model has stronger phylogenetic support, but this region of the RNA is significantly divergent or even absent in some species (see Fig. 1)[Barrick *et al.*, 2005], suggesting the contribution of this region to 6S RNA binding to RNA polymerase may be different in different species. Additional experiments are necessary to define the structure of the RNA in this region and how it is recognized by σ^{70}.

Further information about the importance of individual residues in *E. coli* 6S RNA for binding to σ^{70}-RNA polymerase was provided by an iterative selection strategy that selected 6S RNA mutants that bind to RNA polymerase [Shepard *et al.*, 2010]. 12 critical residues were identified; all were in the upstream stem, demonstrating the essential role of this region of the RNA. Future studies are expected to define the precise role of these nucleotides, such as for direct interaction with protein, or for contributions to an appropriate structure. The selection analysis identified many regions of the 6S RNA where sequence identity is clearly unimportant, and also demonstrated that the region downstream from the central bulge is largely dispensable for binding to σ^{70}-RNA polymerase and pRNA synthesis.

3 6S RNA and Regulation of Transcription

3.1 Regulation of Transcription: *In Vivo* Analysis

6S RNA interacts with the major transcription machinery in the cell, σ^{70}-RNA polymerase, and therefore of particular interest are the effects that this binding has on transcription. Early studies showed that the presence of 6S RNA leads to decreased transcription of a σ^{70}-dependent promoter, but not a σ^{S}-dependent promoter *in vivo* [Wassarman and Storz, 2000]. Downregulation of σ^{70}-dependent transcription *in vivo* during stationary phase also was observed for many promoter-*lacZ* reporters [Trotochaud and Wassarman, 2004; Kim *et al.*, 2004]. It was found that 6S RNA effects are different at different σ^{70}-dependent promoters; some promoters are downregulated 2–5 fold while others remain unchanged even during late stationary phase when 6S RNA levels are maximal and the vast majority of the σ^{70}-RNA polymerase is bound by 6S RNA [Trotochaud and Wassarman, 2004; Cavanagh *et al.*, 2008]. Many of the reporters contain only core promoter sequences that minimize the potential influence of transcription factors, and thus are likely to monitor direct 6S RNA effects on transcription. These observations raise important questions about how binding to the general transcription machinery can lead to promoter specific regulation.

Promoter alignment revealed that tested promoters with strong -35 elements (i. e. nucleotide match to consensus) were found to be insensitive to 6S RNA while promoters with weak -35 elements were sensitive [Cavanagh *et al.*, 2008]. The presence of an extended -10 element (defined by a conserved TG one nucleotide upstream of the -10 element) also correlated with 6S RNA sensitivity. To test whether these two features could determine 6S RNA sensitivity directly, single promoters were systematically mutated to alter the strength of the -35 element or the presence or absence of an extended -10 element without changing flanking or spacer sequences to eliminate the potential contribution of other sequence regions [Cavanagh *et al.*, 2008]. It was found that promoters with strong -35 elements that are insensitive to 6S RNA regulation could be converted to 6S RNA sensitive promoters by mutations that weakened the -35 element. In addition, promoters sensitive to 6S RNA, containing either an extended -10 element or a weak -35 element, could be switched to 6S RNA insensitive promoters by mutating the extended -10 element or strengthening the -35 element, respectively. In contrast, mutations that strengthen or weaken the core -10 element did not have an effect on either sensitive or insensitive promoters indicating the strength of the -10 element does not contribute to 6S RNA sensitivity, and suggesting that overall promoter affinity for σ^{70}-RNA polymerase cannot determine 6S RNA sensitivity.

Microarray experiments comparing expression of genes globally in cells with or without 6S RNA have further explored the question of promoter specificity. In one study, total RNA isolated from cells in late stationary phase when 6S RNA levels are maximal was examined [Cavanagh *et al.*, 2008]. The primary goal of this study was to test further the impact of strength of -35 elements and the presence of extended -10 elements on 6S RNA regulation of transcription. Thus, predictions about 6S RNA sensitivity were made and compared to the microarray results for a subset of genes. Only genes with mapped σ^{70}-dependent promoters [as reported in Mitchell *et al.*, 2003] were included as unambiguous information about promoter location was required to identify core promoter element strength (i. e. match to consensus). Genes with multiple promoters were eliminated unless all reported promoters were predicted to behave similarly in response to 6S RNA (i. e. all sensitive or all insensitive) or there was one promoter that clearly predominated in stationary phase. It is important to note that in microarray experiments all genes with altered expression will be identified, including those that are directly regulated by 6S RNA as well as any that are indirectly regulated. To provide information regarding the mechanism of 6S RNA regulation specifically, only the direct effects are relevant, although for understanding the full physiological response to 6S RNA function, both are important. Even with this caveat, there was a surprising correlation between predictions and observations for both sensitive and insensitive promoters (80 % and 70 %, respectively). These data strongly support the model that 6S RNA sensitivity is driven by the presence of a weak -35 element or an extended -10 element.

A second microarray study used mRNA isolated from mid-log or early stationary phase cells [Neußer *et al.*, 2010]. Interestingly, an abundance of genes were differentially expressed in the presence or absence of 6S RNA, indicating 6S RNA regulates transcription even at earlier times when levels are not maximal. Similar results also have been shown on individually tested promoters [Trotochaud and Wassarman, 2004; 2006]. In contrast to the above report, this study concluded that 6S RNA sensitivity and core promoter element strength do not correlate. The underlying cause of these discrepant conclusions is not clear. However, we suggest any correlation of directly regulated genes might have been overshadowed by genes indirectly regulated in response to 6S RNA when all genes are included in the analysis, including those with multiple or less certain promoters. It also is possible that in early stationary phase indirect regulation might be a larger component of the full 6S RNA dependent response, especially since cells lacking 6S RNA have elevated ppGpp levels and ppGpp is a global regulator itself (see section 4.3) [Neußer *et al.*, 2010; Cavanagh *et al.*, 2010]. ppGpp levels are not altered in a 6S RNA-dependent manner at late stationary phase [Cavanagh *et al.*, 2010].

3.2 Regulation of Transcription: *In Vitro* Approaches

One might expect that the issue of direct versus indirect regulation by 6S RNA could be resolved by *in vitro* transcription assays. The ability of 6S RNA to downregulate transcription has been demonstrated *in vitro*, both in extract and using purified components [Trotochaud and Wassarman, 2005; Gildehaus *et al.*, 2007]. However, 6S RNA effects on transcription and binding of promoter DNA to RNA polymerase *in vitro* do not show the promoter specific effects observed *in vivo*; all σ^{70}-dependent promoters tested are sensitive to 6S RNA including promoters that have been demonstrated to be insensitive to 6S RNA *in vivo* (e. g. RNA1) [Wassarman and Saecker, 2006; Gildehaus *et al.*, 2007]. Of note, nonspecific RNA can have a general inhibitory effect on σ^{70}-RNA polymerase activity, as observed in the presence of tRNA [Krakow and von der Helm, 1970; Spassky, 1979] and the inactive 6S(M5) RNA mutant [Trotochaud and Wassarman, 2005]. Therefore, as mechanistic studies move forward, it becomes increasingly important to design experiments that can differentiate 6S RNA specific from nonspecific RNA effects. In addition, we now know that 6S RNA will be released from σ^{70}-RNA polymerase through the process of pRNA synthesis under transcription conditions (i. e. in the presence of nucleotides) [Wassarman and Saecker, 2006; Wurm *et al.*, 2010], and that the 6S RNA:pRNA hybrid cannot rebind to σ^{70}-RNA polymerase [KMW, unpublished].

It is apparent that *in vitro* assays do not recapitulate the promoter specific regulation observed *in vivo*. One possible explanation is that the physical conditions *in vitro* do not replicate *in vivo* conditions. For example, the salt identity or concentration, or the relative concentrations of 6S RNA, DNA and σ^{70}-RNA polymerase might not allow appropriate competition. It has been observed that 6S RNA binding to σ^{70}-RNA polymerase is very strong in the typical *in vitro* reaction conditions [Trotochaud and Wassarman, 2005; Wassarman and Saecker, 2006; Gildehaus *et al.*, 2007], resulting in negligible dissociation of 6S RNA from σ^{70}-RNA polymerase *in vitro* [Klocko and Wassarman, 2009]. *In vivo* conditions must be more dynamic to allow for the level of transcription observed.

A second possible explanation is that a factor required for the promoter specificity *in vivo* might be missing from the *in vitro* system. Although RNA polymerase subunits were the only proteins found to specifically co-select with 6S RNA from extract [Wassarman and Storz, 2000], such studies cannot eliminate the potential for a sub-stoichiometric population of additional protein(s) or more loosely associated factor(s). Such a "missing factor" is unlikely to be a typical transcription factor that acts through DNA binding. It must act at literally hundreds of genes, including reporter genes that can be switched from sensitive to insensitive or vice versa with 1–3 nucleotide changes in core promoter elements. The "missing factor" would be expected to be a negative regulator of 6S RNA *in vivo* since 6S RNA already inhibits too well *in vitro*. Additionally, it must act after 6S RNA binding to σ^{70}-RNA polymerase as *in vivo* the available σ^{70}-RNA polymerase is essentially saturated with 6S RNA [Wassarman and Storz, 2000]. In any event, further experiments are needed to uncover the mechanism underlying the promoter specificity of 6S RNA regulation.

3.3 6S RNA and Regulation of Transcription: Mechanism

Ultimately, studies of the interaction between 6S RNA and σ^{70}-RNA polymerase, combined with an understanding of what promoter features determine 6S RNA-sensitivity, are aimed to elucidate a molecular understanding of how 6S RNA regulates transcription. At one level, 6S RNA and DNA must compete for binding to σ^{70}-RNA polymerase as both occupy the active site of the enzyme and their binding sites on σ^{70} region 4.2 overlap. *In vitro* binding experiments support a direct competition model; 6S RNA binding to σ^{70}-RNA polymerase blocks DNA binding, and vice versa [Wassarman and Saecker, 2006; Gildehaus *et al.*, 2007]. However, the mechanism for promoter specificity cannot be based simply on overall affinity of promoters for σ^{70}-RNA polymerase, but instead must respond to the strength of -35 element binding [Cavanagh *et al.*, 2008], suggesting the nature of this competition is more complicated. We have hypothesized that the location of competition for binding is on region 4.2 of σ^{70}, facilitated by the overlapping but distinct binding sites for 6S RNA and DNA binding [Cavanagh *et al.*, 2008; Klocko and Wassarman, 2009]. However, it is not clear when this competition takes place *in vivo*. In addition, the potential role of extended -10 elements in 6S RNA sensitivity remains unclear. However, it has been suggested that interaction of the extended -10 element with region 3 of σ^{70} may alter the ability of region 4.2 to bind the -35 element [Minakhin and Severinov, 2003].

3.4 6S RNA and σ^S-Dependent Transcription

Discussion of 6S RNA effects on transcription has focused on σ^{70}-dependent transcription as this is the form of RNA polymerase that specifically and efficiently binds to 6S RNA *in vivo* and *in vitro*. Nevertheless, the presence of 6S RNA leads to altered σ^S-dependent transcription *in vivo*, albeit in the opposite direction as the presence of 6S RNA results in up-regulation of σ^S-dependent transcription [Trotochaud and Wassarman, 2004]. Microarray analyses support changes in many σ^S-transcribed genes [Cavanagh *et al.*, 2008; Neußer *et al.*, 2010]. Once again, not all tested σ^S-dependent promoters are sensitive to 6S RNA, suggesting promoter specificity [Wassarman and Storz, 2000; Trotochaud and Wassarman, 2004; Kim *et al.*, 2004]. *rpoS* transcription is insensitive to 6S RNA and steady state σ^S protein levels are unchanged in cells with and without 6S RNA [Trotochaud and Wassarman, 2004], suggesting the presence of 6S RNA alters σ^S activity rather than production or stability. Given that 6S RNA does not form stable complexes with σ^S-RNA polymerase *in vivo* [Wassarman and Storz, 2000], two models have been proposed for 6S RNA regulation of σ^S-dependent transcription: 1. 6S RNA binding to σ^{70}-RNA polymerase gives an advantage to σ^S for binding to available core RNA polymerase. σ^{70} is considerably more abundant than σ^S, and the affinity of σ^S for core RNA polymerase is much lower than other σ factors, suggesting σ^S-RNA polymerase formation may be a limiting step [Ishihama, 2000]. 2. A trans-acting factor that impedes σ^S activity is down-regulated by 6S RNA. One also could imagine a factor that enhances σ^S activity is upregulated by 6S RNA; however, given that 6S RNA acts directly in decreasing transcription, additional steps would need to be involved for this scenario. Several factors are known to alter σ^S activity [Hengge-Aaronis, 2002; Klauck *et al.*, 2007], but to date none have been found to respond to the presence or absence of 6S RNA in a manner consistent with their involvement (e. g. an inhibitor that is down-regulated by 6S RNA or an activator that is upregulated) (A. T. Cavanagh and KMW, unpublished).

Once again *in vitro* assays examining 6S RNA regulation of σ^S were not particularly enlightening. In fact, σ^S-dependent transcription and DNA binding to σ^S-RNA polymerase are inhibited by the presence of 6S RNA [Gildehaus *et al.*, 2007], in contrast to *in vivo* results where several genes are upregulated in the presence of

6S RNA [Trotochaud and Wassarman, 2004]. In some cases the same promoters were examined (e.g. *bolA*); thus, this discrepancy cannot be due to different effects of 6S RNA on different promoters. Again, further experiments are certainly necessary to sort out the 6S RNA effects on σ^S activity; however, one possibility is that the observed *in vitro* results are nonspecific RNA effects (similar to those that have been reported on σ^{70}-dependent transcription, see above) and that the *in vivo* observations result from indirect 6S RNA effects that would not be observed *in vitro*.

4 Physiological Role of 6S RNA

4.1 6S RNA and Stationary Phase Cell Survival

The presence of 6S RNA leads to changes in expression of literally hundreds of genes by both direct and indirect mechanisms, yet early work suggested that cells lacking 6S RNA grow indistinguishably from wild type through exponential phase and into stationary phase [Lee *et al.*, 1985]. The change in gene expression of any one gene is rather subtle, on the order of 2–5 fold, leading some to question if these changes could be physiologically relevant. However, the magnitude of overall transcriptional changes when amplified by hundreds of genes was harder to overlook. Therefore, researchers turned to competitive growth experiments in an effort to uncover a mutant phenotype, and followed persistence and survival of cells for longer time frames [Trotochaud and Wassarman, 2004]. Indeed, it was found that cells lacking 6S RNA are at a disadvantage for survival when in direct competition (i.e. co-culture) with wild type cells. It took more than 24 hours of co-culture for cells lacking 6S RNA to become underrepresented relative to wild type cells, consistent with observations that 6S RNA levels do not peak until ~24 hours of growth. Cells lacking 6S RNA grew similarly to wild type when not in competition (i.e. mono-culture) during the same time frame. However, it was found that cells lacking 6S RNA do not persist as well as wild type cells at considerably longer times (>20 days)[Trotochaud and Wassarman, 2004]. The precise nature of what leads to either the competitive or long term altered persistence phenotype is not yet understood. It is possible cells lacking 6S RNA are not as efficient at utilizing available resources, perhaps due to the increase in total transcription, or due to changes in expression of a specific gene responsible for altered survival, analogous to the role of *psp* in the altered survival at high pH (see below).

4.2 6S RNA and Stress: Altered Survival at High pH

Surprisingly, it was found that cells lacking 6S RNA survive better than wild type at higher pH (i.e. pH 9.3) when not in competition [Trotochaud and Wassarman, 2006]. Note that changing pH did not alter the competitive phenotype described above. Altered survival at high pH also was only apparent many hours after the transition into stationary phase (>20 hours). In this case, the timing of altered cell survival at high pH was very similar to observed changes in cells lacking *psp* genes [Weiner and Model, 1994], immediately suggesting a potential target for 6S RNA regulation to be tested. It was shown that 6S RNA downregulates expression of *pspF* [Trotochaud and Wassarman, 2006], a transcriptional activator of σ^{54}-RNA polymerase at two promoters (*pspA* operon and *pspG*)[Jovanovic *et al.*, 1996; Green and Darwin, 2004; Lloyd *et al.*, 2004]. 6S RNA dependent changes in *pspF* expression were sufficient to lead to indirect changes in *pspA* and *pspG* expression; 6S RNA does not regulate *pspA* or *pspG* expression directly. *pspF* is required for 6S RNA-dependent changes in survival at high pH, and increasing *pspF* expression in a 6S RNA-independent manner was sufficient to reproduce a similar change in survival. All together, these data definitively demonstrate that the direct 6S RNA dependent change in expression of *pspF* is responsible for the altered growth of cells lacking 6S RNA at high pH. Of note, the 6S RNA-dependent change in *pspF* mRNA levels is only ~2.5 fold, and clearly demonstrates that even this modest change in expression of one direct gene target can be sufficient to lead to altered growth. In addition, these data provided the first connection between a specific change in transcription resulting from 6S RNA function and a specific growth phenotype.

4.3 6S RNA Integration Into Global Pathways

Although the molecular details of how 6S RNA function alters cell survival are not understood, microarray expression studies give hints by revealing which genes change in expression in a 6S RNA-dependent manner [Cavanagh *et al.*, 2008; Neußer *et al.*, 2010]. For example, *relA*, which encodes a ppGpp synthase and is responsible for ppGpp accumulation during the transition from exponential to stationary phase growth, is downregulated by 6S RNA [Cavanagh *et al.*, 2010]. ppGpp is a global regulator itself, binding directly to RNA polymerase and altering transcription of many sensitive genes, including downregulation of ribosomal RNA synthesis and upregulation of genes required for amino acid biosynthesis [Paul *et al.*, 2004; Gralla, 2005; Magnusson *et al.*, 2005; Potrykus and Cashel, 2008]. In cells lacking 6S RNA, ppGpp levels are increased and predicted changes in ppGpp-sensitive transcription have been observed on microarrays, by analysis of reporter genes, and by primer extension [Neußer *et al.*, 2010; Cavanagh *et al.*, 2010]. It is intriguing that 6S RNA, which accumulates in stationary phase and regulates transcription, downregulates *relA*, which is another regulator important during stationary phase. We suggest that one role for 6S RNA may be to limit the extent of induction of stress response genes during nutrient deprivation, as also observed with *pspF* in response to high pH conditions. 6S RNA does not alter the induction of stress genes under appropriate stress conditions (e.g. *pspF* and *relA*), and therefore does not prevent the cellular responses; however, by dampening the stress response, 6S RNA may serve to conserve valuable resources when nutrients are limiting, thereby facilitating longer term survival.

In contrast to *relA*, σ^S activity is enhanced in the presence of 6S RNA [Trotochaud and Wassarman, 2004]. 6S RNA does not alter σ^S levels [Trotochaud and Wassarman, 2004], but instead is likely to alter σ^S protein activity quite differently from the 6S RNA-dependent regulation of transcription of *relA*. σ^S is another important regulator during stationary phase [Hennge-Aronis, 2002; Klauck *et al.*, 2007], and we suggest 6S RNA is one mechanism to coordinate these various factors, leading to enhanced survival of cells in competitive growth and during long term nutrient deprivation. It would appear that many stationary phase regulators function in part to limit expression/activity of other regulators, providing a high degree of control. In addition to 6S RNA down-regulation of *relA* and *pspF* [Cavanagh *et al.*, 2010; Trotochaud and Wassarman, 2006], other stationary phase regulators limit expression of 6S RNA, such as HN-S and Lrp [Neußer *et al.*, 2008]. Interestingly, overexpression of 6S RNA also leads to decreased cell survival during stationary phase, although this phenotype is only apparent in cells lacking σ^S and overexpressing 6S RNA [Wassarman and Storz, 2000]. Further studies testing how the individual 6S RNA-dependent changes in gene expression affect cell survival will be needed to dissect the role of individual gene changes from the cumulative effects of all the changes.

5 Biogenesis of 6S RNA

Other important questions about 6S RNA include how it is generated and what controls its accumulation in stationary phase. A σ^{70}-dependent promoter (P1) initiates transcription 9 nucleotides upstream of the mature 6S RNA 5' end [Hsu *et al.*, 1985; Griffin and Baillie, 1973; Kim and Lee, 2004]. A second promoter (P2) initiates 224 nucleotides upstream and is preferentially utilized by σ^S-RNA polymerase [Kim and Lee, 2004], suggesting 6S RNA accumulation in stationary phase might depend on σ^S. However, 6S RNA accumulates to the same high levels and with similar timing in cells lacking σ^S compared to wild type [Wassarman and Storz, 2000]. Other stationary phase transcriptional regulators have been implicated in direct and indirect regulation of *ssrS* transcription (e. g. HNS, LRP, StpA and Fis) [Neußer *et al.*, 2008]. However, these factors inhibit *ssrS* transcription, suggesting they cannot be responsible for the accumulation of 6S RNA in late stationary phase, but instead appear to limit the extent of expression. Therefore, although it is clear that many factors influence expression of 6S RNA, we do not yet understand what controls the high levels of 6S RNA in stationary phase.

The level of mature 6S RNA also will depend on its processing from precursor RNAs. In addition to the extra nucleotides at the 5' end (9 nt from P1 or 224 nt from P2), transcription proceeds through the downstream gene (*ygfA*), requiring additional RNA processing of the primary transcript to generate the 3' end of 6S RNA [Hsu *et al.*, 1985]. Using a combination of *in vivo* and *in vitro* approaches, it was found that the 5' end of the 6S RNA can be formed by RNase E or RNase G [Kim and Lee, 2004]. Identification of an endonuclease responsible for cleavage between 6S RNA and *ygfA* has been elusive [Kim and Lee, 2004], although several exonucleases can generate the final 3' end of the mature 6S RNA [Li *et al.*, 1998].

In addition to synthesis and processing, a differential RNA stability may contribute to 6S RNA accumulation. The 6S RNA has been reported to be very stable, especially when in a complex with RNA polymerase [Lee *et al.*, 1978]. However, many of the studies examining RNA stability used rifampicin to prevent new RNA synthesis to facilitate analysis. However, one mechanism for release of 6S RNA is through pRNA synthesis, which requires an active RNA polymerase and is inhibited by rifampicin [Wassarman and Saecker, 2006]. It is clear that the stability of 6S RNA is decreased when not bound by RNA polymerase [Lee *et al.*, 1978], and that pRNA synthesis-mediated release of 6S RNA is likely to contribute to decreased 6S RNA levels observed upon outgrowth [Wassarman and Saecker, 2006; Wurm *et al.*, 2010]. It is difficult to significantly overexpress 6S RNA in exponential phase, even when driven from very strong promoters, and it is difficult to underexpress 6S RNA in stationary phase, even when driven from promoters that should be strongly repressed (KMW, unpublished). Therefore, we suggest 6S RNA stability, rather than synthesis rates, is likely to be a primary determinant for 6S RNA accumulation.

6 6S RNAs in Diverse Bacterial Species

6.1 Identification

For many years there has been great interest in how general 6S RNA regulation of transcription is in eubacterial species, both those related and divergent from *E. coli*. One approach used to search for 6S RNAs and/or the genes encoding 6S RNAs has been to search for RNAs with similar properties to *E. coli* 6S RNA (e.g. high abundance, binding to RNA polymerase). Several abundant RNAs have been sequenced, either directly or using cDNA cloning approaches (individual or global RNomics methods) (*Pseudomonas aeruginosa; Synechococcus* PCC6301; *Bacillus subtilis*; *Aquifex aeolicus*; *Boredetella pertussis*)[Vogel *et al.*, 1987; Watanabe *et al.*, 1997; Ando *et al.*, 2002; Suzuma *et al.*, 2002; Willkomm *et al.*, 2005; Trotochaud and Wassarman, 2005] or more recently by deep sequencing (*Helicobacter pylori*) [Sharma *et al.*, 2010]. However, in some cases these RNAs were only predicted to be 6S RNAs later, based primarily on potential secondary structure models (Axmann *et al.*, 2005; Barrick *et al.*, 2005].

By far the majority of predicted 6S RNAs have been identified using computational approaches that either directly searched for candidate 6S RNA encoding genes in genomic sequences [Brösius, 1996; Barrick *et al.*, 2005; Trotochaud and Wassarman, 2005], or came across candidate 6S RNA encoding genes in the course of searches for small RNA genes globally [Axmann *et al.*, 2005; Pánek *et al.*, 2008; del Val *et al.*, 2007; Ulvé *et al.*, 2007; Valverde *et al.*, 2008; Faucher *et al.*, 2010]. Primary sequence searches have been less successful on all but species very highly related to species with identified 6S RNAs (e.g. *E. coli*, *B. subtilis*). In agreement with functional studies that suggest secondary structure is most important for 6S RNA function [Trotochaud and Wassarman, 2005], more divergent candidate 6S

RNAs have been identified using bioinformatic searches for secondary structures similar to known 6S RNAs [Barrick *et al.*, 2005].

Regardless of the approach to identify candidate 6S RNAs, it is important to consider what truly defines a 6S RNA, when a candidate RNA should be called "6S RNA" and what assumptions should be made about its function. In some cases, the similarity in sequence is so high that these questions are trivial. However, as more divergent potential 6S RNAs are identified, it becomes a more thought-provoking discussion. For instance, in α-proteobacteria a number of candidate 6S RNAs have been identified based on overall secondary structure similarity to other 6S RNAs [Barrick *et al.*, 2005]. However, the region analogous to the upstream stem in *E. coli* is significantly shortened and does not contain the elements known to be important for binding to *E. coli* σ^{70}-RNA polymerase. Consistent with these observations, 6S RNA from *A. aeolicus* that also has a shorter, divergent upstream stem (Fig. 1) [Willkolmm *et al.*, 2005; Barrick *et al.*, 2005] does not bind to *E. coli* σ^{70}-RNA polymerase (KMW, unpublished results). Of course it is possible that species specific interactions allow the *A. aeolicus* RNA to bind its cognate RNA polymerase, but lacking experimental data for this RNA, as well as others that share equally divergent structure predictions, it might be prudent to delay assumptions that these RNAs function in a manner similar to the *E. coli* 6S RNA. We suggest the primary determinant for 6S RNA is its ability to bind to RNA polymerase, although a more stringent criteria might be that it binds to the housekeeping form of RNA polymerase (i. e. σ^{70}-RNA polymerase in *E. coli*, σ^{A}-RNA polymerase in *B. subtilis*). Interestingly, one recent report definitively identified 6S RNA in deep sequencing data by the presence of pRNAs encoded by the candidate 6S RNA [Sharma *et al.*, 2010].

Interestingly, *B. subtilis* and some related species have two 6S RNAs, both of which bind to σ^{A}-RNA polymerase (equivalent to the σ^{70}-RNA polymerase in *E. coli*) [Trotochaud and Wassarman, 2005]. Although it is not yet clear whether these two RNAs have distinct or mostly redundant functions, the potential for multiple RNAs certainly adds an additional complication to 6S RNA identification, as well as understanding phenotypic studies lacking individual 6S RNA genes. In the case of *E. coli*, we have searched extensively using biochemical and bioinformatic approaches for any additional 6S-like RNAs without success suggesting *E. coli* only has one 6S RNA (KMW, unpublished results), but one can never eliminate the possibility for multiple RNAs entirely.

6.2 6S RNA Function in Other Species

Many 6S RNAs in divergent species have only recently been identified, and therefore functional and phenotypic studies for most are still emerging. However, in some cases initial characterization of 6S RNAs hint that the accumulation in stationary phase observed for *E. coli* 6S RNA, although common, may not be a global characteristic. For example, in *B. subtilis*, one of the 6S RNAs (called 6S-2, 6Sa, or Bs203 RNA) does not accumulate in stationary phase, although its precise expression profile is less clear as three different studies have observed some variation of RNA levels at different times, presumably due to different growth conditions [Ando *et al.*, 2002; Barrick *et al.*, 2005; Trotochaud and Wassarman, 2005]. In Synechococcus, the 6S RNA (originally called 6Sa RNA) decreases upon entry into stationary phase [Watanabe *et al.*, 1997], while in other cyanobacteria the 6S RNA remains fairly constant through growth [Axmann *et al.*, 2005] or responds to daily light-dark cycles rather than nutrient availability [Axmann *et al.*, 2007]. 6S RNA levels also have been reported to change during stress responses, such as acid or SDS stress in *S. meliloti* [Valverde *et al.*, 2008] and exposure to H_2O_2 in biofilms in *Burkholderia* [Peeters *et al.*, 2010]. Although altered accumulation patterns do not necessarily suggest that 6S RNAs in these species act differently mechanistically, it does suggest that how these RNAs integrate into regulation of biological processes are different.

A recent study provides an exciting view of 6S RNA function in *Legionella pneumophila* [Faucher *et al.*, 2010]. However, there are several perplexing differences from *E. coli* 6S RNA, and we suggest the *L. pneumophila* and *E. coli* 6S RNAs are unlikely to function in a mechanistically similar manner. The *L. pneumophila* 6S RNA is highly abundant and accumulates post-exponentially similarly to *E. coli*. However, microarray experiments revealed that only 8 genes are downregulated in the presence of 6S RNA and 127 are upregulated, hinting that the housekeeping form of RNA polymerase is unlikely to be the target for regulation in a manner analogous to *E. coli* in which hundreds of σ^{70}-dependent promoters are downregulated by 6S RNA. Immunoprecipitation experiments provide evidence that the *L. pneumophila* 6S RNA interacts with RNA polymerase specifically when compared to 5S RNA, but only a small fraction of the 6S RNA co-immunoprecipitates with RNA polymerase suggesting that either the 6S RNA:RNA polymerase complex is in low abundance or that this complex is unstable in *L. pneumophila*. Of note, the *L. pneumophia* 6S RNA upstream region is fairly similar to *E. coli* (see Fig. 1). In contrast, the downstream stem is quite short, although this region is dispensable in *E. coli* [Shepard *et al.*, 2010; KMW, unpublished]. Many of the genes regulated by *L. pneumophila* 6S RNA have functions related to intracellular multiplication. Consistently, cells lacking 6S RNA are reduced for survival intracellularly in host cells, which was even more apparent when cells lacking 6S RNA were in competition with wild type cells [Faucher *et al.*, 2010]. The *L. pneumophila* 6S RNA is required for optimal cell survival and parallels could be made to *E. coli* 6S RNA that also affects survival. More information regarding how and when 6S RNAs from divergent species function is needed to elucidate the full impact of these RNAs on biology, whether they follow the model provided by the *E. coli* 6S RNA or if they provide an expansion of our thinking for how this type of small RNA functions.

7 Concluding Comments

Clearly a lot of progress has been made over the past decade in understanding how and when 6S RNA functions. Many of the specifics of how 6S RNA is recognized by RNA polymerase have been elucidated, although questions still remain regarding the specificity of 6S RNA binding to σ^{70}-RNA polymerase, as opposed to other σ-RNA polymerases, and the intimate details of the interaction. The upstream stem in the *E. coli* 6S RNA is critical for binding to σ^{70}-RNA polymerase. σ^{70} region 4.2 is very well conserved in diverse bacteria, yet the 6S RNA upstream stem is not well conserved in many species, raising interesting questions about how this interaction succeeds in diverse species. One of the most unexpected and exciting discoveries about 6S RNA is the ability of σ^{70}-RNA polymerase to synthesize pRNA using 6S RNA as a template. pRNA synthesis occurs during outgrowth from stationary phase, and is important for the release of σ^{70}-RNA polymerase from 6S RNA. It remains unclear if pRNA synthesis is important at other times of growth. Cells lacking 6S RNA have altered survival, and many genes are regulated in a 6S RNA-dependent manner. However, the connection between changes in gene expression and these phenotypes are still unclear in most cases. There is still much to be learned about this exciting RNA, especially as the studies of 6S RNAs in other organisms hint at more diversity in function than originally suspected from the model *E. coli* system.

References

Ando Y, Asari S, Suzuma S, Yamane K, Nakamura K (2002) Expression of a small RNA, BS203 RNA, from the *yocI-yocJ* intergenic region of *Bacillus subtilis* genome. FEMS Microbiol Lett 207: 29–33

Axmann IM, Holtzendorff J, Voss B, Kensche P, Hess WR (2007) Two distinct types of 6S RNA in *Prochlorococcus*. Gene 406: 69–78

Axmann IM, Kensche P, Vogel J, Kohl S, Herzel H, Hess WR (2005) Identification of cyanobacterial non-coding RNAs by comparative genome analysis. Genome Biol 6: R73

Barrick JE, Sudarsan N, Weinberg Z, Ruzzo WL, Breaker RR (2005) 6S RNA is a widespread regulator of eubacterial RNA polymerase that resembles an open promoter. RNA 11: 774–784

Blattner FR, Plunkett G 3rd, Bloch CA, Perna NT, Burland V, Riley M, Collado-Vides J, Glasner JD, Rode CK, Mayhew GF, Gregor J, Davis NW, Kirkpatrick HA, Goeden MA, Rose DJ, Mau B, Shao Y (1997) The complete genome sequence of *Escherichia coli* K-12. Science 277: 1453–1462

Brösius J (1996) More *Haemophilus* and *Mycoplasma* genes. Science 271: 1302

Brownlee GG (1971) Sequence of 6S RNA of *E. coli*. Nat New Biol 229: 147–149

Campbell EA, Muzzin O, Chlenov M, Sun JL, Olson CA, Weinman O, Trester-Zedlitz ML, Darst SA (2002) Structure of the bacterial RNA polymerase promoter specificity sigma subunit. Mol Cell 9: 527–539

Cavanagh AT, Chandrangsu P, Wassarman KW (2010) 6S RNA regulation of *relA* alters ppGpp levels in early stationary phase. Microbiology 156: 3791–3800

Cavanagh AT, Klocko AD, Liu X, Wassarman KM (2008) Promoter specificity for 6S RNA regulation of transcription is determined by core promoter sequences and competition for region 4.2 of σ^{70}. Mol Microbiol 67: 1242–1256

del Val C, Rivas E, Torres-Quesada O, Toro N, Jiménez-Zurdo JI (2007) Identification of differentially expressed small non-coding RNAs in the legume endosymbiont *Sinorhizobium meliloti* by comparative genomics. Mol Microbiol 66: 1080–1091

Dombroski AJ, Walter WA, Record MT Jr, Siegele DA, Gross CA (1992) Polypeptides containing highly conserved regions of transcription initiation factor σ^{70} exhibit specificity of binding to promoter DNA. Cell 70: 501–512

Dove SL, Darst SA, Hochschild A (2003) Region 4 of sigma as a target for transcription regulation. Mol Microbiol 48: 863–874

Faucher SP, Friedlander G, Livny J, Margalit H, Shuman HA (2010) *Legionella pneumophila* 6S RNA optimizes intracellular multiplication. Proc Natl Acad Sci USA 107: 7533–7538

Gildehaus N, Neußer T, Wurm R, Wagner R (2007) Studies on the function of the riboregulator 6S RNA from *E. coli*: RNA polymerase binding, inhibition of *in vitro* transcription and synthesis of RNA-directed de novo transcripts. Nucleic Acids Res 35: 1885–1896

Gralla JD (2005) *Escherichia coli* ribosomal RNA transcription: regulatory roles for ppGpp, NTPs, architectural proteins and a polymerase-binding protein. Mol Microbiol 55: 973–977

Green RC, Darwin AJ (2004) PspG, a new member of the *Yersinia enterocolitica* phage shock protein regulon. J Bacteriol 186: 4910–4920

Griffin BE, Baillie DL (1973) Precursors of stable RNA accumulated in a mutant of *E. coli*. FEBS Lett 34: 273–279

Gruber TM, Gross CA (2003) Multiple sigma subunits and the partitioning of bacterial transcription space. Annu Rev Microbiol 57: 441–466

Hengge-Aronis R (2002) Recent insights into the general stress response regulatory network in *Escherichia coli*. J Mol Microbiol Biotechnol 4: 341–346

Hindley J (1967) Fractionation of ^{32}P-labeled ribonucleic acids on polyacrylamide gels and their characterization by fingerprinting. J Mol Biol 30: 125–136

Hsu LM, Zagorski J, Wang Z, Fournier MJ (1985) *Escherichia coli* 6S RNA gene is part of a dual-function transcription unit. J Bacteriol 161:1162–1170

Huang DB, Vu D, Cassiday LA, Zimmerman JM, Maher LJ 3rd Ghosh G (2003) Crystal structure of NF-κB $(p50)_2$ complexed to a high-affinity RNA aptamer. Proc Natl Acad Sci USA 100: 9268–9273

Ikemura T, Dahlberg JE (1973) Small ribonucleic acids of *Escherichia coli*. II. Noncoordinate accumulation during stringent control. J Biol Chem 248: 5033–5041

Ishihama A (2000) Functional modulation of *Escherichia coli* RNA polymerase. Annu Rev Microbiol 54: 499–518

Jovanovic G, Weiner L, Model P (1996) Identification, nucleotide sequence, and characterization of PspF, the transcriptional activator of the *Escherichia coli* stress-induced *psp* operon. J Bacteriol 178: 1936–1945

Kim KS, Lee Y (2004) Regulation of 6S RNA biogenesis by switching utilization of both sigma factors and endoribonucleases. Nucleic Acids Res 32: 6057–6068

Kim EY, Shin MS, Rhee JH, Choy HE (2004) Factors influencing preferential utilization of RNA polymerase containing sigma-38 in stationary-phase gene expression in *Escherichia coli*. J Microbiol 42: 103–110

Klauck E, Typas A, Hengge R (2007) The σ^S subunit of RNA polymerase as a signal integrator and network master regulator in the general stress response in *Escherichia coli*. Sci Prog 90: 103–127

Klocko AD, Wassarman KM (2009) 6S RNA binding to $E\sigma^{70}$ requires a positively charged surface of σ^{70} region 4.2. Mol Microbiol 73: 152–164

Krakow JS, von der Helm K (1970) Azotobacter RNA polymerase transitions and the release of sigma. Cold Spring Harbor Symp Quant Biol 35: 73–83

Küster B, Mann M (1998) Identifying proteins and post-translational modifications by mass spectrometry. Curr Opin Struct Biol 8: 393–400

Lee CA, Fournier MJ, Beckwith J (1985) *Escherichia coli* 6S RNA is not essential for growth or protein secretion. J Bacteriol 161: 1156–1161

Lee SY, Bailey SC, Apirion D (1978) Small stable RNAs from *Escherichia coli*: evidence for the existence of new molecules and for a new ribonucleoprotein particle containing 6S RNA. J Bacteriol 133: 1015–1023

Li Z, Pandit S, Deutscher MP (1998) 3' exoribonucleolytic trimming is a common feature of the maturation of small, stable RNAs in *Escherichia coli*. Proc Natl Acad Sci USA 95: 2856–2861

Lloyd LJ, Jones SE, Jovanovic G, Gyaneshwar P, Rolfe MD, Thompson A, Hinton JC, Buck M (2004) Identification of a new member of the phage shock protein response in *Escherichia coli*, the phage shock protein G (PspG). J Biol Chem 279: 55707–55714

Magnusson LU, Farewell A, Nyström T (2005) ppGpp: a global regulator in *Escherichia coli*. Trends Microbiol 13: 236–242
Minakhin L, Severinov K (2003) On the role of the *Escherichia coli* RNA polymerase σ^{70} region 4.2 and α-subunit C-terminal domains in promoter complex formation on the extended -10 *gal*P1 promoter. J Biol Chem 278: 29710–29718
Mitchell JE, Zheng D, Busby SJW, Minchin SD (2003) Identification and analysis of "extended –10" promoters in *Escherichia coli*. Nucleic Acids Res 31: 4689–4695
Murakami KS, Masuda S, Campbell EA, Muzzin O, Darst SA (2002) Structural basis of transcription initiation: an RNA polymerase holoenzyme-DNA complex. Science 296: 1285–1290
Naryshkina T, Kuznedelov K, Severinov K (2006) The role of the largest RNA polymerase subunit lid element in preventing the formation of extended RNA-DNA hybrid. J Mol Biol 361: 634–643
Neußer T, Gildehaus N, Wurm R, Wagner R (2008) Studies on the expression of 6S RNA from *E. coli*: involvement of regulators important for stress and growth adaptation. Biol Chem 389: 285–297
Neußer T, Polen T, Geissen R, Wagner R (2010) Depletion of the non-coding regulatory 6S RNA in *E. coli* causes a surprising reduction in the expression of the translation machinery. BMC Genomics 11: 165–179
Pánek J, Bobek J, Mikulík K, Basler M, Vohradsky J (2008) Biocomputational prediction of small non-coding RNAs in *Streptomyces*. BMC Genomics 9: 217–228
Paul BJ, Ross W, Gaal T, Gourse RL (2004) rRNA transcription in *Escherichia coli*. Annu Rev Genet 38: 749–770
Peeters E, Sass A, Mahenthiralingam E, Nelis H, Coenye T (2010) Transcriptional response of *Burkholderia cenocepacia* J2315 sessile cells to treatments with high doses of hydrogen peroxide and sodium hypochlorite. BMC Genomics 11: 90–108
Poritz MA, Bernstein HD, Strub K, Zopf D, Wilhelm H, Walter P (1990) An *E. coli* ribonucleoprotein containing 4.5S RNA resembles mammalian signal recognition particle. Science 250: 1111–1117
Potrykus K, Cashel M (2008) (p)ppGpp: still magical? Annu Rev Microbiol 62: 35–51
Reiter NJ, Maher LJ 3rd, Butcher SE (2008) DNA mimicry by a high-affinity anti-NF-κB RNA aptamer. Nucleic Acids Res 36: 1227–1236
Ribes V, Römisch K, Giner A, Dobberstein B, Tollervey D (1990) *E. coli* 4.5S RNA is part of a ribonucleoprotein particle that has properties related to signal recognition particle. Cell 63: 591–600
Sharma CM, Hoffmann S, Darfeuille F, Reignier J, Findeiss S, Sittka A, Chabas S, Reiche K, Hackermüller J, Reinhardt R, Stadler PF, Vogel J (2010) The primary transcriptome of the major human pathogen *Helicobacter pylori*. Nature 464: 250–255
Shephard L, Dobson N, Unrau PJ (2010) Binding and release of the 6S transcriptional control RNA. RNA 16: 885–892
Spassky A, Busby SJ, Danchin A, Buc H (1979) On the binding of tRNA to *Escherichia coli* RNA polymerase. Eur J Biochem 99: 187–201
Suzuma S, Asari S, Bunai K, Yoshino K, Ando Y, Kakeshita H, Fujita M, Nakamura K, Yamane K (2002) Identification and characterization of novel small RNAs in the *aspS-yrvM* intergenic region of the *Bacillus subtilis* genome. Microbiol 148: 2591–2598
Toulokhonov I, Landick R (2006) The role of the lid element in transcription by *E. coli* RNA polymerase. J Mol Biol 361: 644–658
Trotochaud AE, Wassarman KM (2004) 6S RNA function enhances long-term cell survival. J Bacteriol 186: 4978–4985
Trotochaud AE, Wassarman KM (2005) A highly conserved 6S RNA structure is required for regulation of transcription. Nat Struct Mol Biol 12: 313–319
Trotochaud AE, Wassarman KM (2006) 6S RNA regulation of *pspF* transcription leads to altered cell survival at high pH. J Bacteriol 188: 3936–3943
Ulvé VM, Sevin EW, Chéron A, Barloy-Hubler F (2007) Identification of chromosomal alphaproteobacterial small RNAs by comparative genome analysis and detection in *Sinorhizobium meliloti* strain 1021. BMC Genomics. 8: 467–483

Valverde C, Livny J, Schlüter JP, Reinkensmeier J, Becker A, Parisi G (2008) Prediction of *Sinorhizobium meliloti* sRNA genes and experimental detection in strain 2011. BMC Genomics 9: 416–440

Vogel DW, Hartmann RK, Struck JC, Ulbrich N, Erdmann VA (1987) The sequence of the 6S RNA gene of *Pseudomonas aeruginosa*. Nucleic Acids Res 15: 4583–4593

Walter P, Blobel G (1983) Disassembly and reconstitution of signal recognition particle. Cell 34: 525–533

Wassarman KM, Saecker RM (2006) Synthesis-mediated release of a small RNA inhibitor of RNA polymerase. Science 314: 1601–1603

Wassarman KM, Storz G (2000) 6S RNA regulates *E. coli* RNA polymerase activity Cell 101: 613–623

Watanabe T, Sugiura M, Sugita M (1997) A novel small stable RNA, 6Sa RNA, from the cyanobacterium *Synechococcus sp.* strain PCC6301. FEBS Lett 416: 302–306

Weiner L, Model P (1994) Role of an *Escherichia coli* stress-response operon in stationary-phase survival. Proc Natl Acad Sci USA 91: 2191–2195

Willkomm DK, Minnerup J, Hüttenhofer A, Hartmann RK (2005) Experimental RNomics in *Aquifex aeolicus*: identification of small non-coding RNAs and the putative 6S RNA homolog. Nucleic Acids Res 33: 1949–1960

Windbichler N, von Pelchrzim F, Mayer O, Csaszar E, Schroeder R (2008) Isolation of small RNA-binding proteins from *E. coli*: evidence for frequent interaction of RNAs with RNA polymerase. RNA Biol 5: 30–40

Wurm R, Neußer T, Wagner R (2010) 6S RNA-dependent inhibition of RNA polymerase is released by RNA-dependent synthesis of small de novo products. Biol Chem 391: 187–196

Zenkin N, Naryshkina T, Kuznedelow K, Severinov K (2006) The mechanism of DNA replication primer synthesis by RNA polymerase. Nature 439: 617–620

Chapter 7
Archaea Employ Small RNAs as Regulators

Ruth Schmitz-Streit, Dominik Jäger, Angelika Jellen-Ritter, Julia Babski, Jörg Soppa, and Anita Marchfelder*

1 Introduction

In addition to well-characterized non-coding RNAs like ribosomal RNAs and transfer RNAs, a plethora of non-coding RNAs exists in organisms of all three domains of life (Hüttenhofer *et al.*, 2005; Storz, 2002). In Eukaryotes the non-coding miRNAs and siRNAs are small RNAs of about 21 nucleotides and they function primarily by base pairing with their target mRNAs, mediating the inhibition of translation or triggering degradation as part of the RISC complex (Krol *et al.*, 2010). The miRNAs act always as part of the RISC complex, where they primarily bind to the Argonaute protein. Homologs to Argonaute proteins have been detected in Archaea but the function of the archaeal Argonaute proteins is not clear yet (Li *et al.*, 2010; Makarova *et al.*, 2009).

In Bacteria, small regulatory RNAs range in size from about 50 to 400 nucleotides and are often required for stress response or virulence gene regulation (Gottesman, 2004a). In *Escherichia coli* more than 70 sRNAs have currently been experimentally identified, of which about 20 have been studied in detail (Storz *et al.*, 2005). In addition, hundreds of sRNAs have been predicted through bioinformatic approaches (Zhang *et al.*, 2004). The majority of these sRNAs are encoded in trans to their target genes and act as antisense RNA, frequently masking sequences close to, or overlapping with, the ribosome binding site of their target transcripts (Gottesman, 2004b). Details about regulation of gene expression by bacterial sRNAs are described in chapters 1–5.

The third domain of life (Figure 1A), represented by the Archaea, resembles in most cellular processes Eukarya rather than Bacteria, as indicated by the discovery of C/D box ribonucleoproteins resembling the eukaryotic C/D box snoRNPs. However, much less is known about the complexity and composition of their sRNA population in general. Neither the sites nor the modes of interaction between regu-

* Biologie II, Universität Ulm, 89069 Ulm, Germany

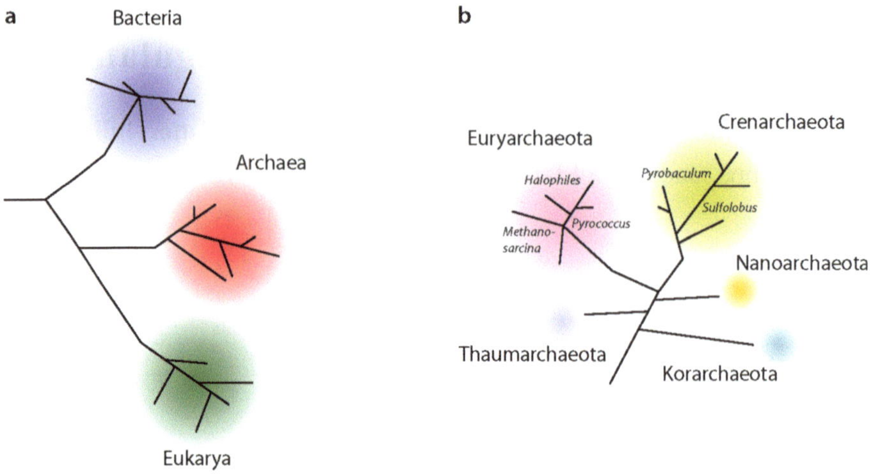

Fig. 1. The domain Archaea
a) The three domains of life: Bacteria, Eukarya, and Archaea.
b) The archaeal domain contains up to five phyla. Whereas Crenarchaeota and Euryarchaeota are accepted as phyla of the archaeal domain, the existence of the other three phyla is still being debated (Brochier-Armanet *et al.*, 2008; Elkins *et al.*, 2008; Huber *et al.*, 2003).

latory RNAs and their targets, which could be RNA as well as protein, have been elucidated in detail so far (a comparison of RNA-mediated regulation between the three domains of life is shown in Figure 2).

The domain of Archaea comprises organisms which are highly diverse and abundant and rather ubiquitous not only in extreme environments as originally proposed. It consists of up to five phyla (Euryarchaeota, Crenarchaeota, Thaumarchaeota, Nanoarchaeota and Korarchaeota) and each of its phyla contains a whole set of divergent organisms (Figure 1B). Only a very small percentage of the archaeal organisms have been studied in detail, mainly due to problems growing them as pure cultures and in sufficient amounts, in the laboratory.

As a result, we have in this book only one chapter about the sRNA population in the whole archaeal domain. Small RNAs have been studied to date in only six organisms of the domain Archaea: *Archaeoglobus fulgidus, Sulfolobus solfataricus, Methanocaldococcus jannaschii, Methanosarcina mazei, Pyrococcus furiosus,* and *Haloferax volcanii.*

In this chapter, we summarize current knowledge about the diversity of sRNA types and their biological functions in Archaea.

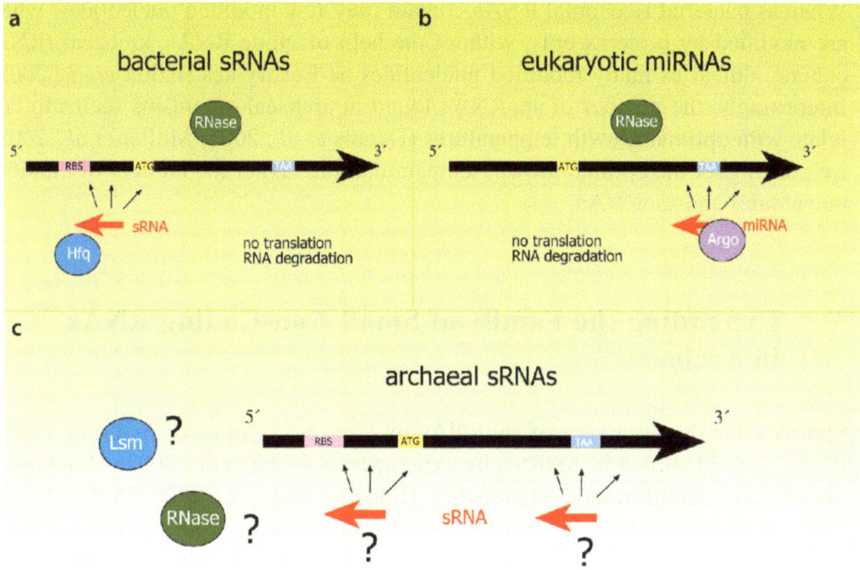

Fig. 2. Different mechanisms of gene expression regulation by small RNAs.
a) In Bacteria, small RNAs characterized to date bind primarily to the 5' end of mRNAs where they often occlude the ribosome binding site (RBS). They pair to the mRNA either with extensive complementarity (cis-encoded sRNAs) or limited complementarity (trans-encoded sRNAs). In addition, bacterial sRNAs have also been shown to activate genes.
b) Eukaryotic miRNAs bind to the 3' end of mRNAs thereby repressing translation and/or triggering mRNA degradation. miRNAs are always active as part of the RISC complex and bind primarily to the Argonaute protein (Argo).
c) The mechanism of regulation by small RNAs in Archaea has not been unravelled yet. Either Archaea use a bacteria-like mechanism or a mechanism similar to the eukaryotic one or they have invented their own mechanism.

2 The Discovery of a New Type of Non-Coding RNA in Archaea: snoRNAs

Up until 2000, the known non-coding RNA population in Archaea consisted of tRNAs, rRNAs, the RNase P RNA and the 7S RNA. The first new type of non-coding RNAs to be discovered in Archaea were the small nucleolar guide RNAs, the snoRNAs (Gaspin *et al.*, 2000; Omer *et al.*, 2000). The discovery of snoRNAs in Archaea was quite surprising since snoRNAs were until then only known from eukaryotic organisms, where they guide the modification of rRNA nucleotides. Eukaryotic ribosomal RNAs contain a high number of methylated and pseudo-uridylated nucleotides. Modification of these nucleotides is guided by snoRNAs.

Whereas bacterial ribosomal RNAs contain only few modified nucleotides, which are modified by proteins only, without the help of guide RNAs, archaeal rRNAs contain almost as many modified nucleotides as Eukaryotes (Noon *et al.*, 2003). Interestingly, the number of snoRNAs found in archaeal organisms seems to correlate with optimal growth temperatures (Dennis *et al.*, 2001; Muller *et al.*, 2008), i.e., the higher the optimal growth temperature, the higher the number of modified nucleotides and snoRNAs.

3 Expanding the Family of Small Non-Coding RNAs in Archaea

Shortly after the discovery of snoRNAs in Archaea, additional small non-coding RNAs were identified in Archaea by experimental (Tang *et al.*, 2002a; Tang *et al.*, 2002b) and bioinformatic approaches (Eddy, 2002). The Hüttenhofer laboratory employed RNomics to identify the sRNA population from the Euryarchaeon *Archaeoglobus fulgidus* (Tang *et al.*, 2002a; Tang *et al.*, 2002b). They identified 86 new small RNAs in *Archaeoglobus*, including 10 snoRNAs, 22 CRISPR-derived RNAs, 4 rRNA fragments, 9 intergenic sRNAs, 33 cis-antisense sRNAs, 5 sRNAs overlapping with ORFs and 3 sense transcripts originating from the sense orientation from ORFs (Table 1). The sRNA population from the Crenarchaeon *Sulfolobus solfataricus* was investigated using three different approaches: RNomics (Hüttenhofer *et al.*, 2005; Tang *et al.*, 2002b), co-immunoprecipitation with the protein L7Ae (Zago *et al.*, 2005) and high throughput sequencing (HTS) (Wurtzel *et al.*, 2009). The HTS approach revealed the highest number of identified sRNAs and also included all sRNAs identified by means of the other two experimental approaches (Wurtzel *et al.*, 2009). Altogether 310 sRNAs were identified, 13 snoRNAs, 18 CRISPR RNAs, 28 sRNAs assigned to transposons and with 185 cis-antisense RNAs – also an unexpectedly high number of such transcripts (Table 1).

Using bioinformatic approaches small RNAs were predicted in *Methanocaldococcus janaschii* and *Pyrococcus furiosus* (Klein *et al.*, 2002; Schattner, 2002) (Table 1). The bioinformatic approaches took advantage of the fact that both organisms are hyperthermophiles and have a high A/T content. Searches for small RNAs concentrated on GC–rich regions, because genes for structural RNAs frequently exhibit an elevated G/C content compared to other regions of the A/T-rich genomes Table 1).

4 Small RNAs in Halophilic Archaea

In order to identify small RNAs in halophilic Archaea, the model organism *Haloferax volcanii* was used. *H. volcanii* belongs to the lineage of Euryarchaeota, it requires about 2.1 M NaCl for optimal growth and adjusts the intracellular salt concentration to similar values to cope with the high salt concentration in the environ-

Table 1. Number of sRNAs identified in Archaea to date. The number of sRNAs identified in the different archaeal organisms is listed. The references for the different identifications are: *A. fulgidus* RNomics (Tang *et al.*, 2002a), *S. solfataricus* RNomics (Tang *et al.*, 2005), CoIP with L7Ae (Zago *et al.*, 2005), HTS (Wurtzel *et al.*, 2009), *M. janaschii* bioinformatics (1) (Klein *et al.*, 2002) and (2) (Schattner, 2002), *P. furiosus* bioinformatics (Klein *et al.*, 2002), *H. volcanii* RNomics (Straub *et al.*, 2009), HTS (Jellen-Ritter, Babski, Soppa and Marchfelder, manuscript in preparation), bioinformatics (1) and (2) (Babski, Tjaden, Voss, Jellen-Ritter, Marchfelder, Hess, Soppa), *M. mazei* HTS (Jäger *et al.*, 2009).

Type of sRNA	***A. fulgid.***	***S. solfataricus***			***M. janaschii***		***P. fur.***	***H. volcanii***				***M. maz.***
Method of identification	RNomics	RNomics	CoIP L7Ae	HTS	bioinf. (1)	bioinf. (2)	bioinf.	RNomics	HTS	bioinf. (1)	bioinf. (2)	HTS
Intergenic sRNA	9	11	3	125	5	18	5	21	145	31	>100	199
cis-antisense	33	19	8	185	–	–	–	18	45	–	–	43
Sense	8	2	13	yes[1]	–	–	–	49	yes[1]	–	–	yes[1]

[1] Sense RNAs were detected, but since additional experiments are required to distinguish sense sRNAs frome stable mRNA degradation products it is diffcult to determine the number of sense RNAs at this point

ment. This fact alone raises interesting questions because the high salinity should have an impact on the formation of biologically active RNA structures as well as on the interaction between different RNA molecules. The genome sequence of *H. volcanii* has been recently annotated, it contains 51 tRNAs, 5S rRNA, 16S rRNA, 23S rRNA, 7S RNA, and RNase P RNA (Hartman *et al.*, 2010).

The sRNA population of *H. volcanii* was thoroughly investigated using three different experimental approaches and two bioinformatic approaches. Candidate sRNAs were further characterized according to their differential expression and biological function (Soppa *et al.*, 2009; Straub *et al.*, 2009).

4.1 Prediction of sRNA Genes

The principle strategy of both bioinformatic approaches for the prediction of sRNA genes in the genome of *H. volcanii* was to compare its intergenic regions with the intergenic regions of related genomes (Babski, Tjaden, Voss, Jellen-Ritter, Marchfelder, Hess, Soppa), inferred from strategies widely established in the prediction of bacterial sRNAs (Backofen und Hess, 2010). The first approach used, in addition to *H. volcanii,* three haloarchaeal species, one crenarchaeal species and one halophilic bacterium. This approach is restricted to the identification of widely conserved sRNAs because species from different phylogenetic groups were included in the analysis. Therefore, it was not surprising that only 31 sRNA genes were predicted.

The second bioinformatic approach concentrated on a comparative analysis within the Haloarchaeales and used the representative species of four genera. This led to the prediction of about 100 sRNA genes that are shared with at least two of the three other species. Unexpectedly and interestingly, only a single sRNA gene was predicted by both bioinformatic approaches. In summary, both approaches led to the prediction of about 130 sRNA genes in the genome of *H. volcanii,* concentrating solely on intergenic regions. Experimental approaches for the verification of the predicted sRNA genes, analysis of their differential expression and characterization of their biological function are described below. Experimental verification is of great importance because the bioinformatic predictions might include some false positives, e. g. the conserved elements might not be small RNA genes but conserved regulatory elements embedded in non-conserved parts of 5'-UTRs or 3'-UTRs of large transcripts of protein-encoding genes. In one case it was indeed found that the predicted "sRNA" is instead part of a long 3'-UTR of the transcript of gene HVO_A0196. Haloarchaeal 5'-UTRs and 3'-UTRs have average lengths of around 30 nt, but in rare cases lengths of up to 150 nt have been determined (Brenneis *et al.*, 2007).

4.2 Experimental Identification of Small RNAs

For the experimental identification of small RNAs (sRNAs) in *H. volcanii,* three different approaches were used, i.e. experimental RNomics, HTS and co-purification with the Lsm protein.

In the RNomics approach, total RNA of *H. volcanii* was size selected to generate a cDNA library representing sRNAs of sizes between 130 and 450 nt. This set-up resulted in the identification of 21 intergenic sRNAs, 18 antisense sRNAs and 49 sRNAs, which originate from the sense orientation of ORFs.

The second experimental approach used HTS of cDNA generated from *Haloferax* RNA ranging in size from 17 to 500 nucleotides. Pooled RNA samples were isolated from *Haloferax* cultures grown at three different growth conditions, each at exponential and stationary phase: optimal conditions, a high temperature and low salt conditions. Altogether 145 intergenic sRNAs, 45 antisense RNAs and 6 snoRNAs were identified by this approach. All sRNAs identified with the RNomics approach were also present in the HTS pool.

In a third approach, sRNAs were identified by affinity isolation of a tagged version of the *H. volcanii* Lsm protein. The analysis showed that 17 different sRNAs could be copurified with the Lsm protein, 10 of the 21 intergenic sRNAs determined in the RNomics approach and 7 bioinformatically-predicted sRNAs (Fischer *et al.*, 2010).

4.3 Expression of Small RNA Genes

In order to verify the expression of experimentally and bioinformatically identified sRNAs, three different experimental approaches were chosen, each exhibiting specific advantages and disadvantages. Northern blot analyses were performed to analyze differential sRNA gene expression and – most importantly – to verify the sizes of the respective sRNAs. To investigate potential differential expression of the sRNAs, RNA from cells grown under ten different conditions was analysed in these northerns, which revealed firstly that most of the selected sRNAs were expressed and secondly that several were clearly differentially expressed.

The HTS approach described above not only provided global information on the diversity of the sRNA population, but moreover, it yielded quantitative information about relative sRNA levels. HTS becomes particularly informative when the numbers of reads are statistically relevant, which is true for many but by far not all sRNAs. While the methodology in essence is very general, until now it has been restricted to the six different conditions that were used for RNA isolation.

As an alternative for the parallel characterization of relative sRNA levels, DNA microarray analyses were used. Probes for 122 selected sRNAs were added to a genome-wide *H. volcanii* microarray, including 62 of the predicted sRNAs. Nearly two-thirds (38) of these predicted sRNA genes were found to be expressed in at least one of the three conditions tested so far ("standard conditions", high temperatures, low osmolarity). Due to the fact that another 17 sRNAs were verified by HTS, the expression of a total of 55 predicted sRNA genes has been experimentally verified so far.

4.4 Functional Analysis

A powerful genetic approach for the elucidation of biological roles of genes with hitherto unknown functions is the construction of deletion mutants and their comparison with the isogenic wild type, e. g. characterization of phenotypes under different conditions or quantification of the differences in the transcriptomes or proteomes. 15 of the 21 intergenic sRNA genes that have been identified by RNomics (see above) have meanwhile been deleted. Phenotypic differences between wild type and mutants could be detected for 9 of the 15 mutants. In each case the mutant exhibited a very specific phenotype and differed from the wild type in only one or very few of the tested conditions, while it behaved indistinguishably from the wild type under most conditions. A very striking example is a mutant which is unable to grow at 51°C, in contrast to the wild type, while both strains grow identically at 48°C (Straub *et al.* 2009).

Whereas the majority of bacterial sRNAs are involved in stress responses, archaeal sRNAs are also involved in metabolic regulation. Four mutants have growth defects on different carbon sources and one mutant is unable to grow via nitrate respiration. The most unexpected findings were behavioural differences between two mutants and the wild type, one mutant had a swarming defect, while the second mutant, in contrast, swarms considerably faster than the wild type. These first analyses strongly indicate that archaeal sRNAs may play many different roles in the cell. This is in congruence with the variety of differential expression patterns of sRNA genes that are beginning to emerge (see above).

4.5 The *Haloferax* Lsm Protein

Proteins of the Lsm family, including eukaryotic Sm proteins and bacterial Hfq, are key players in RNA metabolism. Little is known about the archaeal homologs of these proteins. A detailed review about the Lsm protein is given in chapter 8.

Analysis of the genome of the haloarchaeon *Haloferax volcanii* revealed that *H. volcanii* encodes a single Lsm protein, which belongs to the Lsm1 subfamily. The *lsm* gene is co-transcribed and overlaps with the gene for the ribosomal protein L37e. Northern blot analysis shows that both genes are differentially transcribed (Fischer *et al.*, 2010). The Lsm protein forms homoheptameric complexes and has a copy number of 4000 molecules/cell. *In vitro* analyses using electrophoretic mobility shift assays and ultrasoft mass spectrometry (LILBID) showed a complex formation of the recombinant Lsm protein with oligo(U) RNA, tRNAs and an sRNA

(Fischer *et al.*, 2010). Co-immunoprecipitation with a FLAG-tagged Lsm protein produced *in vivo,* confirmed that the protein binds to small RNAs. Furthermore, the co-immunoprecipitation revealed a plethora of protein interaction partners, suggesting its involvement in several cellular pathways. The deletion of the *lsm* gene is viable, resulting in a pleiotropic phenotype, indicating that the haloarchaeal Lsm is involved in many cellular processes, which is in congruence with the plethora of protein interaction partners (Fischer *et al.*, 2010).

5 Small RNAs in Methanogenic Archaea

Methanogenic Archaea are the largest, phylogenetically most diverse, but also one of the best studied groups within the domain of Archaea. They are strictly anaerobes requiring highly reduced conditions (below -330 mV) for growth (Smith und Hungate, 1958; Thauer, 1998). In accordance with their exceptional diversity, methanogenic Archaea are found in a variety of habitats, ranging from hot springs to the deep ocean floor, sewage treatment plants, landfills, paddy fields and freshwater sediments (Liu und Whitman, 2008). Further, the majority of archaeal strains found in, and on, lower and higher Eukaryotes like ciliates, arthropods, invertebrates and vertebrates belong to this class (Lange *et al.*, 2005; Lepp *et al.*, 2004); e.g. *Methanobrevibacter* species and *Methanosphaera stadtmanae* have been identified as part of the human microbiota (Belay *et al.*, 1990; Eckburg *et al.*, 2005; Miller and Wolin, 1982; Miller and Wolin, 1985). Though restricted in their energy metabolisms to the formation of methane from CO_2 and H_2, formate, methanol, methylmamines, and acetate, methanogenic Archaea play a significant role in the global carbon cycle, forming approximately 1 billion tons of methane per year in various anaerobic habitats on Earth (Thauer, 1998). The most widespread catabolic reaction is the reduction of CO_2 to CH_4, using molecular hydrogen as the reductant. Members of the order *Methanosarcinales* have the most versatile substrate spectrum within methanogenic Archaea and are able to grow on acetate or methylotrophic substrates such as methanol or methylamines as the sole carbon and energy source (Ferry, 1999; Thauer, 1998; Thauer *et al.*, 2008), in addition to CO_2 and H_2. The biochemical basis of methanogenesis has been analyzed in considerable detail, frequently resulting in the discovery of unique reactions and cofactors (Deppenmeier, 2002; Ferry, 1999; Thauer *et al.*, 2008).

Currently, the genome sequences of 33 methanogenic Archaea have become available (http://www.ncbi.nlm.nih.gov/genomes/genlist.cgi?taxid=2157&type=0&name =Complete%20Archaea). Many representatives of this group of Archaea are able to survive extended periods of nutrient starvation or stress resulting from drastic changes in the environment. Whereas it is known in great detail how certain bacteria cope with such situations, only very little insight has been gained so far into the global regulatory networks of methanogenic Archaea. Considerable numbers of potential transcriptional regulators have been predicted from the available genome sequences (including members of the LrpR, ArsR, AscN and MarR families) but only a few global transcriptional regulators have been identified and characterized (Brinkman *et al.*, 2003; Lee *et al.*, 2008; Lie and Leigh, 2003; Weidenbach *et al.*, 2008). Only very few examples of post-transcriptional regulation have been reported in methanogenic Archaea so far. These will be discussed in the following.

5.1 Un-translated Regions of mRNAs

The presence and potentially regulatory function of un-translated regions of mRNAs (5′-UTR or 3′-UTR) has until recently never been systematically studied for methanogenic Archaea. First experimental implications for regulatory 5'-leader sequences have been observed for leaders of a DEAD-box RNA helicase of *Methanococcus burtonii* (Lim *et al.*, 2000) and the methanol-specific methyltransferase 2 genes (*mtaA1* and *mtaA2*) of *Methanosarcina acetivorans* (Bose *et al.*, 2008; Opulencia *et al.*, 2009), however the potential role of the respective leader regions in regulation has not been further elucidated or confirmed. Recently, Sowers and collaborators reported a potential cis-acting element of the CO dehydrogenase / acetyl coenzymeA synthase (CODH/ACS) mRNA of *Methanosarcina acetivorans,* indicating a regulatory role of the 5′-UTR (Anderson *et al.*, 2009), potentially terminating transcriptional elongation during methylotrophic growth.

Very recently, a pyrosequencing approach of cDNA libraries using the newly developed differential RNA-seq method selective for newly initiated transcripts (Sharma *et al.*, 2010), discovered 876 transcriptional start sites across the *Methanosarcina mazei* strain Göl genome (Jäger *et al.*, 2009). Unlike other Archaea such as *Sulfolobus* and Haloarchaea, where leaderless mRNAs are prevalent (Brenneis *et al.*, 2007), the majority of the detected mRNAs in *M. mazei* carry long un-translated 5'-regions of up to 500 nucleotides (with an average size of 150–200 nucleotides). The high number of these long 5′-UTRs in *M. mazei* argues for extensive post-transcriptional regulation at 5′-UTRs in methanogenic Archaea, e.g. at the level of transcript stability, by regulatory proteins or RNAs, or through riboswitches, which has to be demonstrated experimentally. Despite the experimental implications for potentially regulatory 5′-UTRs in *M. mazei*, only a single, highly conserved, riboswitch candidate present in Bacteria and Archaea was predicted by *in silico* analysis via comparative (meta)genomics in *M. acetivorans*, namely *crcB*, which is located upstream of the *crcB* gene and supposed to bind a yet unknown metabolite (Weinberg *et al.*, 2010).

5.2 Small RNAs in *M. mazei*

Until very recently, no small non-coding RNAs have been reported for methanogenic Archaea besides (i) *in silico* predicted snoRNAs (UCSC Archaeal Genome Browser (Ellis *et al.*, 2010; Schneider *et al.*, 2006)) and (ii) four stable transcripts with a high degree of secondary structures in the hyperthermophile *Methanocaldococcus jannashii* (previously *Methanococcus jannashii*), which were identified by computational screens for GC-rich regions in the AT-rich genome and experimen-

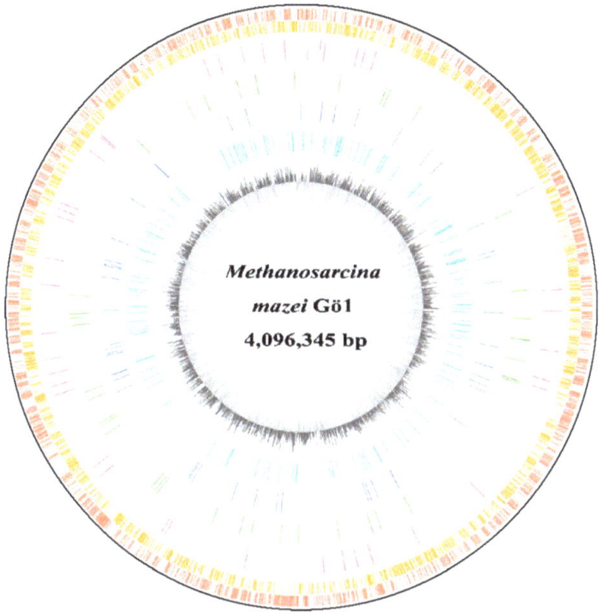

Fig. 3. Newly identified transcripts aligned to the *M. mazei* genome. Each transcript derived from cDNA sequencing was mapped onto the *M. mazei* chromosome and visualized with a DNA plotter from the Sanger Institute (Carver *et al.*, 2009). The two outer rings show the ORFs of *M. mazei* identified by Deppenmeier *et al.* 2002 (Deppenmeier *et al.*, 2002). The following four cycles display the genomic localization of newly identified transcripts, obtained by massive parallel sequencing. New ORFs (red), spRNAs (green), asRNAs (violet) and sRNAs (blue) are indicated. The inner ring indicates the GC content

tally verified (Klein *et al.*, 2002). However, genetic and biochemical studies have yet to follow to confirm and elucidate the function of those small RNA candidates.

At the end of 2009, a high number of non-coding RNAs were reported, which were identified by a massively parallel pyrosequencing approach of cDNA libraries derived from *M. mazei* growing under different nitrogen sources (Jäger *et al.*, 2009). A high number of small RNA (sRNA) candidates were identified, mostly located within intergenic regions but also in antisense orientation to 5' and 3' regions of mRNAs (asRNA), scattered across the genome (208 candidates) (Fig. 3). Genome-wide comparative screens for the presence of those sRNA candidates in the *Methanosarcina bakeri* and *M. acetivorans* genomes, both representing a similar low percentage of coding regions in the original annotation (Galagan *et al.*, 2002; Maeder *et al.*, 2006), demonstrated that 30 % of the identified asRNAs and 21 % of sRNAs originating from intergenic regions were conserved in all three *Methanosarcina* species. Moreover, 40 sRNA candidates containing potential ORFs encoding oligopeptides smaller than 30 amino acids were identified, the majority of which is conserved in the three *Methanosarcina* strains, indicating that some of those sRNAs in fact are short mRNAs that have remained non-annotated so far or might have a dual function as mRNA and regulatory sRNA. However, at present, no such small peptide has been demonstrated in *M. mazei*. 57 out of 77 randomly selected sRNA candidates have been verified on northern blots, allowing an estimate of ~180 for the total number of sRNAs in this organism. Confirming the expression of a high number of sRNA genes by Northern blot analysis further verified differential expression in response to nitrogen availability for 18 small RNAs, representing the first regulatory RNAs potentially involved in nitrogen stress response in prokaryotes. Overall, the strong conservation of the identified sRNA candidates and the high number of differentially transcribed sRNAs in response to nitrogen argue for a prominent regulatory function of sRNAs in the nitrogen and general stress response of *M. mazei* and potentially in methanoarchaea in general.

The finding that a considerably high number of asRNA candidates has been identified in antisense orientation to transposase transcripts in *M. mazei,* indicates that transposon mobility might be regulated by an RNA-antisense mechanism on the post-transcriptional level, which has been demonstrated for Tn*10* and Tn*30* in *Escherichia coli* (Arini *et al.*, 1997; Ma and Simons, 1990) as well as for insertion elements in *S. solfataricus* (Tang *et al.*, 2005). Several of those asRNAs identified in *M. mazei* show differential transcript levels in response to nitrogen availability (Jäger *et al.*, 2009), predicting a relationship between nitrogen availability and transposition events in *M. mazei*.

6 Conclusion

In all archaeal organisms investigated so far, a complex sRNA population has been identified, encompassing sRNAs derived from intergenic regions, cis-antisense sRNAs and sense RNAs. Functional analyses of these sRNAs suggest their role in the regulation of gene expression. However, up to date experimental studies to identify sRNA interaction partners have not been performed and thus no targets are known for these sRNAs. Therefore, it is unclear where these sRNAs and their target molecules interact (e. g. mRNA upstream region or mRNA 3' end) and how the stabilities and/or translation efficiencies of the target mRNAs are influenced. While it seems reasonable to assume that the antisense sRNAs interact with the mRNA encoded on the opposite strand, this has not been experimentally tested yet. Further research is needed to identify the direct intracellular targets of archaeal sRNAs and to unravel their molecular mechanisms of action. Given the extreme growth conditions such as high temperatures, salinities and pH to which many Archaea are adapted, fundamental insight into the mechanisms of RNA-RNA interaction and the involvement of additional protein factors can be expected. Therefore, the investigation of the archaeal sRNAs constitutes a fascinating topic and the future will show what kind of mechanisms Archaea use for gene expression regulation with sRNAs. One may expect mechanisms similar to the bacterial sRNA pathways, but it is also possible that the fundamental principles are related more closely to one of the eukaryotic non-coding RNA pathways. It also appears feasible that unique ways of regulating gene expression with sRNAs have been invented within the archaeal domain.

References

Anderson KL, Apolinario EE, MacAuley SR, Sowers KR (2009) A 5' leader sequence regulates expression of methanosarcinal CO dehydrogenase/acetyl coenzyme A synthase. *Journal of bacteriology* 191: 7123–7128

Arini A, Keller MP, Arber W (1997) An antisense RNA in IS*30* regulates the translational expression of the transposase. *Biol Chem* 378: 1421–1431

Backofen R, Hess WR (2010) Computational prediction of sRNAs and their targets in bacteria. *RNA (New York, NY* 7: 33–42. Epub 2010 Jan 2013.

Belay N, Mukhopadhyay B, Conway de Macario E, Galask R, Daniels L (1990) Methanogenic bacteria in human vaginal samples. *Journal of clinical microbiology* 28: 1666–1668

Bose A, Pritchett MA, Metcalf WW (2008) Genetic analysis of the methanol- and methylamine-specific methyltransferase 2 genes of Methanosarcina acetivorans C2A. *Journal of bacteriology* 190: 4017–4026

Brenneis M, Hering O, Lange C, Soppa J (2007) Experimental characterization of Cis-acting elements important for translation and transcription in halophilic archaea. *PLoS Genet* 3: e229.

Brinkman AB, Ettema TJ, de Vos WM, van der Oost J (2003) The Lrp family of transcriptional regulators. *Mol Microbiol* 48: 287–294

Brochier-Armanet C, Boussau B, Gribaldo S, Forterre P (2008) Mesophilic Crenarchaeota: proposal for a third archaeal phylum, the Thaumarchaeota. *Nat Rev Microbiol* 6: 245–252.

Carver T, Thomson N, Bleasby A, Berriman M, Parkhill J (2009) DNAPlotter: circular and linear interactive genome visualization. *Bioinformatics (Oxford, England)* 25: 119–120

Dennis PP, Omer A, Lowe T (2001) A guided tour: small RNA function in Archaea. *Mol Microbiol* 40: 509–519.

Deppenmeier U (2002) The unique biochemistry of methanogenesis. *Prog Nucleic Acid Res Mol Biol* 71: 223–283

Deppenmeier U, Johann A, Hartsch T, Merkl R, Schmitz RA, Martinez-Arias R, Henne A, Wiezer A, Baumer S, Jacobi C, Bruggemann H, Lienard T, Christmann A, Bomeke M, Steckel S, Bhattacharyya A, Lykidis A, Overbeek R, Klenk HP, Gunsalus RP, Fritz HJ, Gottschalk G (2002) The genome of *Methanosarcina mazei*: evidence for lateral gene transfer between bacteria and archaea. *J Mol Microbiol Biotechnol* 4: 453–461

Eckburg PB, Bik EM, Bernstein CN, Purdom E, Dethlefsen L, Sargent M, Gill SR, Nelson KE, Relman DA (2005) Diversity of the human intestinal microbial flora. *Science (New York, NY* 308: 1635–1638

Eddy SR (2002) Computational genomics of noncoding RNA genes. *Cell* 109: 137–140

Elkins JG, Podar M, Graham DE, Makarova KS, Wolf Y, Randau L, Hedlund BP, Brochier-Armanet C, Kunin V, Anderson I, Lapidus A, Goltsman E, Barry K, Koonin EV, Hugenholtz P, Kyrpides N, Wanner G, Richardson P, Keller M, Stetter KO (2008) A korarchaeal genome reveals insights into the evolution of the Archaea. *Proc Natl Acad Sci U S A* 105: 8102–8107. Epub 2008 Jun 8105.

Ellis JC, Brown DD, Brown JW (2010) The small nucleolar ribonucleoprotein (snoRNP) database. *RNA (New York, NY* 16: 664–666

Ferry JG (1999) Enzymology of one-carbon metabolism in methanogenic pathways. *FEMS Microbiol Rev* 23: 13–38

Fischer S, Benz J, Späth B, Maier L-K, Straub J, Granzow M, Raabe M, Urlaub H, Hoffmann J, Brutschy B, Allers T, Soppa J, Marchfelder A (2010) The Archaeal Lsm protein binds to small RNAs. *J Biol Chem* 285: 34429–34438

Galagan JE, Nusbaum C, Roy A, Endrizzi MG, Macdonald P, FitzHugh W, Calvo S, Engels R, Smirnov S, Atnoor D, Brown A, Allen N, Naylor J, Stange-Thomann N, DeArellano K, Johnson R, Linton L, McEwan P, McKernan K, Talamas J, Tirrell A, Ye W, Zimmer A, Barber RD, Cann I, Graham DE, Grahame DA, Guss AM, Hedderich R, Ingram-Smith C, Kuettner HC, Krzycki JA, Leigh JA, Li W, Liu J, Mukhopadhyay B, Reeve JN, Smith K, Springer TA, Umayam LA, White O, White RH, Conway de Macario E, Ferry JG, Jarrell KF, Jing H, Macario AJ, Paulsen I, Pritchett M, Sowers KR, Swanson RV, Zinder SH, Lander E, Metcalf

WW, Birren B (2002) The genome of *M. acetivorans* reveals extensive metabolic and physiological diversity. *Genome Res* 12: 532–542
Gaspin C, Cavaille J, Erauso G, Bachellerie JP (2000) Archaeal homologs of eukaryotic methylation guide small nucleolar RNAs: lessons from the Pyrococcus genomes. *J Mol Biol* 297: 895–906.
Gottesman S (2004a) Small RNAs shed some light. *Cell* 118: 1–2.
Gottesman S (2004b) The small RNA regulators of Escherichia coli: roles and mechanisms. *Annu Rev Microbiol* 58: 303–328.
Hartman AL, Norais Cd, Badger JH, Delmas Sp, Haldenby S, Madupu R, Robinson J, Khouri H, Ren Q, Lowe TM, Maupin-Furlow J, Pohlschroder M, Daniels C, Pfeiffer F, Allers T, Eisen JA (2010) The Complete Genome Sequence of Haloferax volcanii DS2, a Model Archaeon. *PLoS ONE* 5: e9605
Huber H, Hohn MJ, Stetter KO, Rachel R (2003) The phylum Nanoarchaeota: present knowledge and future perspectives of a unique form of life. *Res Microbiol* 154: 165–171.
Hüttenhofer A, Schattner P, Polacek N (2005) Non-coding RNAs: hope or hype? *Trends Genet* 21: 289–297.
Jäger D, Sharma CM, Thomsen J, Ehlers C, Vogel J, Schmitz RA (2009) Deep sequencing analysis of the Methanosarcina mazei Go1 transcriptome in response to nitrogen availability. *Proceedings of the National Academy of Sciences of the United States of America* 106: 21878–21882
Klein RJ, Misulovin Z, Eddy SR (2002) Noncoding RNA genes identified in AT-rich hyperthermophiles. *Proceedings of the National Academy of Sciences of the United States of America* 99: 7542–7547
Krol J, Loedige I, Filipowicz W (2010) The widespread regulation of microRNA biogenesis, function and decay. *Nat* 11: 597–610. Epub 2010 Jul 2027.
Lange M, Westermann P, Ahring BK (2005) Archaea in protozoa and metazoa. *Applied microbiology and biotechnology* 66: 465–474
Lee SJ, Surma M, Hausner W, Thomm M, Boos W (2008) The role of TrmB and TrmB-like transcriptional regulators for sugar transport and metabolism in the hyperthermophilic archaeon Pyrococcus furiosus. *Archives of microbiology* 190: 247–256
Lepp PW, Brinig MM, Ouverney CC, Palm K, Armitage GC, Relman DA (2004) Methanogenic Archaea and human periodontal disease. *Proceedings of the National Academy of Sciences of the United States of America* 101: 6176–6181
Li Y, Liu X, Huang L, Guo H, Wang XJ (2010) Potential coexistence of both bacterial and eukaryotic small RNA biogenesis and functional related protein homologs in Archaea. *J* 37: 493–503.
Lie TJ, Leigh JA (2003) A novel repressor of *nif* and *glnA* expression in the methanogenic archaeon *Methanococcus maripaludis*. *Mol Microbiol* 47: 235–246
Lim J, Thomas T, Cavicchioli R (2000) Low temperature regulated DEAD-box RNA helicase from the Antarctic archaeon, Methanococcoides burtonii. *Journal of molecular biology* 297: 553–567
Liu Y, Whitman WB (2008) Metabolic, phylogenetic, and ecological diversity of the methanogenic archaea. *Annals of the New York Academy of Sciences* 1125: 171–189
Ma C, Simons RW (1990) The IS*10* antisense RNA blocks ribosome binding at the transposase translation initiation site. *The EMBO journal* 9: 1267–1274
Maeder DL, Anderson I, Brettin TS, Bruce DC, Gilna P, Han CS, Lapidus A, Metcalf WW, Saunders E, Tapia R, Sowers KR (2006) The *Methanosarcina barkeri* genome: comparative analysis with *Methanosarcina acetivorans* and *Methanosarcina mazei* reveals extensive rearrangement within methanosarcinal genomes. *Journal of bacteriology* 188: 7922–7931
Makarova KS, Wolf YI, van der Oost J, Koonin EV (2009) Prokaryotic homologs of Argonaute proteins are predicted to function as key components of a novel system of defense against mobile genetic elements. *Biol Direct* 4: 29.
Miller TL, Wolin MJ (1982) Enumeration of *Methanobrevibacter smithii* in human feces. *Archives of microbiology* 131: 14–18
Miller TL, Wolin MJ (1985) *Methanosphaera stadtmaniae* gen. nov., sp. nov.: a species that forms methane by reducing methanol with hydrogen. *Archives of microbiology* 141: 116–122

Muller S, Leclerc F, Behm-Ansmant I, Fourmann JB, Charpentier B, Branlant C (2008) Combined in silico and experimental identification of the Pyrococcus abyssi H/ACA sRNAs and their target sites in ribosomal RNAs. *Nucleic Acids Res* 36: 2459–2475. Epub 2008 Feb 2427.

Noon KR, Guymon R, Crain PF, McCloskey JA, Thomm M, Lim J, Cavicchioli R (2003) Influence of temperature on tRNA modification in archaea: Methanococcoides burtonii (optimum growth temperature [Topt], 23 degrees C) and Stetteria hydrogenophila (Topt, 95 degrees C). *Journal of bacteriology* 185: 5483–5490

Omer AD, Lowe TM, Russell AG, Ebhardt H, Eddy SR, Dennis PP (2000) Homologs of small nucleolar RNAs in Archaea. *Science* 288: 517–522.

Opulencia RB, Bose A, Metcalf WW (2009) Physiology and posttranscriptional regulation of methanol:coenzyme M methyltransferase isozymes in Methanosarcina acetivorans C2A. *Journal of bacteriology* 191: 6928–6935

Schattner P (2002) Searching for RNA genes using base-composition statistics. *Nucleic Acids Res* 30: 2076–2082.

Schneider KL, Pollard KS, Baertsch R, Pohl A, Lowe TM (2006) The UCSC Archaeal Genome Browser. *Nucleic acids research* 34: D407–410

Sharma CM, Hoffmann S, Darfeuille F, Reignier J, Findeiss S, Sittka A, Chabas S, Reiche K, Hackermuller J, Reinhardt R, Stadler PF, Vogel J (2010) The primary transcriptome of the major human pathogen Helicobacter pylori. *Nature* 464: 250–255

Smith PH, Hungate RE (1958) Isolation and characterization of Methanobacterium ruminantium n. sp. *Journal of bacteriology* 75: 713–718

Soppa J, Straub J, Brenneis M, Jellen-Ritter A, Heyer R, Fischer S, Granzow M, Voss B, Hess WR, Tjaden B, Marchfelder A (2009) Small RNAs of the halophilic archaeon Haloferax volcanii. *Biochem Soc Trans* 37: 133–136.

Storz G (2002) An expanding universe of noncoding RNAs. *Science* 296: 1260–1263.

Storz G, Altuvia S, Wassarman KM (2005) An abundance of RNA regulators. *Annu Rev Biochem* 74: 199–217

Straub J, Brenneis M, Jellen-Ritter A, Heyer R, Soppa J, Marchfelder A (2009) Small RNAs in haloarchaea: Identification, differential expression and biological function. *RNA Biol* 6: 281–292

Tang TH, Bachellerie JP, Rozhdestvensky T, Bortolin ML, Huber H, Drungowski M, Elge T, Brosius J, Hüttenhofer A (2002a) Identification of 86 candidates for small non-messenger RNAs from the archaeon Archaeoglobus fulgidus. *Proc Natl Acad Sci U S A* 99: 7536–7541.

Tang TH, Polacek N, Zywicki M, Huber H, Brugger K, Garrett R, Bachellerie JP, Huttenhofer A (2005) Identification of novel non-coding RNAs as potential antisense regulators in the archaeon *Sulfolobus solfataricus*. *Mol Microbiol* 55: 469–481

Tang TH, Rozhdestvensky TS, d'Orval BC, Bortolin ML, Huber H, Charpentier B, Branlant C, Bachellerie JP, Brosius J, Hüttenhofer A (2002b) RNomics in Archaea reveals a further link between splicing of archaeal introns and rRNA processing. *Nucleic Acids Res* 30: 921–930.

Thauer RK (1998) Biochemistry of methanogenesis: a tribute to Marjory Stephenson. 1998 Marjory Stephenson Prize Lecture. *Microbiology* 144 (Pt 9): 2377–2406

Thauer RK, Kaster AK, Seedorf H, Buckel W, Hedderich R (2008) Methanogenic archaea: ecologically relevant differences in energy conservation. *Nature reviews* 6: 579–591

Weidenbach K, Ehlers C, Kock J, Ehrenreich A, Schmitz RA (2008) Insights into the NrpR regulon in *Methanosarcina mazei* Go1. *Archives of microbiology* 190: 319–332

Weinberg Z, Wang JX, Bogue J, Yang J, Corbino K, Moy RH, Breaker RR (2010) Comparative genomics reveals 104 candidate structured RNAs from bacteria, archaea, and their metagenomes. *Genome biology* 11: R31

Wurtzel O, Sapra R, Chen F, Zhu Y, Simmons BA, Sorek R (2009) A single-base resolution map of an archaeal transcriptome. *Genome Res* 20: 133–141

Zago MA, Dennis PP, Omer AD (2005) The expanding world of small RNAs in the hyperthermophilic archaeon Sulfolobus solfataricus. *Mol Microbiol* 55: 1812–1828

Zhang Y, Zhang Z, Ling L, Shi B, Chen R (2004) Conservation analysis of small RNA genes in Escherichia coli. *Bioinformatics* 20: 599–603. Epub 2004 Jan 2022.

Chapter 8
Structure, Function and RNA Binding Mechanisms of the Prokaryotic Sm-like Protein Hfq

Poul Valentin-Hansen*

1 Introduction

Over the last decade it has become increasingly clear that the Sm-like (Lsm) protein Hfq, initially identified as a host factor required for Qβ bacteriophage replication, acts as a global post-transcriptional regulator in enterobacteria (reviewed in Valentin-Hansen *et al.* 2004; Storz *et al.* 2004; Gottesman *et al.* 2006; Vogel 2009a). Most of our current knowledge derives from detailed studies of small regulatory RNAs (sRNAs) in *Escherichia coli* and *Salmonella*. In particular, Hfq has been implicated in the regulation of translation and stability of an increasing number of mRNAs and its action has been shown to be intimately linked with the activity of a family of chromosomally encoded sRNAs that function in post-transcriptional control by base-pairing with cognate target mRNAs. Hfq associates with both regulatory RNAs and target RNA(s) and is required for proper riboregulation in enterobacteria. The importance of Hfq is further underscored by its multiple roles in bacterial physiology, including stress responses, bacteriocin production, nitrogen fixation, symbiosis, motility, quorum sensing and virulence of various bacterial pathogens (for reviews of Hfq activities see Vassilieva and Garber 2002; Valentin-Hansen *et al.* 2004; Dienst *et al.* 2008; Fröhlich and Vogel 2009). Here I will focus on novel findings about Hfq, its structure, RNA-binding mechanisms and mode of action in riboregulation and mRNA decay.

* Department of Biochemistry and Molecular Biology, University of Southern Denmark, Campusvej 55, DK-5230 Odense M

2 Prevalence of the Sm Fold

Sm and Lsm proteins are members of a growing family of factors known to affect different aspects of RNA metabolism (for reviews see Beggs 2005; Wilusz and Wilusz 2005). These proteins, which are generally found in eukaryotes and archaea, are characterized by the presence of two relatively conserved motifs, Sm1 and Sm2, which are separated by a region of variable length and sequence. The bipartite sequence motif constitutes an autonomously folded domain (the Sm fold) composed of an N-terminal α-helix followed by a twisted five-stranded β-sheet, responsible for both protein oligomerization and RNA binding (Séraphin 1995; Cooper *et al.* 1995; Hermann *et al.* 1995). The evolutionary connection between the Sm/Lsm proteins and Hfq was established in 2002 by the finding that Hfq possesses certain structural and functional properties in common with the Sm proteins (Møller *et al.* 2002a; Zhang *et al.* 2002). However, although the Sm1 motif is clearly present in the Hfq proteins, no obvious Sm2 motif can be identified. Rather, these proteins contain a different but conserved motif in the corresponding region, with the sequence signature [Y/F]KHAI. Like its eukaryotic counterparts, Hfq oligomerizes into ring-shaped structures, but its structure is homohexameric rather than heteroheptameric like the eukaryotic Sm/Lsm complexes (Møller *et al.* 2002a; Zhang *et al.* 2002; Beggs 2005; Brennan and Link 2007). Crystal structures of Hfq are currently solved for five bacterial species: *Staphylococcus aureus* (PDB code: 1KQ1), *E. coli* (PDB code: 1HK9), *Pseudomonas aeruginosa* (PDB code: 1U1S and 1U1T), *Anabaena* PCC 7120 (PDB code: 3HFN) and *Synechocystis* sp. PCC 6803 (PDB code: 3HFO), and one archaeon, *Methanococcus jannaschii* (PDB code: 2QTX) (Schumacher *et al.* 2002; Sauter *et al.* 2003; Nikulin 2005; Bøggild *et al.* 2009; Nielsen *et al.* 2007). Furthermore, the crystal structures of *S. aureus* Hfq bound to the U-rich oligo-nucleotide, 5′-AUUUUUG-3′ (PDB code: 1KQ2), and a C-terminal truncation mutant of *E. coli* Hfq bound to a poly(A) tail (PDB code: 3GIB), revealed details of RNA binding (Schumacher *et al.* 2002; Link *et al.* 2009).

3 Biochemical and Genetic Analysis of Hfq

Initial studies revealed that Hfq binds to polyuridylate, and especially polyriboadenylate [poly(A)] with high affinity (Carmichael 1975). In line with this observation, binding to the Qβ and R17 phage RNAs indicated that Hfq preferentially binds to segments rich in adenylate residues (Table 1) (Senear and Steitz 1976). In a more detailed study of the binding properties of linear and cyclic oligoadenylates, de Haseth and Uhlenbeck suggested that the binding site of Hfq has a circular spatial orientation and that each hexamer can accommodate 16-18 contiguous nucleotides (de Haseth and Uhlenbeck 1980b). These authors also observed that below 0.3 M NaCl, Hfq hexamers aggregate to form dodecamers or larger aggregates without affecting the binding properties to poly(A). In contrast, increasing the ionic strength results in a drastic reduction of the affinity for Qβ RNA, suggesting that Hfq aggregates, when formed at low ionic strengths, can interact cooperatively to several low affinity sites on the phage RNA (de Haseth and Uhlenbeck 1980a). More recently, Hfq-binding sites have been pinpointed on five sRNAs and three target mRNAs: OxyS (Zhang *et al.* 2002), Spot 42 (Møller *et al.* 2002a), DsrA (Brescia *et al.* 2003; Lease and Woodson 2004), RyhB and *ompA* mRNA (Moll *et al.* 2003), *sodB* mRNA (Geissmann and Touati 2004), MicA (Rasmussen *et al.* 2005), *rpoS* leader (Soper and Woodson 2008) and MicM RNA (Figueroa-Bossi *et al.* 2009; P.V-H unpublished). Furthermore, *in vitro* selection experiments have been used to define the RNA-binding characteristics of Hfq (Lorenz *et al.* 2010). The various binding sites have been identified principally by protection assays, mutagenesis, deletion mapping and EMSAs. The assays indicate that Hfq does not bind with high sequence specificity; rather the protein is characterized by a relaxed specificity for unpaired A-rich or A/U-rich sequences (Table 1). However, the protein may interact with more than one position on an RNA molecule, and adjacent secondary structures seem to be as important for Hfq binding as an unpaired A/U-rich sequence (Møller *et al.* 2002a; Zhang *et al.* 2002; Brescia *et al.* 2003; Soper and Woodson 2008; Lorenz *et al.* 2010). Specifically, an upstream binding site in the *rpoS* leader, consisting of four consecutive (AAN) triplets, recruits Hfq to facilitate pairing with regulatory sRNAs (Soper and Woodson 2008; Updegrove *et al.* 2008). In addition, mutational analyses of *E. coli* Hfq have identified two independent RNA binding faces (Mikulecky *et al.* 2004; Sun and Wartell 2006), and suggested that other positions may be involved in target docking (Sonnleitner *et al.* 2004; Sun and Wartell 2006; Ziolkowska *et al.* 2006; Updegrove *et al.* 2008). Finally, Hfq has been shown to interact specifically with tRNAs despite their lack of A/U-rich single-stranded regions (Lee and Feig 2008).

Table 1. Small RNAs with a potential role in the oxidative stress response of bacteria

R17 (1):	(AAG)(AAU)(AAU)(AAA)AUA
Qβ-1 (1):	ACC(AAU)ACU(AAA)(AAG)
Qβ-2 (1):	(AAU)(AAA)UUAUCAC(AAU)UACUCUUACG
OxyS:	SL-G**AAUAACUAAAG**CCAACGUGAACUUUU-SL
DsrA:	SL-AACG**AAUUUUUU**AAG-SL
RyhB:	SL-CCAG**UAUUA**CUU-SL
Spot 42:	SL-AUCGG**AUUU**GGCUG**AAUAUUU**AGCC-SL
MicA:	SL-G**AAAUUUU**-SL
MicM:	SL-CAU(**AAU**)(**AAU**)(**AAA**)(**AAA**)AUG(**AAA**) UUCCUCUUUGACG-SL
sodB mRNA:	SL-GCA(AAUU)(AAU)(AAU)(AAA)GGAGAGUAG- CA
ompA mRNA:	SL-CG**UAUUUU**GG
rpoS leader unpaired:	**(AAA)**$_2$
rpoS leader AAYAA element:	**(AAN)**$_4$
Selected Hfq sites:	**(AAN)**$_{3\text{-}4}$

[1] Hfq-bound fragments of RNase T_1 digested phage RNA. SL denotes stem-loop structure. Bases protected from attack by ribonucleases, hydroxyl radicals or DMS are shown in bold. Consecutive AAN triples and AANN quartets are in brackets.

3.1 The RNA-binding Modes of Hfq

Structural studies of Hfq-RNA complexes have provided invaluable insight into the RNA- binding mechanisms of Hfq (for a review see Brennan and Link 2007). One major RNA- binding site, first identified in X-ray crystallographic studies of *S. aureus* Hfq in complex with the hepta-oligoribonucleotide, 5′-AUUUUUG, is situated on the "proximal side" and consists of six essentially identical binding pockets that can accommodate U or A-nucleotides (Schumacher *et al.* 2002). In this complex the RNA is bound in a circular, unwound manner around the pore of the Hfq hexamer, within a basic patch. A second, independent RNA- binding site, which was discovered through studies of a series of structure-based site-directed mutants, is present on the "distal" side and is responsible for the high-affinity binding of poly(A) tails by Hfq (Mikulecky *et al.* 2004; Sun and Wartell 2006). In the structure of a C-terminally truncated mutant of *E. coli* Hfq, Hfq_{69} (lacking residues 70-102), bound to the oligoribonucleotide, A_{15}, the poly(A) tract also binds in a circular manner utilizing six identical tripartite binding motifs (Link *et al.* 2009). Significantly, the structure suggested that this RNA binding site was degenerate and the distal face could potentially accommodate $(A\text{-}R\text{-}N)_i$ or $(A\text{-}R\text{-}N\text{-}N')_i$ tracts, where R is a purine nucleotide and N is any nucleotide. Such binding promiscuity would increase the number and type of RNA molecules that might interact functionally with Hfq; and most importantly, this RNA-binding mode would be consistent with the interaction with various RNAs, such as the R17 phage, MicM, *sodB mRNA*, *rpoS* leader and selected high affine sites (Table 1).

3.2 Hfq-mediated sRNA-mRNA Annealing

The importance of Hfq for trans-acting sRNA was first recognized in studies of OxyS RNA, a riboregulator of the oxidative stress response (Zhang *et al.* 1998). Subsequently, numerous interactions between Hfq and sRNAs have been identified and several of these have been studied in great detail (Wassarman *et al.* 2001; Zhang *et al.* 2003; Sittka *et al.* 2009; Waters and Storz 2009). Hfq carries out its regulatory functions mainly by stabilizing most of the sRNA (see below) and, as discovered in studies with OxyS and Spot 42, by directly promoting sRNA-mRNA duplex formation (Møller *et al.* 2002b; Zhang *et al.* 2002). More recent data have shown that Hfq strongly enhances association rates of sRNAs and mRNAs, a property that would seem crucial for the inhibitory efficiency of sRNAs that have to compete with ribosome binding (Kawamoto *et al.* 2006; Aiba 2007; Arluison *et al.* 2007; Rajkowitsch and Schroeder 2007; Soper and Woodson 2008; Rasmussen *et al.* 2009). Therefore, it seems likely that binding (i.e. regulation) of the trans-acting regulatory sRNAs is controlled by association rates, rather than determined by the thermodynamic stability of the complex between antisense and target RNA. It has been consistently established that the regulatory effect of conventional cis-encoded antisense RNAs on their target RNAs is proportional to the rate of complex formation, and that such Hfq-independent antisense RNAs, or their targets, typically contain YUNR motifs (where Y is a pyrimidine nucleotide, R a purine nucleotide and N is any nucleotide) that specify a dynamic U-turn loop structure (Nordström and Wagner 1994; Franch and Gerdes 2000). This structure creates a sharp bend in the RNA-phosphate backbone and presents the following three to four bases in a solvent-exposed, stacked conformation providing a scaffold for rapid interaction with complementary RNA.

The mechanism by which Hfq promotes RNA-RNA annealing is not fully understood but it seems likely that the protein, via distinct binding surfaces for A/U-rich and (ARN)-rich tracks of RNAs, serves as a docking platform to bring partners in close proximity, thereby physically increasing the likelihood of annealing. However, it is also possible that Hfq could increase the local concentrations of the RNA molecules involved in pairing via interaction between two RNA-bound Hfq hexamers (de Haseth and Uhlenbeck 1980a). Alternatively, or additionally, the ability of Hfq to bind RNA molecules in a circular, unwound manner could promote annealing by refolding sRNA/target mRNA, thereby permitting RNA-RNA interactions. (Schumacher *et al.* 2002; Geissmann and Touati 2004; Storz *et al.* 2004).

A recent study of positive control of *rpoS* expression by DsrA RNA questioned the idea that the main role of Hfq is to enhance the rate of pairing and proposed instead that Hfq's role might be to stabilize sRNA complexes with the *rpoS* leader (Soper *et al.* 2010). In part, the model is based on the finding that the requirement for Hfq can be bypassed by over-expressing DsrA and that Hfq *in vitro* only had a modest effect on the rate of complex formation. However, other evidence showed that Hfq enhances the rate of DsrA binding to a full length *rpoS* leader RNA as much as 50 fold and stabilizes the final complex *in vitro* (Soper and Woodson 2008). Clearly, such a dual role of Hfq would be ideally suited for a base-paring-based activation of translation, but the exact role of Hfq in positive control of *rpoS* translation still awaits experimental clarification.

4 Hfq in RNA decay

Hfq is an important player in regulating the stability of mRNA and sRNA and the protein can stabilize or destabilize RNA substrates. In Gram-negative bacteria, such as *E. coli* and *Salmonella*, mRNAs are typically degraded by endonucleolytic cleavage, followed by 3′→5′exonucleolytic decay. For the majority of mRNAs, the rate-limiting step is considered to be cleavage by the essential single-strand specific endonuclease RNase E, with minor roles attributed to more specialized endonucleases such as RNase III, which has specificity for double-stranded regions of 20–40 bps on RNA. The main exonucleases acting in *E. coli* are RNase II, PNPase and RNase R of which the two former can get through secondary structure in RNA, as long as a 3′ unpaired tail is available. In the absence of endonucleolytic cleavage of RNAs with a structured 3′ end (e.g. a rho-independent terminator), another pathway of decay, the polyadenylation-dependent pathway, becomes activated. The main enzyme responsible for adding poly(A) tails on RNAs is Poly(A) polymerase I (PAPI), encoded by the *pcnB* gene (for reviews, see Viegas and Arraiano 2008; Anderson and Dunman 2009).

Hfq has been shown to form complexes - presumably RNA-mediated - with PNPase, PAPI and RNase E and has been implicated in all of the above-mentioned steps (Hajnsdorf and Régnier 2000; Folichon *et al.* 2003; Mohanty *et al.* 2004; Morita *et al.* 2005; Viegas and Arraiano 2008; Worrall *et al.* 2008). Firstly, most of the trans-acting sRNAs in *E. coli* and *Salmonella* are less stable in the absence of Hfq, probably because Hfq binding protects against RNase E cleavage. Accordingly, binding sites for Hfq and the cleavage sites for RNase E share sequence similarity and Hfq binding may occlude the cleavage site. This has been demonstrated directly or indirectly for the *sodB* and *ompA* messages and for the sRNAs DsrA and RybB (Moll *et al.* 2003). However, sRNA cleavage by RNase E can also result in stabilization as exemplified by the MicX RNA of *Vibrio cholerae* (Davis and Waldor 2007). Here the primary transcript is processed in an RNase E and Hfq-dependent fashion to a shorter, still active and much more stable form of the sRNA. Furthermore, base-pairing between sRNA and mRNA can alter processing and increase the stability of the target RNA (Opdyke *et al.* 2004; Fröhlich and Vogel 2009).

Secondly, PAPI requires Hfq to polyadenylate transcripts containing Rho-independent transcription terminators (Hajnsdorf and Régnier 2000; Mohanty *et al.* 2004). Most probably Hfq binding partially melts the stem-loop structure so that PAPI can bind and polyadenylate the 3′end. Moreover, Hfq changes PAPI to a processive enzyme, but only after the addition of ~20 adenosine nucleotides to the transcripts, thereby generating substrates that are readily degraded by PNPase and RNase II. The Hfq-poly(A) binding mechanism suggests how this happens: simply, initial Hfq binding interferes with PAPI interaction until the distal Hfq poly(A) binding site has been completely occupied and, when fulfilled, the tighter binding of Hfq might prevent backtracking of PAPI.

Thirdly, Hfq-mediated complex formation between regulatory RNA and target mRNA typically takes place at the ribosome binding site, interfering with the formation of translational initiation complexes. Decreased ribosome loading might, in turn, promote degradation of an mRNA by increasing the access for degradation factors to recognition sites normally masked by translating ribosomes (Deana and Belasco 2005). In most studied cases a rapid sRNA-induced mRNA decay is observed, but the prevailing view is that translational repression is the primary event in negative regulation. This has been demonstrated directly for down-regulation of protein synthesis by the sugar-stress induced SgrS RNA (Morita *et al.* 2008). Furthermore, translational repression without a significant increase in mRNA decay has been described for Spot 42 regulation in the *gal* operon (Møller *et al.* 2002b). However, sRNA-induced mRNA decay can play an important or essential role as observed for the conventional stress-induced sRNAs, such as the iron-stress induced RyhB, the anaerobically induced FnrS and the SgrS sRNAs, where pairing results in translational repression and a concomitant rapid coupled degradation of the regulatory RNA and target mRNA in an RNase E-dependent manner (Massé *et al.* 2003; Morita *et al.* 2005; Boysen *et al.* 2010). Here, the co-degradation of the regulatory RNA provides a rapid and efficient OFF-switch for sRNA regulation upon adaptation to non-stressed conditions (Gottesman *et al.* 2006).

A second striking example is provided by *Salmonella* MicC RNA, which, in addition to OmpC, a major outer membrane protein, down-regulates the synthesis of the OmpD porin (Pfeiffer *et al.* 2009). While the regulation of OmpC follows the conventional pattern, the biochemically validated target for *ompD* is situated within the coding region about 70 nt downstream of the start codon. *In vitro* binding of MicC to the message does not repress translational initiation and it would also seem unlikely that the short duplex formed with MicC could interfere with elongating ribosomes *in vivo* as these can get through stable secondary structures in mRNA. Rather, the primary mechanism of MicC regulation relies on rapid Hfq and RNase E-dependent mRNA degradation (for a review see Wagner 2009).

Another interesting outcome of base-pairing concerns the silencing of the YbfM chitosugar porin by the Hfq-dependent MicM sRNA (ChiX and ChiP, respectively in *Salmonella*) (Figueroa-Bossi *et al.* 2009; Overgaard *et al.* 2009; Rasmussen *et al.* 2009). Here, pairing of MicM to the translational initiation region (TIR) of *ybfM* mRNA leads to an almost complete elimination of the target RNA. However, MicM RNA is not co-degraded with its target RNA, but rather the sRNA is recycled and can participate in the decay of many target molecules (Overgaard *et al.* 2009). Moreover, it seems likely that the *micM* gene is expressed constitutively (Mandin and Gottesman 2009a). Genetic screens uncovered that transcription of the first intergenic region (IGR) of the chitobiose operon relieved the tight repression of *ybfM* expression by promoting MicM degradation (Figueroa-Bossi *et al.* 2009; Overgaard *et al.* 2009). This region of the *chb* transcript exhibits short complementarity to the same region of MicM that pairs with the *ybfM* TIR, but the pairing is slightly longer and is interrupted by a central mismatched loop. Thus, the *chbBCARFG* mRNA, encoding the transport and degradation system for chitobiose, the main degradation product of chitin, functions as a *bona fide* riboregulator that traps MicM and unleashes the synthesis of the chitosugar porin (for reviews see Mandin and Gottesman 2009b; Vogel, 2009b). The mRNA-based regulation makes biological sense and is reminiscent of target mimicry for regulation of microRNA activity in plants by non-coding RNAs (Franco-Zorrilla *et al.* 2007). In summary, the effect of Hfq-mediated base-pairing ranges from translational blocking to mRNA degradation, to sRNA killing and to coupled degradation of regulatory and target RNAs. The molecular details of how pairing stimulates degradation of one or more of the interacting RNAs and what exactly happens at the site of pairing remain to be fully understood.

5 The Role of the C-terminus of Hfq Proteins

The N-terminal portion of Hfq proteins consists of the evolutionarily conserved Sm motif, corresponding to residues 7-66 in *E. coli* Hfq. In contrast, the C-termini of the Hfq proteins vary greatly in length and sequence, with the β- and γ-proteobacteria having the longest tails (Sun *et al.* 2002; Attia *et al.* 2008; Schilling and Gerischer 2009). These features suggest that the C-terminal region might not play a significant role in the major functions of Hfq. This suggestion was supported by a study showing that the insertion of an Ω cassette in the *hfq* gene after codon 78 did not affect the activity of the protein (Tsui *et al.* 1994). Furthermore,

Hfq variants with short C-terminal extensions, such as *E. coli* Hfq-75, consisting of the first 75 N-terminal amino acid residues, and the Hfq homologue of *Pseudomonas aeruginosa* (82 aa residues) can functionally replace *E. coli* Hfq for phage Qβ replication and for the *rpoS* expression regulation (Sonnleitner *et al.* 2002; Sonnleitner *et al.* 2004). Also, an Hfq variant, comprising the first 65 N-terminal amino acid residues (Hfq-65), displayed no great defect in binding to the RyhB and DsrA sRNAs and retained the capacity to support phage Qβ replication. The mutant protein, however, was impaired in binding to some mRNAs, in promoting sRNA/target interaction and in the stimulation of *rpoS* mRNA translation (Vecerek *et al.* 2008). Based on these latter findings, the authors hypothesized that the C-terminal extension of *E. coli* Hfq constitutes a hitherto unrecognized RNA interaction surface with specificity for mRNAs. Further, these authors proposed that Hfq proteins with short C-terminal extensions (e.g. those of Gram-positive bacteria) might phenotypically behave like the *E. coli* Hfq_{65} variant, despite the fact that they, in previous work, had provided evidence that the variant Hfq-66 was proficient in supporting down-regulation of *ompA* translation (Sonnleitner *et al.* 2004; Vecerek *et al.* 2008). This leaves a confusing picture of the role of the C-terminal tail for riboregulation, and highlights the need for further direct experimentation. In this regard, we tested the hypothesis in an *hfq* knock-out background by expressing C-terminal truncation mutants to levels that match those seen for full-length Hfq in a wild-type strain. We could show that residues 66-102 of *E. coli* Hfq can be deleted without affecting the ability to support *rpoS* expression upon entry into stationary growth phase as well as riboregulation by various well-characterized sRNAs, including CyaR, FnrS, MicA, MicM, RybB and RyhB (Olsen *et al.* 2010). Collectively, these data support the view that the C-terminal tail of *E. coli* Hfq plays little or no role in riboregulation The dispensability of this part of the protein for sRNA-regulation is further supported by studies of Hfq homologues from *Listeria monocytogenes* and *Methanococcus jannaschii* (see below). As yet, knowledge of a definitive biological role of the C-terminus of *E. coli* is lacking, but the extension may possibly have an effect on the stability of the protein (Arluison *et al.* 2004). Alternatively, it could play a role in protein-protein communication, binding to DNA (Updegrove *et al.* 2010) or in the cellular localization of the Hfq protein (Diestra *et al.* 2009).

6 Role of Hfq in Low GC Gram-positive Bacteria

The finding that the vast majority of chromosomally trans-encoded antisense RNAs in enterobacteria require Hfq for *in vivo* regulatory efficiency led to the early suggestion that a need for an RNA chaperone may be common for regulatory RNAs that have to interact through short stretches of complementarity. However, the lack of corroborating data from low GC Gram-positive bacteria has upset this view. For example, in *S. aureus* and *B. subtilis* the trans-encoded regulatory sRNAs studied so far all function without the need for Hfq (Jousselin *et al.* 2009). In addition, the *hfq* gene was shown to be weakly expressed in three *S. aureus* backgrounds and no significant differences between wild-type and *hfq* mutant strains were observed in comparative studies using Phenotype Microarray Technology (Bohn *et al.* 2007). On this basis, it was hypothesized that the Hfq protein is dispensable for sRNA-mediated antisense regulation in Gram positives (Jousselin *et al.* 2009; Repoila and Darfeuille 2009; Romby and Charpentier 2010). However, in some *S. aureus* strains the *hfq* gene is well expressed and novel data suggest that also *S. aureus* Hfq acts as a global post-transcriptional regulator involved in stress and virulence (Liu *et al.* 2010). Furthermore, Hfq in the pathogen *Listeria monocytogenes* contributes to stress tolerance and interacts with at least three sRNAs, suggesting that the protein participates in their function (Christiansen *et al.* 2004; Christiansen *et al.* 2006). Accordingly, one of the sRNAs, LhrA, is strongly destabilized in strains lacking Hfq and a recent study of this sRNA provided the first experimental evidence for Hfq-dependent antisense regulation in a Gram-positive species (Nielsen *et al.* 2010). In addition, the study showed that *Listeria* and *E. coli* Hfqs display very similar biochemical and biological properties.

It is possible that additional RNA chaperones could mediate fast RNA pairing in low GC Gram-positive species or substitute for Hfq in bacteria lacking this protein. Alternatively, or additionally, the trans-encoded antisense-regulated systems in these bacteria may have evolved structural features akin to U-turn loop structures in conventional cis-encoded antisense RNAs that facilitate rapid bi-molecular RNA-RNA interaction (Franch and Gerdes 2000). Curiously, the YURN U-turn motif is also found in several of the sRNAs recently identified in low GC Gram-positive bacteria, including the LhrC RNAs of *L. monocytogenes, S. aureus* RsaE and RsaH RNAs and mRNA target sites for RNAIII (Christiansen *et al.* 2006; Geissmann *et al.* 2009; Toledo-Arana *et al.* 2009; Chevalier *et al.* 2010; Romby and Charpentier 2010). In addition, a second conserved and unpaired sequence motif, UCCC, is found in many of the sRNAs, and the *S. aureus* RNAIII also carries this motif in three hairpin loops that are known to pair with the TIR of target mRNAs, suggesting that sRNAs carrying unpaired UCCC elements constitute a class of riboregulators that target mRNAs by a shared mechanism (Geissmann *et al.* 2009).

6.1 Hfq in Cyanobacteria

Standard BLAST searches initially identified Hfq candidates from about half of the completed or nearly completed bacterial genomes, approximately 140 at the time (Sun *et al.* 2002). Based on the apparent lack of Hfq-type proteins in several bacterial phyla, in some proteobacteria and in a number of Gram-positive bacteria, it was suggested that the Hfq protein was of ancient origin, and subsequently lost in some branches of evolution. However, the combined use of motif and pattern sequence searches led to the identification of an Hfq orthologue in some of the organisms from which it was thought to be missing, including the cyanobacterium *Anabaena* PCC 7120, and subsequently in a wide variety of unicellular and filamentous cyanobacteria, among these *Synechocystis* sp. PCC 6803 (Valentin-Hansen *et al.* 2004). Furthermore, bioinformatics and experimental studies revealed the existence of a large number of putative regulatory RNAs in cyanobacteria (Georg *et al.* 2009 and references herein).

The cyanobacterial proteins constitute a new group of Hfq proteins, quite separate from other bacterial Hfqs. Particularly, the Hfq signature sequence (Y/F)KHAI is significantly different and substitutes for RLAAI in *Synechocystis* and WKQAI in *Anabaena.* In *Synechocystis* sp. PCC 6803 the *hfq* gene can be deleted without any detrimental effects on cell growth, but mutant cells are not naturally transformable and exhibit a non-phototactic phenotype (Dienst *et al.* 2008). Transcriptomics consistently revealed that the most strongly affected genes upon *hfq* inactivation belong to two operons controlled by the cAMP receptor protein, which had previously been shown to be involved in motility of *Synechocystis* (Dienst *et al.* 2008).

The crystal structures of *Anabaena* and *Synechocystis* Hfq revealed that they retained the classical Sm fold despite low sequence conservation (Bøggild *et al.* 2009). However, the proteins possess variant RNA binding sites and their binding affinity for enterobacterial small regulatory sRNAs is lowered by several orders of magnitude *in vitro*. Furthermore, the cyanobacterial Hfqs cannot mediate Hfq-dependent regulation in *E. coli* suggesting that they may function in a different way than the enterobacterial orthologues.

6.2 Archaeal Hfq Protein

The prevalence of Hfq is not restricted to the bacterial domain. Thus, database searches identified a potential homologue in the thermophilic archaeon *Methanococcus jannaschii* (Sauter *et al.* 2003; Valentin-Hansen *et al.* 2004). Archaeal genomes generally encode a single or a few small Sm-related proteins. These form homo-hexameric or homo-heptameric rings and share oligomerization behaviour and RNA-binding properties with their eukaryotic counterparts (Achsel *et al.* 2001; Beggs 2005; Khusial *et al.* 2005). Though the structure and RNA-binding activity of archaeal Sm rings have been investigated in detail, their function has remained elusive. However, the *Archaeoglobus fulgidus* Sm1- and Sm2-type proteins associate with RNase P, suggesting a potential role in post-transcriptional processing of stable RNAs (Törö *et al.* 2001).

Intriguingly, the genome of *M. jannaschii* does not contain any gene encoding a conventional archaeal Sm-type protein, suggesting that Hfq proteins and archaeal Sm-like homologues can complement each other functionally. The archaeal Hfq protein consists of an isolated Sm domain without the flanking C-terminal region found in many of the bacterial Hfq proteins. Within the Sm domain, the protein contains the conserved FKHAI motif, and the residues that form the A/U-binding pockets in *S. aureus* Hfq and the poly(A) binding determinants of *E. coli* Hfq are also well conserved in the archaeal protein. Structural analysis of the protein established that it forms a hexameric ring whose overall structure is similar to that found in bacterial Hfq proteins (Nielsen *et al.* 2007). Furthermore, functional analysis revealed that *E. coli* and *M. jannaschii* Hfqs display very similar biochemical and biological properties. Specifically, the archaeal Hfq is capable of modulating translation and facilitating small RNA function in *E. coli*; and in *Salmonella* it is functional in mediating Hfq-dependent porin mRNA repression and Hfq-dependent regulation of OppA synthesis by the small GcvB regulatory RNA (Nielsen *et al.* 2007; Sittka *et al.* 2009). In addition, the protein can complement, at least partly, a number of phenotypes displayed by an *E. coli hfq* knockout strain (Nielsen *et al.* 2007). Collectively, these findings indicate that the archaeal and enterobacterial Hfq proteins are largely functionally interchangeable. It therefore seems probable that *M. jannaschii* Hfq plays a role *in vivo* that is comparable to that of *E. coli* Hfq.

The exact evolutionary relationship between Hfq and Sm-like proteins is not clear, but it is believed that archaea and eukaryotes evolved from a common predecessor that diverged earlier from the bacterial domain (Woese *et al.* 1990). In keeping with this view, the Sm domains of archaea and eukaryotes are homologous at both the sequence and structural levels, and the proteins also share similar RNA binding properties and oligomerization behaviour. In contrast, Hfq has a divergent sequence compared to the eukaryotic and archaeal Sm/Lsm proteins. Furthermore, the conventional enterobacterial Hfq protein possesses rather relaxed RNA-binding specificities that enable it to interact with many RNA molecules and to participate in various stages of RNA metabolism. Hence Hfq may represent an ancient, less specialized form of the Sm family of proteins. Consistent with this view, phylogenetic analyses provided no evidence for lateral transfer of the *hfq* gene from bacteria to archaeal species (Sun *et al.* 2002).

7 Concluding Remarks

There has been substantial progress in recent years in understanding the role of Hfq in post-transcriptional control of gene expression in enterobacteria but the crucial question remains: How exactly does Hfq work? Thus, whereas a rough sketch is now possible of how Hfq may promote RNA-RNA annealing, a number of aspects of the mechanism remain to be explained; among these, does Hfq possess additional RNA binding sites and does the formation of aggregates of Hfq hexamers take place *in vivo*? Obviously, additional mutational analysis combined with structural studies of Hfq in complex with target RNAs, or portions of these molecules, should shed some light on this issue. A second important issue concerns the outcome of Hfq-mediated pairing which can lead to mRNA and/or sRNA degradation in some cases but not others. What exactly happens is still an open question and whether Hfq is present or leaves the complex is not clear. A deeper understanding will require an integration of existing knowledge with knowledge to be acquired about the organization and structure of various nucleoprotein complexes, the RNA decay pathway(s) and how ribonucleases are guided to cleave one or more of the interacting RNA molecules. A third issue is whether Hfq can act independently of sRNAs in post-transcriptional control of translation. In any case this will be very difficult to prove experimentally *in vivo* because of the large number of chromosomally encoded sRNAs, and conclusions based solely on *in vitro* results may be misleading. In addition, the study of Hfq proteins in most prokaryotes is in its infancy and work on Hfq associated molecules in, for example, Gram-positives, archaea and cyanobacteria could give us valuable examples to enable further understanding of the role of Hfq in prokaryotic physiology. In this regard, it is clear that paradigms based on *E. coli* and *Salmonella* Hfqs do not necessarily hold true for other bacteria and with the advent of high-throughput sequencing and proteomics, novel classes of Hfq-associated RNAs and proteins may be uncovered, and there are likely to be many surprises in store. Finally, it has become obvious that Hfq and Lsm complexes have much in common and lessons learnt from the study of Hfq may open new perspectives in the understanding of Lsm functions in archaea and eukaryotes (Wilusz and Wilusz 2005; Scheller *et al.* 2009; Galão *et al.* 2010)

Acknowledgements

I am grateful to Birgitte Kallipolitis, Jakob Møller-Jensen and Mikkel Girke Jørgense for helpful comments. The work in the laboratory was supported by the Danish National Research Foundation (Danmarks Grundforskningsfond) Centre for mRNP Biogenesis and Metabolism, and grants from The Danish Natural Science Research Council. I apologize to those whose work was not discussed due to space limitations.

References

Achsel T, Stark H, Lührmann R (2001) The Sm domain is an ancient RNA-binding motif with oligo(U) specificity. Proc Natl Acad Sci U S A 98:3685–3689

Aiba H (2007) Mechanism of RNA silencing by Hfq-binding small RNAs. Curr Opin Microbiol 10:134–139

Anderson KL, Dunman PM (2009) Messenger RNA Turnover Processes in *Escherichia coli*, *Bacillus subtilis*, and Emerging Studies in *Staphylococcus aureus*. Int J Microbiol 2009:525491

Arluison V, Folichon M, Marco S, Derreumaux P, Pellegrini O, Seguin J, Hajnsdorf E, Regnier P (2004) The C-terminal domain of *Escherichia coli* Hfq increases the stability of the hexamer. Eur J Biochem 271:1258–1265

Arluison V, Hohng S, Roy R, Pellegrini O, Regnier P, Ha T (2007) Spectroscopic observation of RNA chaperone activities of Hfq in post-transcriptional regulation by a small non-coding RNA. Nucleic Acids Res 35:999–1006

Attia AS, Sedillo JL, Wang W, Liu W, Brautigam CA, Winkler W, Hansen EJ (2008) *Moraxella catarrhalis* expresses an unusual Hfq protein. Infect Immun 76:2520–2530

Beggs J (2005) Lsm proteins and RNA processing. Biochem Soc Trans 33:433–438

Bøggild A, Overgaard M, Valentin-Hansen P, Brodersen DE (2009) Cyanobacteria contain a structural homologue of the Hfq protein with altered RNA-binding properties. FEBS J 276:3904–3915

Bohn C, Rigoulay C, Bouloc P (2007) No detectable effect of RNA-binding protein Hfq absence in *Staphylococcus aureus*. BMC Microbiol 7:10

Boysen A, Møller-Jensen J, Kallipolitis B, Valentin-Hansen P, Overgaard M (2010) Translational regulation of gene expression by an anaerobically induced small non-coding RNA in *Escherichia coli*. J Biol Chem 285:10690–10702

Brennan RG, Link TM (2007) Hfq structure, function and ligand binding. Curr Opin Microbiol 10:125–133

Brescia CC, Mikulecky PJ, Feig AL, Sledjeski DD (2003) Identification of the Hfq-binding site on DsrA RNA: Hfq binds without altering DsrA secondary structure. RNA 9:33–43

Carmichael GG (1975) Isolation of bacterial and phage proteins by homopolymer RNA-cellulose chromatography. J Biol Chem 250:6160–6167

Chevalier C, Boisset S, Romilly C, Masquida B, Fechter P, Geissmann T, Vandenesch F, Romby P (2010) *Staphylococcus aureus* RNAIII binds to two distant regions of coa mRNA to arrest translation and promote mRNA degradation. PLoS Pathog 6:e1000809

Christiansen JK, Larsen MH, Ingmer H, Søgaard-Andersen L, Kallipolitis BH (2004) The RNA-binding protein Hfq of *Listeria monocytogenes*: role in stress tolerance and virulence. J Bacteriol 186:3355–3362

Christiansen JK, Nielsen JS, Ebersbach T, Valentin-Hansen P, Søgaard-Andersen L, Kallipolitis BH (2006) Identification of small Hfq-binding RNAs in *Listeria monocytogenes*. RNA 12:1383–1396

Cooper M, Johnston LH, Beggs JD (1995) Identification and characterization of Uss1p (Sdb23p): a novel U6 snRNA-associated protein with significant similarity to core proteins of small nuclear ribonucleoproteins. EMBO J 14:2066–2075

Davis BM, Waldor MK (2007) RNase E-dependent processing stabilizes MicX, a *Vibrio cholerae* sRNA. Mol Microbiol 65:373–385

de Haseth PL, Uhlenbeck OC (1980a) Interaction of *Escherichia coli* host factor protein with Q beta ribonucleic acid. Biochemistry 19:6146–6151

de Haseth PL, Uhlenbeck OC (1980b) Interaction of *Escherichia coli* host factor protein with oligoriboadenylates. Biochemistry 19:6138–6146

Deana A, Belasco JG (2005) Lost in translation: the influence of ribosomes on bacterial mRNA decay. Genes Dev 19:2526–2533

Dienst D, Dühring U, Mollenkopf HJ, Vogel J, Golecki J, Hess WR, Wilde A (2008) The cyanobacterial homologue of the RNA chaperone Hfq is essential for motility of *Synechocystis* sp. PCC 6803. Microbiology 154:3134–3143

Diestra E, Cayrol B, Arluison V, Risco C (2009) Cellular electron microscopy imaging reveals the localization of the Hfq protein close to the bacterial membrane. PLoS One 4:e8301

Figueroa-Bossi N, Valentini M, Malleret L, Fiorini F, Bossi L (2009) Caught at its own game: regulatory small RNA inactivated by an inducible transcript mimicking its target. Genes Dev 23:2004–2015

Folichon M, Arluison V, Pellegrini O, Huntzinger E, Regnier P, Hajnsdorf E (2003) The poly(A) binding protein Hfq protects RNA from RNase E and exoribonucleolytic degradation. Nucleic Acids Res 31:7302–7310

Franch T, Gerdes K (2000) U-turns and regulatory RNAs. Curr Opin Microbiol 3:159–164

Franco-Zorrilla JM, Valli A, Todesco M, Mateos I, Puga MI, Rubio-Somoza I, Leyva A, Weigel D, Garcia JA, Paz-Ares J (2007) Target mimicry provides a new mechanism for regulation of microRNA activity. Nat Genet 39:1033–1037

Fröhlich KS, Vogel J (2009) Activation of gene expression by small RNA. Curr Opin Microbiol 12:674–682

Galão RP, Chari A, ves-Rodrigues I, Lobao D, Mas A, Kambach C, Fischer U, Diez J (2010) LSm1-7 complexes bind to specific sites in viral RNA genomes and regulate their translation and replication. RNA 16:817–827

Geissmann T, Chevalier C, Cros MJ, Boisset S, Fechter P, Noirot C, Schrenzel J, Francois P, Vandenesch F, Gaspin C, Romby P (2009) A search for small noncoding RNAs in *Staphylococcus aureus* reveals a conserved sequence motif for regulation. Nucleic Acids Res 37:7239–7257

Geissmann TA, Touati D (2004) Hfq, a new chaperoning role: binding to messenger RNA determines access for small RNA regulator. EMBO J 23:396–405

Georg J, Vosz B, Scholz I, Mitschke J, Wilde A, Hess WR (2009) Evidence for a major role of antisense RNAs in cyanobacterial gene regulation. Mol Syst Biol 5:

Gottesman S, McCullen CA, Guillier M, Vanderpool CK, Majdalani N, Benhammou J, Thompson KM, FitzGerald PC, Sowa NA, FitzGerald DJ (2006) Small RNA regulators and the bacterial response to stress. Cold Spring Harb Symp Quant Biol 71:1–11

Hajnsdorf E, Régnier P (2000) Host factor Hfq of *Escherichia coli* stimulates elongation of poly(A) tails by poly(A) polymerase I. Proc Natl Acad Sci U S A 97:1501–1505

Hermann H, Fabrizio P, Raker VA, Foulaki K, Hornig H, Brahms K, Lührmann H (1995) snRNP Sm proteins share two evolutionarily conserved sequence motifs which are involved in Sm protein-protein interactions. EMBO J 14:2076–2088
Jousselin A, Metzinger L, Felden B (2009) On the facultative requirement of the bacterial RNA chaperone, Hfq. Trends Microbiol 17:399–405
Kawamoto H, Koide Y, Morita T, Aiba H (2006) Base-pairing requirement for RNA silencing by a bacterial small RNA and acceleration of duplex formation by Hfq. Mol Microbiol 61:1013–1022
Khusial P, Plaag R, Zieve GW (2005) LSm proteins form heptameric rings that bind to RNA via repeating motifs. Trends Biochem Sci 30:522–528
Lease RA, Woodson SA (2004) Cycling of the Sm-like protein Hfq on the DsrA small regulatory RNA. J Mol Biol 344:1211–1223
Lee T, Feig AL (2008) The RNA binding protein Hfq interacts specifically with tRNAs. RNA 14:514–523
Link TM, Valentin-Hansen P, Brennan RG (2009) Structure of *Escherichia coli* Hfq bound to polyriboadenylate RNA. Proc Natl Acad Sci U S A 106:19292–19297
Liu Y, Wu N, Dong J, Gao Y, Zhang X, Mu C, Shao N, Yang G (2010) Hfq is a global regulator that controls the pathogenicity of *Staphylococcus aureus*. PLoS One 5:e13069
Lorenz C, Gesell T, Zimmermann B, Schoeberl U, Bilusic I, Rajkowitsch L, Waldsich C, von HA, Schroeder R (2010) Genomic SELEX for Hfq-binding RNAs identifies genomic aptamers predominantly in antisense transcripts. Nucleic Acids Res
Mandin P, Gottesman S (2009a) A genetic approach for finding small RNAs regulators of genes of interest identifies RybC as regulating the DpiA/DpiB two-component system. Mol Microbiol 72:551–565
Mandin P, Gottesman S (2009b) Regulating the regulator: an RNA decoy acts as an OFF switch for the regulation of an sRNA. Genes Dev 23:1981–1985
Massé E, Escorcia FE, Gottesman S (2003) Coupled degradation of a small regulatory RNA and its mRNA targets in *Escherichia coli*. Genes Dev 17:2374–2383
Mikulecky PJ, Kaw MK, Brescia CC, Takach JC, Sledjeski DD, Feig AL (2004) *Escherichia coli* Hfq has distinct interaction surfaces for DsrA, *rpoS* and poly(A) RNAs. Nat Struct Mol Biol 11: 1206–1214.
Mohanty BK, Maples VF, Kushner SR (2004) The Sm-like protein Hfq regulates polyadenylation dependent mRNA decay in *Escherichia coli*. Mol Microbiol 54:905–920
Moll I, Afonyushkin T, Vytvytska O, Kaberdin VR, Bläsi U (2003) Coincident Hfq binding and RNase E cleavage sites on mRNA and small regulatory RNAs. RNA 9:1308–1314
Møller T, Franch T, Højrup P, Keene DR, Bächinger HP, Brennan RG, Valentin-Hansen P (2002a) Hfq: a bacterial Sm-like protein that mediates RNA-RNA interaction. Mol Cell 9:23–30
Møller T, Franch T, Udesen C, Gerdes K, Valentin-Hansen P (2002b) Spot 42 RNA mediates discoordinate expression of the *E. coli* galactose operon. Genes Dev 16:1696–1706
Morita T, Maki K, Aiba H (2005) RNase E-based ribonucleoprotein complexes: mechanical basis of mRNA destabilization mediated by bacterial noncoding RNAs. Genes Dev 19:2176–2186
Morita T, Maki K, Yagi M, Aiba H (2008) Analyses of mRNA destabilization and translational inhibition mediated by Hfq-binding small RNAs. Methods Enzymol 447:359–378
Nielsen JS, Bøggild A, Andersen CB, Nielsen G, Boysen A, Brodersen DE, Valentin-Hansen P (2007) An Hfq-like protein in archaea: crystal structure and functional characterization of the Sm protein from *Methanococcus jannaschii*. RNA 13:2213–2223
Nielsen JS, Lei LK, Ebersbach T, Olsen AS, Klitgaard JK, Valentin-Hansen P, Kallipolitis BH (2010) Defining a role for Hfq in Gram-positive bacteria: evidence for Hfq-dependent antisense regulation in *Listeria monocytogenes*. Nucleic Acids Res 38:907–919
Nikulin A (2005) Structure of *Pseudomonas aeruginosa* Hfq protein. Acta Crystallogr D 61:141–146
Nordström K, Wagner EG (1994) Kinetic aspects of control of plasmid replication by antisense RNA. Trends Biochem Sci 19:294–300
Olsen AS, Møller-Jensen J, Brennan RG, Valentin-Hansen P (2010) C-terminally truncated derivatives of *Escherichia coli* Hfq are proficient in riboregulation. J Mol Biol 404:173–82

Opdyke JA, Kang JG, Storz G (2004) GadY, a small-RNA regulator of acid response genes in *Escherichia coli*. J Bacteriol 186:6698–6705

Overgaard M, Johansen J, Moller-Jensen J, Valentin-Hansen P (2009) Switching off small RNA regulation with trap-mRNA. Mol Microbiol 73:790–800

Pfeiffer V, Papenfort K, Lucchini S, Hinton JC, Vogel J (2009) Coding sequence targeting by MicC RNA reveals bacterial mRNA silencing downstream of translational initiation. Nat Struct Mol Biol 16:840–846

Rajkowitsch L, Schroeder R (2007) Dissecting RNA chaperone activity. RNA 13:2053–2060

Rasmussen AA, Johansen J, Nielsen JS, Overgaard M, Kallipolitis B, Valentin-Hansen P (2009) A conserved small RNA promotes silencing of the outer membrane protein YbfM. Mol Microbiol 72:566–577

Rasmussen AA, Eriksen M, Gilany K, Udesen C, Franch T, Petersen C, Valentin-Hansen P (2005) Regulation of *ompA* mRNA stability: the role of a small regulatory RNA in growth phase-dependent control. Mol Microbiol 58:1421–1429

Repoila F, Darfeuille F (2009) Small regulatory non-coding RNAs in bacteria: physiology and mechanistic aspects. Biol Cell 101:117–131

Romby P, Charpentier E (2010) An overview of RNAs with regulatory functions in gram-positive bacteria. Cell Mol Life Sci 67:217–237

Sauter C, Basquin J, Suck D (2003) Sm-like proteins in Eubacteria: the crystal structure of the Hfq protein from *Escherichia coli*. Nucleic Acids Res 31:4091–4098

Scheller N, Mina LB, Galao RP, Chari A, Gimenez-Barcons M, Noueiry A, Fischer U, Meyerhans A, Diez J (2009) Translation and replication of hepatitis C virus genomic RNA depends on ancient cellular proteins that control mRNA fates. Proc Natl Acad Sci U S A 106:13517–13522

Schilling D, Gerischer U (2009) The *Acinetobacter baylyi* Hfq gene encodes a large protein with an unusual C terminus. J Bacteriol 191:5553–5562

Schumacher MA, Pearson RF, Møller T, Valentin-Hansen P, Brennan RG (2002) Structures of the pleiotropic translational regulator Hfq and an Hfq-RNA complex: a bacterial Sm-like protein. EMBO J 21:3546–3556

Senear AW, Steitz JA (1976) Site-specific interaction of Qbeta host factor and ribosomal protein S1 with Qbeta and R17 bacteriophage RNAs. J Biol Chem 251:1902–1912

Séraphin B (1995) Sm and Sm-like proteins belong to a large family: identification of proteins of the U6 as well as the U1, U2, U4 and U5 snRNPs. EMBO J 14:2089–2098

Sittka A, Sharma CM, Rolle K, Vogel J (2009) Deep sequencing of *Salmonella* RNA associated with heterologous Hfq proteins in vivo reveals small RNAs as a major target class and identifies RNA processing phenotypes. RNA Biol 6:266–275

Sonnleitner E, Moll I, Bläsi U (2002) Functional replacement of the *Escherichia coli hfq* gene by the homologue of *Pseudomonas aeruginosa*. Microbiology-Sgm 148:883–891

Sonnleitner E, Napetschnig J, Afonyushkin T, Ecker K, Vecerek B, Moll I, Kaberdin VR, Blasi U (2004) Functional effects of variants of the RNA chaperone Hfq. Biochem Biophys Res Commun 323:1017–1023

Soper T, Mandin P, Majdalani N, Gottesman S, Woodson SA (2010) Positive regulation by small RNAs and the role of Hfq. Proc Natl Acad Sci U S A

Soper TJ, Woodson SA (2008) The *rpoS* mRNA leader recruits Hfq to facilitate annealing with DsrA sRNA. RNA 14:1907–1917

Storz G, Opdyke JA, Zhang A (2004) Controlling mRNA stability and translation with small, non-coding RNAs. Curr Opin Microbiol 7:140–144

Sun X, Wartell RM (2006) *Escherichia coli* Hfq binds A18 and DsrA domain II with similar 2:1 Hfq6/RNA stoichiometry using different surface sites. Biochemistry 45:4875–4887

Sun XG, Zhulin I, Wartell RM (2002) Predicted structure and phyletic distribution of the RNA-binding protein Hfq. Nucl Acids Res 30:3662–3671

Toledo-Arana A, Dussurget O, Nikitas G, Sesto N, Guet-Revillet H, Balestrino D, Loh E, Gripenland J, Tiensuu T, Vaitkevicius K, Barthelemy M, Vergassola M, Nahori MA, Soubigou G, Regnault B, Coppee JY, Lecuit M, Johansson J, Cossart P (2009) The *Listeria* transcriptional landscape from saprophytism to virulence. Nature 459:950–956

Törö I, Thore S, Mayer C, Basquin J, Séraphin B, Suck D (2001) RNA binding in an Sm core domain: X-ray structure and functional analysis of an archaeal Sm protein complex. EMBO J 20:2293–2303

Tsui HCT, Leung HCE, Winkler ME (1994) Characterization of broadly pleiotropic phenotypes caused by an Hfq insertion mutation in *Escherichia-coli*-K-12. Mol Microbiol 13:35–49

Updegrove T, Wilf N, Sun X, Wartell RM (2008) Effect of Hfq on RprA-*rpoS* mRNA pairing: Hfq-RNA binding and the influence of the 5' *rpoS* mRNA leader region. Biochemistry 47:11184–11195

Updegrove TB, Correia JJ, Galletto R, Bujalowski W, Wartell RM (2010) *E. coli* DNA associated with isolated Hfq interacts with Hfq's distal surface and C-terminal domain. Biochim Biophys Acta 1799:588–96

Valentin-Hansen P, Eriksen M, Udesen C (2004) The bacterial Sm-like protein Hfq: a key player in RNA transactions. Mol Microbiol 51:1525–1533

Vassilieva IM, Garber MB (2002) The regulatory role of the Hfq protein in bacterial cells. Mol Biol 36:970–977

Vecerek B, Rajkowitsch L, Sonnleitner E, Schroeder R, Bläsi U (2008) The C-terminal domain of *Escherichia coli* Hfq is required for regulation. Nucleic Acids Res 36:133–143

Viegas SC, Arraiano CM (2008) Regulating the regulators: How ribonucleases dictate the rules in the control of small non-coding RNAs. RNA Biol 5:230–243

Vogel J (2009a) A rough guide to the non-coding RNA world of *Salmonella*. Mol Microbiol 71:1–11

Vogel J (2009b) An RNA trap helps bacteria get the most out of chitosugars. Mol Microbiol 73:737–741

Wagner EG (2009) Kill the messenger: bacterial antisense RNA promotes mRNA decay. Nat Struct Mol Biol 16:804–806

Wassarman KM, Repoila F, Rosenow C, Storz G, Gottesman S (2001) Identification of novel small RNAs using comparative genomics and microarrays. Genes Dev 15:1637–1651

Waters LS, Storz G (2009) Regulatory RNAs in bacteria. Cell 136:615–628

Wilusz CJ, Wilusz J (2005) Eukaryotic Lsm proteins: lessons from bacteria. Nat Struct Mol Biol 12:1031–1036

Woese CR, Kandler O, Wheelis ML (1990) Towards a natural system of organisms: proposal for the domains Archaea, Bacteria, and Eucarya. Proc Natl Acad Sci U S A 87:4576–4579

Worrall JA, Gorna M, Crump NT, Phillips LG, Tuck AC, Price AJ, Bavro VN, Luisi BF (2008) Reconstitution and analysis of the multienzyme *Escherichia coli* RNA degradosome. J Mol Biol 382:870–883

Zhang A, Wassarman KM, Ortega J, Steven AC, Storz G (2002) The Sm-like Hfq protein increases OxyS RNA interaction with target mRNAs. Mol Cell 9:11–22

Zhang A, Wassarman KM, Rosenow C, Tjaden BC, Storz G, Gottesman S (2003) Global analysis of small RNA and mRNA targets of Hfq. Mol Microbiol 50:1111–1124

Ziolkowska K, Derreumaux P, Folichon M, Pellegrini O, Regnier P, Boni IV, Hajnsdorf E (2006) Hfq variant with altered RNA binding functions. Nucleic Acids Res 34:709–720

Chapter 9
CRISPR/Cas and CRISPR/Cmr Immune Systems of Archaea

Shiraz A. Shah, Gisle Vestergaard, and Roger A. Garrett*

1 Introduction

The CRISPR/Cas (Clustered Regularly Interspaced Short Palindromic Repeats/ CRISPR-Associated Genes) and CRISPR/Cmr systems (Cmr: Cas module-RAMP (Repeat-Associated Mysterious Proteins)) provide the basis for adaptive and hereditable immune responses directed against the DNA and RNA, respectively, of invading elements. The former consists of CRISPR loci physically linked to a cassette of *cas* genes which together appear to constitute integral genetic modules. *cmr* genes, clustered in Cmr modules, are sometimes physically linked to CRISPR/Cas modules. The CRISPR/Cas immune system occurs in almost all archaea and about 40 % of bacteria. Cmr modules are less common, occurring in only about one third of genomes carrying CRISPR/Cas modules. An outline of how the CRISPR/Cas and CRISPR/Cmr systems function is indicated in Figure 1 where the former targets DNA and the latter RNA (mRNA and/or viral RNA) of the genetic elements.

Archaeal CRISPR loci consist of clusters of spacer-repeat units varying in size from one to more than one hundred spacer-repeat units where each unit is about 60 – 90 bp with repeats and spacers of, on average, 30 bp and 40 bp, respectively (Lillestøl et al., 2006; Grissa et al., 2008). CRISPR loci are preceded by a non protein coding leader region which varies in size from about 150 to 550 bp and are invariably physically linked to a *cas* gene cassette (Jansen et al., 2002; Haft et al., 2005; Makarova et al., 2006; Lillestøl et al., 2006; Lillestøl et al., 2009). Cas and Cmr proteins, involved in the two different targeting pathways, are functionally and phylogenetically diverse. The CRISPR/Cas system specifically targets DNA elements (Marraffini and Sontheimer, 2008; Shah et al., 2009) while the CRISPR/Cmr system targets RNA, although whether mRNA and/or viral RNA remains unclear (Hale et al., 2009). CRISPR/Cas modules have been classified into families on the basis of sequences of their *cas* genes, leaders and repeats. Although these modules

* Archaea Centre, Department of Biology, Copenhagen University, Copenhagen, Denmark

show a capacity for transfer between phyla of the archaeal and bacterial Domains, and supposedly rarely across Domain boundaries, archaea-specific features are nevertheless apparent.

Crucial for the functioning of the immune systems are the spacer sequences which derive from foreign invading elements (Mojica et al., 2005; Pourcel et al., 2005; Bolotin et al., 2005; Lillestøl et al., 2006; Barrangou et al., 2007). The CRISPR loci generate whole transcripts which initiate within the leader sequence adjacent to the first repeat (Lillestøl et al., 2009). These are subsequently processed in their repeat regions yielding end-products that constitute single spacer-containing crRNAs (Tang et al., 2002; Tang et al., 2005; Lillestøl et al., 2006; Lillestøl et al., 2009). Processing is effected by specific Cas or Cmr proteins and, at least for the

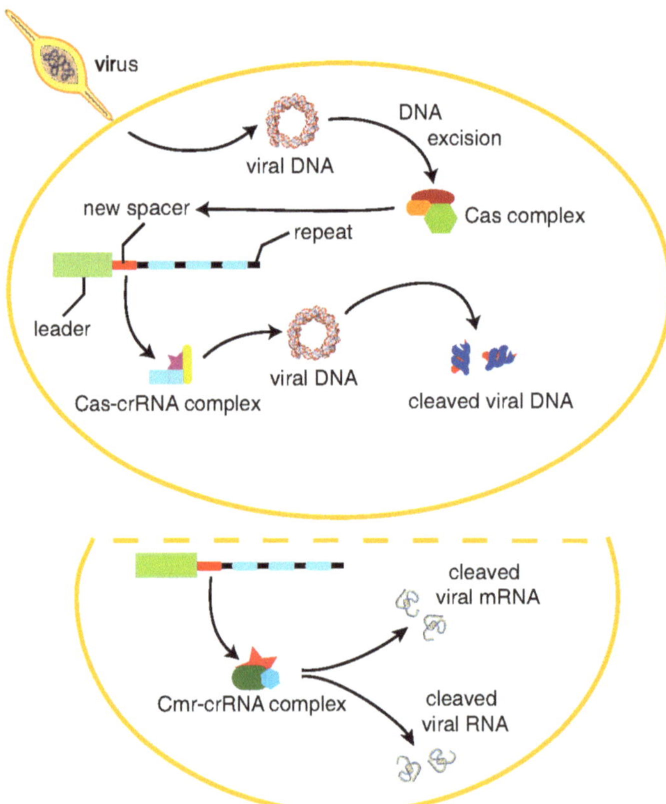

Fig. 1. Diagram illustrating how CRISPR/Cas and CRISPR/Cmr systems target genetic elements invading a host cell. crRNAs are processed from whole transcripts of CRISPR loci. For the CRISPR/Cas system, Cas proteins complex with the crRNA and guide it to the complementary protospacer sequence in the invading DNA element where they anneal prior to DNA degradation. Cmr proteins also complex with crRNA and guide them to either mRNA or viral RNA, targeting them for degradation

latter, two discrete archaeal crRNAs are produced each carrying 8 nt of the repeat at the 5'-end and lacking 5 nt or 11 nt from the 3'-end of each spacer (Carte et al., 2008; Hale et al., 2009). Complexes of Cas or Cmr proteins transport the processed crRNAs to target, and inactivate, DNA or RNA, respectively, of invading genetic elements (Brouns et al., 2008; Hale et al., 2009). Base pairing mismatches occurring between the 5' 8 nt repeat sequence of the crRNA and the Protospacer-Associated Motif (PAM) sequence adjacent to the targeted protospacer of the invading DNA are essential for subsequent degradation of the latter and for ensuring that the chromosomal CRISPR locus, itself, is not targeted (Horvath and Barrangou, 2010; Marraffini and Sontheimer, 2010; Lillestøl *et al.,* 2009; Gudbergsdottir et al., 2010).

2 Archaeal Viruses and Plasmids and Chromosomal Evolution

Although few comprehensive studies have been performed on the relative abundance of different virus-like particle (VLP) morphotypes in archaea-rich environments, available results indicate that spindles, filaments, rods and spheres predominate in terrestial hot springs and hydrothermal vents, while spindle-shaped and spherical virus-like particles (VLPs) prevail in hypersaline environments (Rachel et al., 2002; Porter et al., 2007; Bize et al., 2008). Bacteriophage-like head-tail VLPs are found infrequently, although their proviruses have been detected in a few halo- and methanoarchaeal genomes (Porter et al., 2007; Krupovic et al., 2010).

Several viruses, mainly from terrestial hot springs have been classified into eight new archaeal viral families and examples of their diverse morphotypes are illustrated in Figure 2. Other viruses including several haloarchaeal viruses remain to be classified (Porter et al., 2007). The latter process is complicated by the absence of a consistent relationship between morphology and genomic properties for euryarchaeal and crenarchaeal viruses. Overall these discoveries underline the major differences between the archaeal and bacterial virospheres (Prangishvili et al., 2006a; Lawrence et al., 2009).

Archaeal viral genomes fall in the size range 15 to 75 kb dsDNA and are circular or linear. Some linear genomes have free ends whereas others, including those of rudiviruses and some lipothrixviruses, have modified ends or are covalently closed and some genomes carry base-specific modifications (Zillig et al., 1998; Peng et al., 2001). Consistent with the unusual and sometimes unique viral morphologies (Figure 2), the viral genomes yielded very few significant sequence matches with genes in public sequence databases (Prangishvili et al., 2006b). These results are summarised in histograms of the major hyperthermophilic crenarchaeal viruses in Figure 3 where a large percentage of the genes are classified as unique for each virus. The most extreme case was for genes of the thermoneutrophilic virus PSV which yielded almost no significant sequence matches in the original study (Bettstetter et al., 2003).

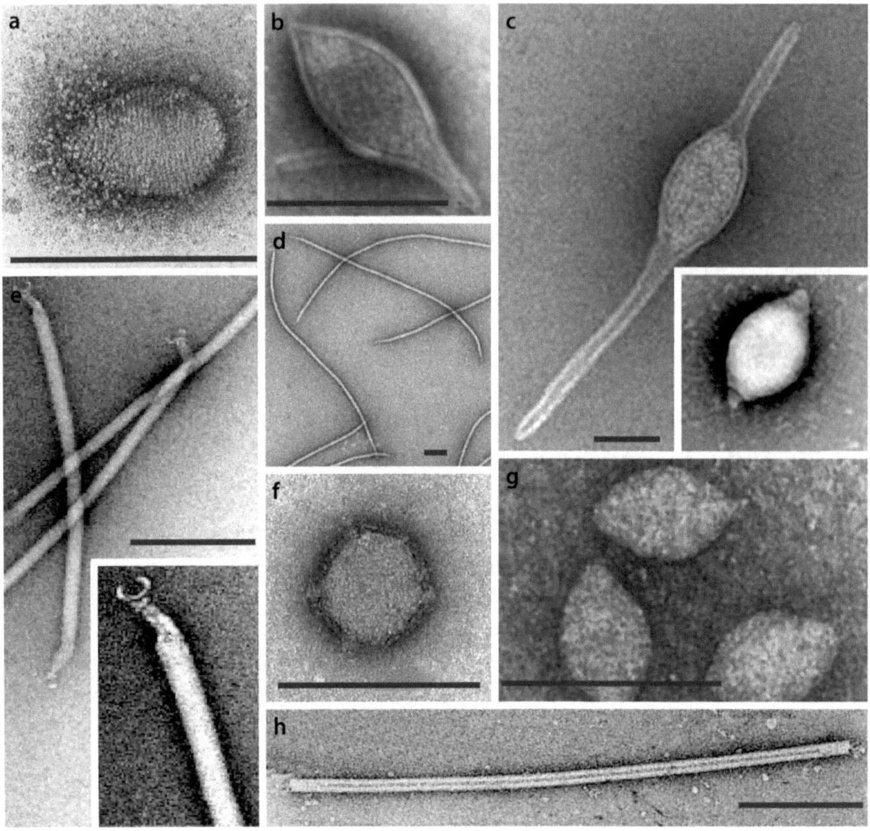

Fig. 2. Typical morphologies of representatives of different families of archaeal viral families. **a**, SNDV; **b**, STSV1; **c**, ATV; **d**, SIFV; **e**, AFV1; **f**, PSV; **g**, SSV4; **h**, ARV1. Bars are 100 nm

With the availability of an increasing number of archaeal genome sequences, it has become clear that archaeal viruses and plasmids have played a major role in the evolution of host genomes. This process has apparently been fuelled by the entrapment of foreign DNA elements in host chromosomes via an archaea-specific integrative process. Many archaeal integrase genes partition on integration such that, if the free form of the element is lost, the integrase will not be expressed and cannot effect excision of the genetic element from the chromosome (She et al., 2001). Many of the encaptured elements are recognisable as intact or degenerate genetic entities and Markov-model analyses of whole archaeal genomes suggest that such genes of viral or plasmid origin contribute disproportionately to the genes of unknown function in archaeal chromosomes (Cortez et al., 2009).

Archaeal viruses and plasmids have also evolved complex relationships as dependents or antagonists. Thus, in the presence of a fusellovirus, pRN family plasmids

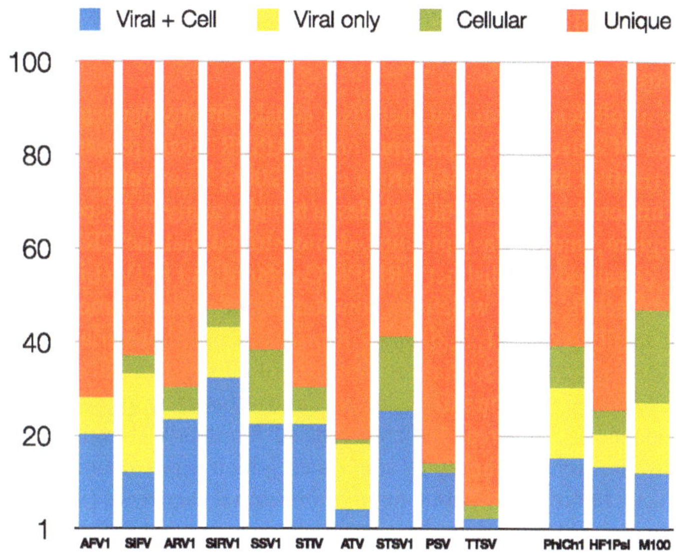

Fig. 3. Histogram showing a summary of archaeal viral gene homologies to other viruses (virus only genes) and cellular chromosomes (cellular); unique indicates no detectable homologs. Homologs in closely related viruses, including the rudiviruses ARV1 and SIRV1 and the spherical viruses PSV and TTSV1 are not included (Prangishvili *et al.*, 2006b)

pSSVx and pSSVi are packaged into fusellovirus-like particles and spread through *Sulfolobus* host cultures as satellite viruses (Arnold et al., 1999; Wang et al., 2007). In contrast, when a strain of *Acidianus hospitalis* carrying the conjugative plasmid pAH1 was infected with the lipothrixvirus AFV1, plasmid replication appeared to be inhibited (Basta et al., 2009). Moreover, as mentioned below, *Sulfolobus* conjugative plasmids pNOB8 and pKEF9 carry CRISPR loci which may directly target and inactivate archaeal viruses (She et al., 1998; Greve et al., 2004).

3 Diversity of Archaeal CRISPR/Cas and CRISPR/Cmr Immune Systems

Bioinformatic analyses have demonstrated that homologs of a few core Cas proteins occur widely throughout the archaeal and bacterial domains while others occur less commonly and some are predominantly archaeal or bacterial in character. Core gene sets typify the *cas* and *cmr* gene cassettes (Figure 4). For the former, the *cas* genes fall into groups 1 and 2. This division is based on different factors including co-occurrence, co-regulation and synteny of the genes and, possibly, functional differences for the groups of proteins (see below). The *cas6* gene can occur in either group and is likely to be cofunctional with both CRISPR/Cas and CRISPR/Cmr systems (Hale et al., 2009). For the *cmr* cassette, the two most conserved genes *cmr2* and *cmr5* are interspersed with diverse RAMP-motif containing proteins (Figure 4B).

It has also been shown that there is a consistent phylogenetic linkage between sequences of selected Cas proteins and CRISPR locus repeats for archaea and bacteria (Haft et al., 2005; Makarova et al., 2006; Kunin et al., 2007; Shah et al., 2009). Furthermore for the Sulfolobales, a broader analysis of sequences of repeats, leader regions, and of Cas1 proteins, demonstrated that the CRISPR/Cas modules could be classified into distinct CRISPR/Cas families I to IV (Lillestøl et al., 2009; Shah and Garrett, 2011) which are components of an earlier more broadly defined group of families CASS1, 5, 6, and 7 from archaea and bacteria (Haft et al., 2005; Makarova et al., 2006). Spatial distributions of all the archaeal and bacterial families are illustrated in Figure 5A using a Markov clustering approach based on Cas1 protein sequences. Whereas the crenarchaeal families I, II and III tend to cluster separately, the archaeal family IV sequences, which derive mainly from mesophilic euryarchaea, fall together with a family of bacterial sequences (in green). A closely similar spatial distribution is also observed when crenarchaeal families I to IV are

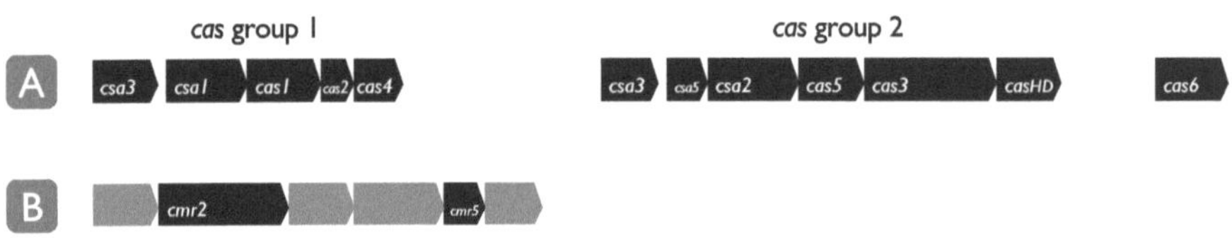

Fig. 4. Gene maps of A. *cas cassettes* and B. a Cmr module showing only conserved core genes. Many other genes that occur less frequently are not included. The *cas* genes are divided into two groups 1 and 2 (see text). The Cmr module contains the highly conserved *cmr2* and *cmr5* genes and genes shaded grey, which correspond to different genes encoding RAMP motif-containing proteins that are present in 3 to 5 copies in the different Cmr module families (Garrett *et al.*, 2011b)

---

next page:

▸ **Fig. 5.** CRISPR/Cas modules can be divided into families based on their unique characteristics, including the Cas1 protein sequence and nucleotide sequences of the repeat and leader regions.

a) Spheres represent Cas1 protein sequences from different organisms. Small distances between spheres reflects higher sequence similarity between them. All Cas1 sequences that are currently publicly available are represented. Markov clustering reveals that all the sequences fall within about 20 families (each coloured differently), 5 of which are very large. Strongly coloured spheres represent archaeal Cas1 sequences while bacterial sequences are shown in faded colours. It is evident that some families are specific to bacteria, whereas others are archaea-specific. A few CRISPR/Cas families are shared between both archaea and bacteria. Sulfolobales families I – IV are marked (Lillestøl *et al.* 2009) and others remain to be formally classified. An earlier broader classification, CASS1 to 7, is also included (Haft *et al.* 2005; Makarova *et al.* 2006).

b) Leaders from the Sulfolobales are clustered based on their sequence similarities and they fall into the same group of families (I–IV) as those found for the Cas1 proteins, and a similar result is obtained when repeat sequences are clustered (Lillestøl *et al.* 2009) clustered on the basis of their leader sequences (Figure 5b). Clearly, there are other archaea-specific families (strongly coloured in Figure 5a) which remain to be analysed and classified.

Family I CRISPR/Cas modules are the most common amongst the Sulfolobales and other crenarchaea, and the most conserved in structural organisation. The two conserved groups of *cas* genes are located between a pair of CRISPR loci and externally at one end of the module. The separation may be functionally significant with the former involved in processing and insertion of DNA spacer-repeat units and the latter encoding RNA processing and effector proteins (Shah and Garrett, 2011).

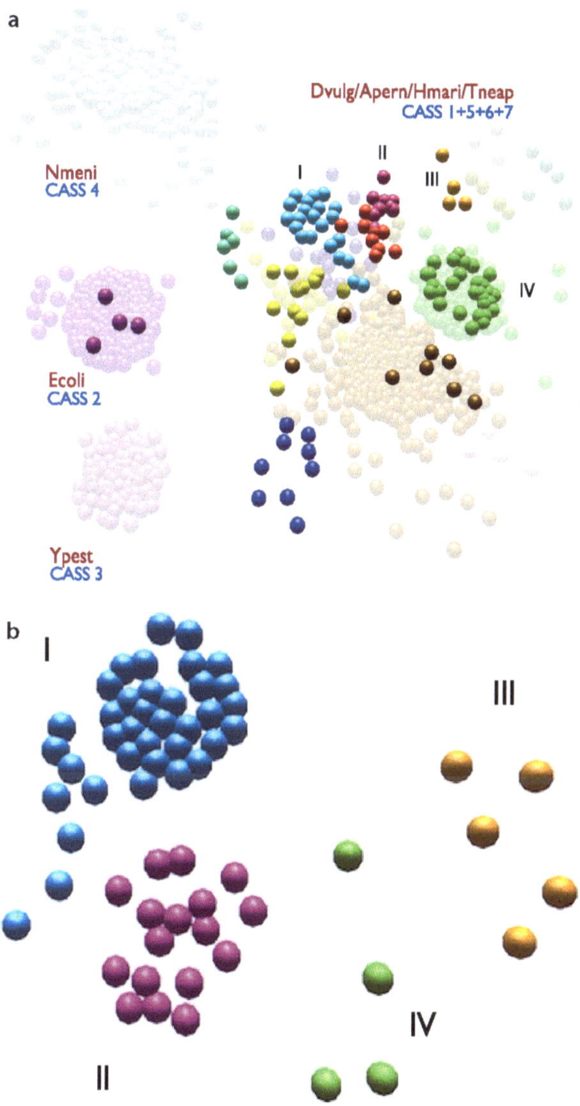

The methanoarchaea and haloarchaea, which carry the majority of the family IV CRISPR/Cas modules show the least conservation in their *cas* gene contents. In particular, their group 2 *cas* genes range from those typical of crenarchaea to those common amongst bacteria. Putative genetic exchange between archaea and bacteria has generally been attributed to the methanoarchaea and haloarchaea thriving in environments rich in bacteria.

Cmr modules invariably coexist with, and are sometimes physically linked to, CRISPR/Cas modules but they occur less widely than the latter. For archaea they are found in about 70 % of genomes carrying CRISPR/Cas modules, more prevalent than for CRISPR/Cas-carrying bacterial genomes (about 30 %). Both CRISPR/Cas modules and Cmr modules frequently occur in multiple copies in a given archaeal genome. *cmr* genes are mainly co-transcribed and their protein products have been implicated in processing of crRNAs and in the guiding of crRNAs to target RNA of invading genetic elements, whether viral RNA, transcripts, or both, remains unclear (Hale et al., 2009).

Comparison of phylogenetic trees for the CRISPR/Cas and Cmr modules, based on archaeal and bacterial sequences of Cas1 and the Cmr2 protein, and its homologs Csm1 and Csx11, revealed five major families of Cmr modules, named A to E, showing distinctive gene syntenies (Garrett et al., 2011b; Guo *et al.*, 2011).

Given that Cmr and CRISPR/Cas modules are sometimes physically linked and can potentially be mobilised as a unit, and that they have to recognise CRISPR repeat sequences of similar sequence, it is likely that some degree of coevolution has occurred. In support of this idea, there are many examples of family II CRISPR/Cas modules coexisting with family D Cmr modules amongst the Sulfolobales and this relationship extends to other archaea including for example, the euryarchaeon *Methanospirillum hungatei.*

Sizes of CRISPR loci vary from a single spacer bordered by repeats to more than 100 spacer-repeat units (Lillestøl et al., 2006; Grissa et al., 2008). New spacer-repeat units are added at the leader-repeat junction and the CRISPR loci also undergo deletions of spacer-repeat units, probably via recombination at the direct repeats, without impairing the overall CRISPR/Cas functionality, and the deletions can range from one to several spacer-repeat units. Moreover, there are also putative examples of duplications of spacer-repeat units, or small groups thereof, occurring and exchange between CRISPR loci within a genome (Lillestøl et al., 2006; Lillestøl et al., 2009; Shah and Garrett, 2011; Gudbergsdottir et al., 2011).

4 Development and Stability of CRISPR Loci

CRISPR loci generally appear to be quite stable, gradually adding spacer-repeat units at the junction with the leader, albeit at different rates for different loci within an organism. There is also a compensatory mechanism for gradual loss of internal spacers which probably involves recombination between the identical direct repeats of a given locus, and occasionally between different loci carrying identical repeats (Lillestøl *et al.*, 2009; Shah and Garrett, 2011). A specific example of such changes is illustrated in Figure 6, showing the pairwise alignments of CRISPR locus A of *Sulfolobus solfataricus* strains P1, P2 and 98/2 where shared spacers are shaded, as well as spacers added adjacent to the leader region after these strains diverged. The pattern of shared spacers for each pair of organisms demonstrate that strain 98/2 separated prior to the divergence of strains P1 and P2 which carry more common spacers. Those spacers which show significant matches to known genetic elements are also colour-coded (Figure 6A,B) indicating a wide variety of matches especially to rudiviruses, bicaudaviruses and conjugative plasmids (Lillestøl et al., 2009).

Earlier evidence suggested that CRISPR loci were strongly resistant to integrative events (Lillestøl et al., 2006). For example, three strains of *S. solfataricus* P1, P2 and 98/2, which carry multiple large CRISPR loci, in addition to locus A in Figure 6. They are also extremely rich in active transposable elements (about 350 in strain P2) which have contributed to extensive genome shuffling (Brügger et al., 2004) but no IS insertions were detected in the extensive CRISPR loci (Lillestøl et al., 2009; Shah and Garrett, 2011). Thus, although they do occasionally occur intergenically in the *cas* and *cmr* gene clusters, there appears to be a strong selective pressure to maintain the integrity of CRISPR loci which are essential for the function of both CRISPR/Cas and CRISPR/Cmr systems. Whether this is a general rule for archaea or is dependent on environmental conditions, including the levels of viruses and plasmids present, is unclear. A different picture has emerged from bacterial studies. For example, in a biofilm carrying acidophilic *Leptospirillum* group II bacteria, about 20 % of the partially sequenced CRISPR loci contained IS elements (Tyson and Banfield, 2008).

Many archaeal and bacterial chromosomes, with or without CRISPR/Cas modules, carry short CRISPR-like clusters lacking associated leader regions and *cas* genes (Grissa et al., 2008). Although their origin(s) remain unknown, they may have separated from intact CRISPR/Cas modules, possibly via transposable elements. If preceded by promoters their transcripts can, in principle, be processed and activated. Such CRISPR loci are present in *Sulfolobus* conjugative plasmids pNOB8 and pKEF9 (She et al., 1998; Greve et al., 2004) and at least for the latter,

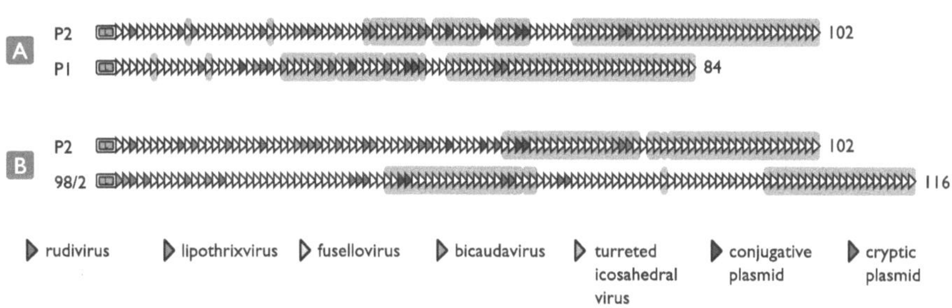

Fig. 6. Pairwise comparison of the spacer-repeat units of CRISPR A locus of three closely related strains of *S. solfataricus* P1, P2 and 98/2. Shaded regions indicate identical spacer-repeat units shared by two CRISPR loci. Colour-coded spacer-repeat units indicate that spacers have significant sequence matches to the viruses or plasmid families indicated on the Figure

the spacer-repeat cluster is transcribed and RNA processed in a *S. solfataricus* host, suggesting that at least some of these small clusters can be activated and functional if complementary Cas or Cmr proteins are present (Lillestøl et al., 2009; Shah and Garrett, 2011).

5 Mobility of CRISPR/Cas and Cmr Modules

Genomic analyses of closely related *Sulfolobus* species have provided strong evidence for CRISPR/Cas modules being mobilised given that they occur at different genomic positions even when there is high level of genomic gene synteny present and they are generally confined to the variable genetic regions (Shah and Garrett, 2011). Their ability to transfer between organisms is also supported by the different combinations of CRISPR/Cas families found in closely related organisms (Lillestøl *et al.*, 2009; Shah and Garrett, 2011). For example, in *S. islandicus* strains HVE10/4 and REY15A, the former carries family I and III CRISPR/Cas modules and one Cmr module, while the latter exhibits a family I CRISPR/Cas module and two family B Cmr modules (Shah and Garrett, 2011; Guo *et al.*, 2011). Further support for such transfer was provided by analysis of the *Pyrococcus furiosus* genome where a 155 kb fragment bordered by a CRISPR locus and a repeat showing significantly different properties of G+C content, third codon position and codon usage from the rest of the genome (Portillo and Gonzalez, 2009).

Evidence for gene exchange within the CRISPR/Cas modules derived from examination of the structural integrities of the paired family I CRISPR/Cas modules of several closely related *Sulfolobus* strains. The results indicated that the internal group 1 *cas gene*s, which are functionally implicated in spacer addition at the leader-repeat junction (Figure 4) seem to coevolve, and be mobilised, with the CRISPR locus whereas the group 2 *cas* genes, putatively involved in RNA processing and crRNA mobility (Figure 4), were retained within the strains, suggesting that some exchange within *cas* gene cassettes can occur (Shah and Garrett, 2011).

The mechanism(s) of transfer of CRISPR/Cas modules, varying in size from about 7 kb to 25 kb, remains unclear. The larger CRISPR/Cas modules, at least, may be too large to be borne on plasmids as has been proposed for bacteria (Godde and Bickerton, 2006). At least for the crenarchaea, genetic elements are relatively small and, although small CRISPR loci have been detected in crenarchaeal conjugative plasmids, transfer is more likely to result from chromosomal conjugation which may well be facilitated by integrated conjugative plasmids (Lillestøl et al., 2009).

6 Targets of the CRISPR/Cas and CRISPR/Cmr Systems

Bioinformatic evidence indicated that the spacer crRNAs carrying significant sequence matches to the protospacer sequence were complementary to either strand of genes implying that they were not exclusively targeting mRNAs (Lillestøl *et al.*, 2006). Moreover, extensive analyses of significant matches to the many known viruses and plasmids of the Sulfolobales revealed several matches to protospacers lying between genes. They demonstrated, further, that the locations of the protospacers were randomly distributed along, and on either strand of, the genetic elements. This is illustrated in Figure 7 for five crenarchaeal viruses and two plasmids, where the positions of the significant matches are shown in relation to the annotated gene locations. A similar conclusion that DNA, and not mRNA, was targeted by the

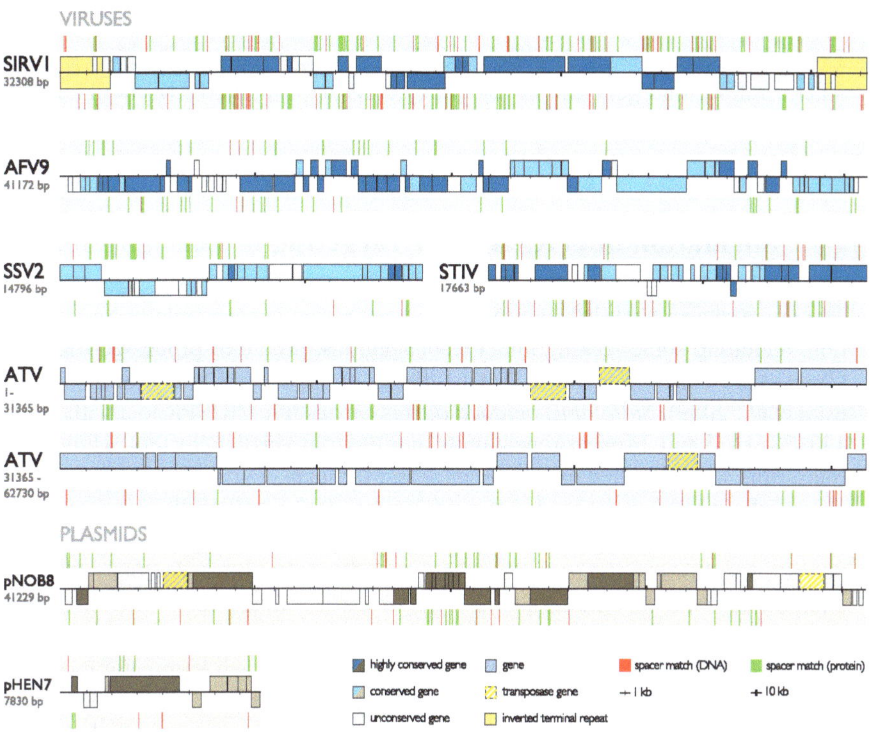

Fig. 7. Significant CRISPR spacer matches to protospacer sequences are superimposed on genomes of the following representative viruses and plasmids: SIRV1 – rudiviruses, AFV9 – betalipothrixviruses, SSV2 – fuselloviruses, STIV – turreted icosahedral viruses, ATV – bicaudavirus, pNOB8 – conjugative plasmids. and pHEN7 – cryptic plasmids where circular genomes (SSV2, STIV, ATV, pNOB8 and pHEN7) are presented in a linear form. Protein coding regions are boxed and shaded, as indicated on the Figure, according to their levels of conservation for those genomes. No comparative genomic data were used for ATV. Spacer sequence matches are indicated by lines above and below the genomes for the two DNA strands and they are colour-coded according to whether they occur exclusively at a nucleotide level (red) or at an amino acid level (green). Significant spacer matches were found by setting an e-value cut off corresponding to a 10 % false positive ratio, which was estimated by using the genome of *S. acidocaldarius* as a negative control (Chen *et al.*, 2005). These data are updated from an earlier study (Shah *et al.*, 2009)

CRISPR/Cas system of the bacterium *Staphylococcus epidermidis* was achieved experimentally (Marraffini and Sontheimer, 2008).

However, more recently it was demonstrated that crRNAs complexed with Cmr proteins target RNA carrying matching protospacers (Hale *et al.*, 2009) but it is still unclear whether this includes both mRNAs and viral RNAs. For archaea, this will only be resolved when the first archaeal RNA viruses have been characterised.

A few sequence matches have been detected between archaeal CRISPR spacers and IS elements suggesting that CRISPR/Cas system can target transposable elements (Lillestøl et al., 2006; Held and Whitaker, 2009; Mojica et al., 2009; Shah et al., 2009). However, most of those reported can be attributed to transposase genes carried on viral genomes or plasmids, including, for example, spacer matches to each of the four *orfB* transposable elements of the bicaudavirus ATV (Figure 7). These transposase genes/IS elements are presumably indistinguishable from any other viral/plasmid genomic target if they carry appropriate PAM motifs adjacent to protospacer sites.

7 Formation of crRNAs and Targeting of Foreign Elements

The few archaeal CRISPR loci that have been tested experimentally for transcription, including some lacking intact leader regions, produced processed transcripts (Tang et al., 2002; Tang et al., 2005; Lillestøl et al., 2006; Carte et al., 2008; Lillestøl et al., 2009). *Sulfolobus acidocaldarius* carries five CRISPR loci with sizes of 133, 78, 11, 5 and 2 spacer-repeat units. For the four smaller clusters, whole length transcripts were detected experimentally and for locus-78, the maximum transcript size of about 5000 nt, exceeded the size of the 4930 bp CRISPR locus, consistent with the whole transcript extending from within the leader region and terminating downstream from the locus (Lillestøl *et al.*, 2006; Lillestøl *et al.*, 2009). However, a large fraction of the transcripts also fell in the size range 3000–3500 nt suggesting that endogenous degradation, premature termination or processing had occurred towards the 3'-end of the transcript. Given that promoter and terminator motifs will be randomly taken up in spacers of CRISPR loci (Shah et al., 2009), there must be some form of transcriptional regulation to ensure the formation of whole CRISPR transcripts, possibly involving the *Sulfolobus* CRISPR repeat binding protein (Peng et al., 2003).

In the euryarchaeon *P. furiosus* and in *Escherichia coli* RNA transcripts are processed within repeats, 8 nt from the spacer start by the Cas6-type endonuclease. The processing of the 3'-end is less clear but for *P. furiosus* it occurs at two sites within the spacer, at 5 nt and 11 nt from the 3'-end of the spacer sequence. Complexes of Cas or Cmr proteins guide the mature crRNAs to their targets (Brouns et al., 2008; Hale et al., 2009). Annealing of the spacer sequence of the crRNA to the protospacer of the invading element is crucial for the recognition and inactivation of the target. For the bacterium *Streptococcus thermophilus* it was claimed that 100% sequence matching between the crRNAs and protospacer RNAs was essential for target inactivation (Barrangou et al., 2007; Horvath and Barrangou, 2010).

However, for *S. solfataricus* and *Sulfolobus islandicus* the requirements appear to be much less stringent because even with 3 mismatches between crRNA and protospacer targeting was still effective (Gudbergsdottir et al., 2011).

There may also be differences between some archaea and bacteria in the role of the family specific Protospacer-Associated Motif (PAM) complementary to part of the 5'-repeat sequence of the crRNA which, in *Sulfolobus* species constitutes a conserved dinucleotide (Lillestøl et al., 2009). For *S. islandicus* it was shown that altering the PAM motif inhibited protospacer targeting (Gudbergsdottir et al., 2011) whereas for the bacterium *Staphylococcus epidermidis* it was concluded that any sequence mismatch with the 5'-end of the crRNA ensured protospacer targeting and that sequence complementarity to the PAM motif was not essential (Marraffini and Sontheimer, 2010).

The CRISPR-like locus of the conjugative plasmid pKEF9 lacks an associated *cas* gene cassette and leader region but when transformed into *S. solfataricus* P2 it produced transcripts covering the whole CRISPR locus initiating 32 bp upstream from the first repeat and these were found to be processed. Processing sites were detected within each repeat spacer unit but some of the sites occurred within the spacer. At the time it was presumed that some inaccurate processing had occurred, possibly reflecting mismatches occurring between the plasmid repeat sequence and the host Cas proteins (Lillestøl *et al.*, 2009), but it was not known then that Cmr proteins process within the 3'-ends of spacers (Hale *et al.*, 2009).

In contrast to reports on the euryarchaeal CRISPR transcripts (Carte *et al.*, 2008) and a bacterium (Brouns *et al.*, 2008) transcripts were detected from both DNA strands of each of the five CRISPR loci of *S. acidocaldarius* (Lillestøl *et al.*, 2006; Lillestøl *et al.*, 2009). The largest CRISPR locus Saci-133 was probed against spacer sequences distributed along the cluster and each yielded clear signals in Northern analyses. The smallest processed products in the size range 55–60 nt were larger than those of leader strand crRNAs and were less regularly processed. These small RNAs were observed for all five *S. acidocaldarius* repeat-clusters and must contain most or all of the spacer sequence because the corresponding band was not detected when the spacer probe was replaced by a repeat probe. It remains to be shown which other archaea produce these antisense crRNAs and whether they have a functional role in protecting the mature crRNAs when there are no invading elements present remains unclear.

8 Anti CRISPR/Cas and CRISPR/Cmr Systems

Examples have been recorded of archaeal CRISPR/Cas modules being lost from genomes. For example, a variant strain of *S. solfataricus* P2 (strain P2A) was characterised that had lost four closely linked CRISPR/Cas modules, A to D, apparently via a single recombination event between bordering IS elements (Redder and Garrett, 2006). Bordering IS elements also have the potential to generate transposons carrying whole CRISPR/Cas or Cmr modules. Possibly this loss reflects *S. solfataricus* P2A being a laboratory strain where the immune system had become an unnecessary burden on the cell's energy resources in the absence of invading genetic elements and this may be analogous to the many bacterial endosymbionts which lack functional CRISPR/Cas systems (Grissa et al., 2008; Mojica et al., 2009).

There are also examples of viruses, which circumvent or interfere with the CRISPR systems. Some members of the viral families *Rudiviridae* and *Lipothrixviridae*, carry 12 bp indels, probably deletions, in their genomes often lying within, but not disrupting, open reading frames (Peng et al., 2004; Vestergaard et al., 2008). Although the function of these elements is unknown they may be generated in response to the CRISPR-based immune systems to avoid crRNA targeting. The presence of multiple recombination sites in some archaeal viruses and conjugative plasmids may also facilitate genomic rearrangements and sequence changes (Greve et al., 2004; Garrett et al., 2010).

Analysis of the genome of *S. islandicus* strain M.16.4 isolated in Kamchatka, Russia (Reno et al., 2009), revealed the presence of a more direct viral interference where an M164 provirus 1, has integrated into, and disrupted, the *csa3* gene encoding a putative transcriptional regulator of the group 1 *cas* genes (Figure 8). The insertion event seems to be recent since the truncated parts of the *csa3* gene show high sequence similarity to genes of closely related species, and it may be reversible. The closely related strain M.16.27 carries and intact *csa3* gene (Figure 8A) but also, unlike strain M.16.4, carries a CRISPR spacer sequence perfectly matching the provirus.

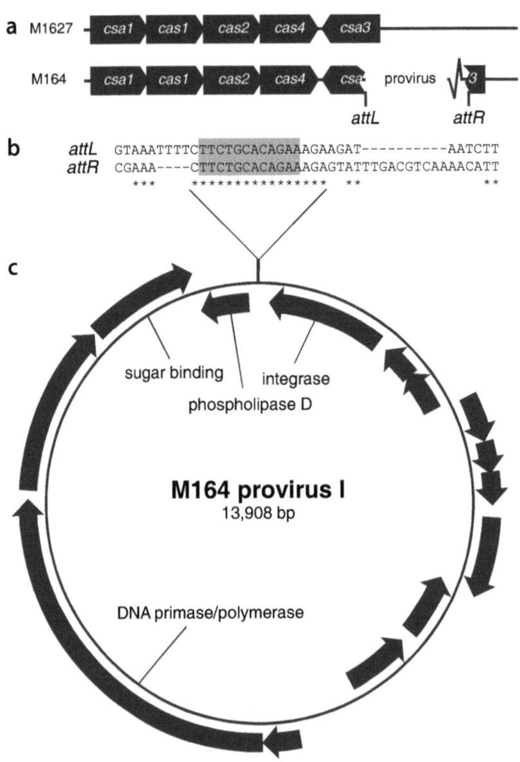

Fig. 8. An example of a *cas* gene cassette that has been inactivated in the gene for the putative transcriptional regulator *csa3* of *S. islandicus* M.16.4 by the integration of an M164 provirus 1.
a) Strain M.16.27, lacking the integrated provirus, carries a CRISPR spacer with a perfect match to the provirus whereas *S. islandicus* M.16.4, carrying the integrated provirus, contains no spacer sequence matching the proviral sequence.
b) The integration *att* site in the *csa3* gene.
c) Gene map of the integrated provirus showing some predicted functional assignments

9 Evolutionary Considerations

The view that archaeal and bacterial CRISPR/Cas systems are closely related has prevailed since their discovery and was underpinned by the similar ordering of spacer-repeat units in the CRISPR loci and by extensive sequence similarities between Cas proteins (Haft et al., 2005; Godde and Bickerton, 2006; Makarova et al., 2006). This view has been reinforced by the shared mechanism of elongation of CRISPR loci at the leader-repeat junction as well as similarities in the processing mechanisms of crRNAs in both Domains (Tang et al., 2002; Tang et al., 2005; Brouns et al., 2008; Hale et al., 2008; Hale et al., 2009). Nevertheless, there are distinctive features. CRISPR/Cas modules are more common amongst archaea and tend to be larger, structurally more complex and more labile (Lillestøl et al., 2006; Grissa et al., 2008; Shah and Garrett, 2011). Many repeat sequences show a bias to archaea or bacteria CRISPR loci, and many archaeal repeats lack inverted repeats common to those of bacteria suggesting that different RNA processing signals occur within transcript repeats (Lillestøl et al., 2006; Kunin et al., 2007). Moreover, many crenarchaea encode the CRISPR repeat binding protein of elusive function (Peng et al., 2003).

Phylogenetic analyses imply that periodic inter-Domain exchange of CRISPR/Cas modules has occurred (Haft et al., 2005; Godde and Bickerton, 2006; Makarova et al., 2006). Clearly, crossing Domain boundaries would be a very complex process given the basic differences in the transcriptional and translational mechanisms of archaea and bacteria (Torarinsson et al., 2005; Santangelo et al., 2009). Moreover, conjugal DNA transfer would also have to overcome the major barriers of different membrane and cell wall structures, and different conjugative systems, of archaea and bacteria (Greve et al., 2004; Veith et al., 2009). Nevertheless, coevolution of archaeal and bacterial CRISPR/Cas systems would only require cross Domain events to succeed rarely. The more archaea-specific components may be associated with systems that have evolved in environments of high temperature, extremes of pH, or hypersaline conditions where levels of bacteria are relatively low, which is also supported by the *cas* gene compositions of different CRISPR/Cas families. Other mechanistic differences may surface as the different systems are studied in more depth. Importantly, however, crenarchaeal viruses have radically different virus-host relationships from those of bacteria that may require altered responses from the immune systems (Prangishvili et al., 2006a; Bize et al., 2009) and it is likely that the CRISPR-based immune systems have maintained and/or undergone Domain-specific adaptations during evolution.

Small interference RNA systems (siRNA) are widespread in eukarya where they have multiple roles including the discrimination and targeting of "foreign" genetic elements such as viruses and transposons (Hannon 2002; Jinek and Doudna, 2009). There are broad mechanistic parallels between these eukaryal siRNA systems and the DNA- and RNA-targeting CRISPR systems. They all have to distinguish foreign DNA from self-DNA, and target nucleic acids which show little sequence similarity and can undergo continual sequence change. However, whereas the CRISPR systems employ ssRNAs for targeting foreign elements, the eukaryal anti-viral systems generate small 21–22 bp dsRNAs for targeting viruses which are subsequently converted to ssRNAs by an Argonaute protein-RISC complex.

The closest parallel to the crRNAs and CRISPR loci amongst the eukaryal siRNA systems are the Argonaute Piwi-interacting RNAs (piRNAs) directly processed from large transcripts of piRNA clusters which are rich in transposons and repeat-sequence elements and, as for the CRISPR loci, occur at specific chromosomal sites (Lillestøl et al., 2009; Karginov and Hannon, 2010). This eukaryal system probably plays a role in maintaining germline integrity and development (Aravin et al., 2007; Klattenhoff and Theurkauf, 2008). As for CRISPR loci, the piRNA clusters increase their informational capacity by the insertion of transposon sequences which provide novel sequence content and are maintained in the piRNA clusters by selection. Thus, continual expansion of piRNA clusters occurs, as for CRISPR loci, but the process is passive rather than directed. Moreover, as for the CRISPR/Cas system, the newly incorporated DNA derives exclusively from genetic elements that are to be targeted. No homologous proteins have been detected from sequence analyses between proteins of the eukaryal siRNA systems and those of the CRISPR system, although similarities may appear at a tertiary structural level.

10 Conclusions

The CRISPR/Cas and CRISPR/Cmr immune machinery provide an effective defence against foreign genetic elements in archaea and some bacteria. The system is dynamic and hereditable, although the benefit for the cell in evolutionary terms is transitional because DNA from extra chromosomal elements taken up as spacers in CRISPR loci, have a rapid turnover and are lost again via recombination at repeats and/or transpositional events. Current evidence suggests that CRISPR/Cas and Cmr modules can behave like integral genetic elements. They tend to be located in the most variable regions of chromosomes, sometimes physically linked, and are frequently displaced as a result of genome shuffling, including possibly transposition of whole modules. CRISPR loci may be broken up, and dispersed, in chromosomes with the potential for creating genetic novelty. Small leaderless CRISPR-like loci are commonly found in chromosomes, and in plasmids, and some can be transcribed and processed and therefore constitute potentially functional accessories to the CRISPR-based immune systems. Both CRISPR/Cas and Cmr modules appear to exchange readily between closely related organisms, possibly via chromosomal conjugation, where they may be subjected to strong selective pressure. While universal phylogenetic trees based on the Cas1 and Cmr2 proteins of the CRISPR/Cas and Cmr modules, respectively, suggest that transfers between archaea and bacteria have occurred, the relatively large number of archaea-specific Cas/Cmr proteins suggests that these may have been very rare events, consistent with the incompatibility of the transcriptional, translational and conjugative systems of the two Domains (Shah and Garrett, 2011). Parallels to the eukaryal siRNAs exist, and especially germ cell piRNAs which are also directed by effector proteins to silence or destroy invading foreign DNA and transposons.

References

Aravin AA, Hannon GJ, Brennecke J (2007) The Piwi-piRNA pathway provides an adaptive defense in the transposon arms race. Science 318: 761–764

Arnold HP, She Q, Phan H, Stedman K, Prangishvili D, Holz I et al. (1999) The genetic element pSSVx of the extremely thermophilic crenarchaeon *Sulfolobus* is a hybrid between a plasmid and a virus. Mol Microbiol 34: 217–226

Barrangou R, Fremaux C, Deveau H, Richards M, Boyaval P, Moineau S et al. (2007) CRISPR provides acquired resistance against viruses in prokaryotes. Science 315: 1709–1712

Basta T, Smyth J, Forterre P, Prangishvili D, Peng X (2009) Novel archaeal plasmid pAH1 and its interactions with the lipothrixvirus AFV1. Mol Microbiol 71: 23–34

Bettstetter M, Peng X, Garrett RA, Prangishvili D (2003) AFV1, a novel virus infecting hyperthermophilic archaea of the genus *Acidianus*. Virology 315: 68–79

Bize A, Karlsson EA, Ekefjard K, Quax TE, Pina M, Prevost MC et al. (2009) A unique virus release mechanism in the Archaea. Proc Natl Acad Sci U S A 106: 11306–11311

Bize A, Peng X, Prokofeva M, Maclellan K, Lucas S, Forterre P et al. (2008) Viruses in acidic geothermal environments of the Kamchatka Peninsula. Res Microbiol 159: 358–366

Bolotin A, Quinquis B, Sorokin A, Ehrlich SD (2005) Clustered regularly interspaced short palindrome repeats (CRISPRs) have spacers of extrachromosomal origin. Microbiology 151: 2551–2561

Brouns SJ, Jore MM, Lundgren M, Westra ER, Slijkhuis RJ, Snijders AP et al. (2008) Small CRISPR RNAs guide antiviral defense in prokaryotes. Science 321: 960–964

Brügger K, Torarinsson E, Redder P, Chen L, Garrett RA (2004) Shuffling of *Sulfolobus* genomes by autonomous and non-autonomous mobile elements. Biochem Soc Trans 32: 179–183

Carte J, Wang R, Li H, Terns RM, Terns MP (2008) Cas6 is an endoribonuclease that generates guide RNAs for invader defense in prokaryotes. Genes Dev 22: 3489–3496

Chen L, Brugger K, Skovgaard M, Redder P, She Q, Torarinsson E et al. (2005) The genome of *Sulfolobus acidocaldarius*, a model organism of the Crenarchaeota. J Bacteriol 187: 4992–4999

Cortez D, Forterre P, Gribaldo S (2009) A hidden reservoir of integrative elements is the major source of recently acquired foreign genes and ORFans in archaeal and bacterial genomes. Genome Biol 10: R65

Garrett RA, Prangishvili D, Shah SA, Reuter M, Stetter KO, Peng X (2010) Metagenomic analyses of novel viruses and plasmids from a cultured environmental sample of hyperthermophilic neutrophiles. Environ Microbiol 12: 2918–2930

Garrett RA, Shah SA, Vestergaard G, Deng L, Gudbergsdottir S, Kenchappa CS et al. (2011) CRISPR-based immune systems of the Sulfolobales – complexity and diversity. Biochem Soc Trans 39: 51–57

Godde JS and Bickerton A (2006) The repetitive DNA elements called CRISPRs and their associated genes: evidence of horizontal transfer among prokaryotes. J Mol Evol 62: 718–729

Greve B, Jensen S, Brügger K, Zillig W, Garrett RA (2004) Genomic comparison of archaeal conjugative plasmids from *Sulfolobus*. Archaea 1: 231–239

Grissa I, Vergnaud G, Pourcel C (2008) CRISPRcompar: a website to compare clustered regularly interspaced short palindromic repeats. Nucleic Acids Res 36: W145-W148

Gudbergsdottir S, Deng L, Chen Z, Jensen JVK, Jensen LR, She Q *et al.* (2011) Dynamic properties of the *Sulfolobus* CRISPR/Cas and CRISPR/Cmr systems when challenged with vector-borne viral and plasmid genes and protospacers. Mol Microbiol 79: 35–49

Guo L, Brügger K, Chao Liu C, Shah SA, Zheng H, Zhu Y et al. (2011) Genome analyses of Icelandic strains of *Sulfolobus islandicus*, model organisms for genetic and virus-host interaction studies. J Bacteriol in press

Haft DH, Selengut J, Mongodin EF, Nelson KE (2005) A guild of 45 CRISPR-associated (Cas) protein families and multiple CRISPR/Cas subtypes exist in prokaryotic genomes. PLoS Comput Biol 1: e60

Hale C, Kleppe K, Terns RM, Terns MP (2008) Prokaryotic silencing (psi)RNAs in *Pyrococcus furiosus*. RNA 14: 2572–2579

Hale CR, Zhao P, Olson S, Duff MO, Graveley BR, Wells L et al. (2009) RNA-guided RNA cleavage by a CRISPR RNA-Cas protein complex. Cell 139: 945–956

Hannon GJ (2002) RNA interference. Nature 418: 244–251

Held NL and Whitaker RJ (2009) Viral biogeography revealed by signatures in *Sulfolobus islandicus* genomes. Environ Microbiol 11: 457–466

Horvath P and Barrangou R (2010) CRISPR/Cas, the immune system of bacteria and archaea. Science 327: 167–170

Jansen R, Embden JD, Gaastra W, Schouls LM (2002) Identification of genes that are associated with DNA repeats in prokaryotes. Mol Microbiol 43: 1565–1575

Jinek M and Doudna JA (2009) A three-dimensional view of the molecular machinery of RNA interference. Nature 457: 405–412

Karginov FV and Hannon GJ (2010) The CRISPR system: small RNA-guided defense in bacteria and archaea. Mol Cell 37: 7–19

Klattenhoff C and Theurkauf W (2008) Biogenesis and germline functions of piRNAs. Development 135: 3–9

Krupovic M, Forterre P, Bamford DH (2010) Comparative analysis of the mosaic genomes of tailed archaeal viruses and proviruses suggests common themes for virion architecture and assembly with tailed viruses of bacteria. J Mol Biol 397: 144–160

Kunin V, Sorek R, Hugenholtz P (2007) Evolutionary conservation of sequence and secondary structures in CRISPR repeats. Genome Biol 8: R61

Lawrence CM, Menon S, Eilers BJ, Bothner B, Khayat R, Douglas T et al. (2009) Structural and functional studies of archaeal viruses. J Biol Chem 284: 12599–12603

Lillestøl RK, Redder P, Garrett RA, Brugger K (2006) A putative viral defence mechanism in archaeal cells. Archaea 2: 59–72

Lillestøl RK, Shah SA, Brugger K, Redder P, Phan H, Christiansen J et al. (2009) CRISPR families of the crenarchaeal genus *Sulfolobus*: bidirectional transcription and dynamic properties. Mol Microbiol 72: 259–272

Makarova KS, Grishin NV, Shabalina SA, Wolf YI, Koonin EV (2006) A putative RNA-interference-based immune system in prokaryotes: computational analysis of the predicted enzymatic machinery, functional analogies with eukaryotic RNAi, and hypothetical mechanisms of action. Biol Direct 1: 7

Marraffini LA and Sontheimer EJ (2008) CRISPR interference limits horizontal gene transfer in staphylococci by targeting DNA. Science 322: 1843–1845

Marraffini LA and Sontheimer EJ (2010) Self versus non-self discrimination during CRISPR RNA-directed immunity. Nature 463: 568–571

Mojica FJ, Diez-Villasenor C, Garcia-Martinez J, Almendros C (2009) Short motif sequences determine the targets of the prokaryotic CRISPR defence system. Microbiology 155: 733–740

Mojica FJ, ez-Villasenor C, Garcia-Martinez J, Soria E (2005) Intervening sequences of regularly spaced prokaryotic repeats derive from foreign genetic elements. J Mol Evol 60: 174–182

Peng X, Blum H, She Q, Mallok S, Brügger K, Garrett RA et al. (2001) Sequences and replication of genomes of the archaeal rudiviruses SIRV1 and SIRV2: relationships to the archaeal lipothrixvirus SIFV and some eukaryal viruses. Virology 291: 226–234

Peng X, Brügger K, Shen B, Chen L, She Q, Garrett RA (2003) Genus-specific protein binding to the large clusters of DNA repeats (short regularly spaced repeats) present in *Sulfolobus* genomes. J Bacteriol 185: 2410–2417

Peng X, Kessler A, Phan H, Garrett RA, Prangishvili D (2004) Multiple variants of the archaeal DNA rudivirus SIRV1 in a single host and a novel mechanism of genomic variation. Mol Microbiol 54: 366–375

Porter K, Russ BE, Dyall-Smith ML (2007) Virus-host interactions in salt lakes. Curr Opin Microbiol 10: 418–424

Portillo MC and Gonzalez JM (2009) CRISPR elements in the Thermococcales: evidence for associated horizontal gene transfer in *Pyrococcus furiosus*. J Appl Genet 50: 421–430

Pourcel C, Salvignol G, Vergnaud G (2005) CRISPR elements in *Yersinia pestis* acquire new repeats by preferential uptake of bacteriophage DNA, and provide additional tools for evolutionary studies. Microbiology 151: 653–663

Prangishvili D, Forterre P, Garrett RA (2006a) Viruses of the Archaea: a unifying view. Nat Rev Microbiol 4: 837–848

Prangishvili D, Garrett RA, Koonin EV (2006b) Evolutionary genomics of archaeal viruses: unique viral genomes in the third domain of life. Virus Res 117: 52–67

Rachel R, Bettstetter M, Hedlund BP, Haring M, Kessler A, Stetter KO et al. (2002) Remarkable morphological diversity of viruses and virus-like particles in hot terrestrial environments. Arch Virol 147: 2419–2429

Redder P and Garrett RA (2006) Mutations and rearrangements in the genome of *Sulfolobus solfataricus* P2. J Bacteriol 188: 4198–4206

Reno ML, Held NL, Fields CJ, Burke PV, Whitaker RJ (2009) Biogeography of the *Sulfolobus islandicus* pan-genome. Proc Natl Acad Sci U S A 106: 8605–8610

Santangelo TJ, Cubonova L, Skinner KM, Reeve JN (2009) Archaeal intrinsic transcription termination in vivo. J Bacteriol 191: 7102–7108

Shah SA and Garrett RA (2011) CRISPR/Cas and Cmr modules, mobility and evolution of adaptive immune systems. Res Microbiol 162: 27–38

Shah SA, Hansen NR, Garrett RA (2009) Distribution of CRISPR spacer matches in viruses and plasmids of crenarchaeal acidothermophiles and implications for their inhibitory mechanism. Biochem Soc Trans 37: 23–28

She Q, Peng X, Zillig W, Garrett RA (2001) Gene capture in archaeal chromosomes. Nature 409: 478

She Q, Phan H, Garrett RA, Albers SV, Stedman KM, Zillig W (1998) Genetic profile of pNOB8 from *Sulfolobus*: the first conjugative plasmid from an archaeon. Extremophiles 2: 417–425

Tang TH, Bachellerie JP, Rozhdestvensky T, Bortolin ML, Huber H, Drungowski M et al. (2002) Identification of 86 candidates for small non-messenger RNAs from the archaeon *Archaeoglobus fulgidus*. Proc Natl Acad Sci U S A 99: 7536–7541

Tang TH, Polacek N, Zywicki M, Huber H, Brügger K, Garrett R et al. (2005) Identification of novel non-coding RNAs as potential antisense regulators in the archaeon *Sulfolobus solfataricus*. Mol Microbiol 55: 469–481

Torarinsson E, Klenk HP, Garrett RA (2005) Divergent transcriptional and translational signals in Archaea. Environ Microbiol 7: 47–54

Tyson GW and Banfield JF (2008) Rapidly evolving CRISPRs implicated in acquired resistance of microorganisms to viruses. Environ Microbiol 10: 200–207

Veith A, Klingl A, Zolghadr B, Lauber K, Mentele R, Lottspeich F et al. (2009) *Acidianus*, *Sulfolobus* and *Metallosphaera* surface layers: structure, composition and gene expression. Mol Microbiol 73: 58–72

Vestergaard G, Shah SA, Bize A, Reitberger W, Reuter M, Phan H et al. (2008) *Stygiolobus* rod-shaped virus and the interplay of crenarchaeal rudiviruses with the CRISPR antiviral system. J Bacteriol 190: 6837–6845

Wang Y, Duan Z, Zhu H, Guo X, Wang Z, Zhou J et al. (2007) A novel *Sulfolobus* non-conjugative extrachromosomal genetic element capable of integration into the host genome and spreading in the presence of a fusellovirus. Virology 363: 124–133

Zillig W, Arnold HP, Holz I, Prangishvili D, Schweier A, Stedman K et al. (1998) Genetic elements in the extremely thermophilic archaeon *Sulfolobus*. Extremophiles 2: 131–140

Chapter 10
Control of Bacterial Heat Shock and Virulence Genes by RNA Thermometers

Franz Narberhaus*

Bacteria have developed numerous strategies in order to sense and respond to environmental challenges. Apart from well-known, protein-based mechanisms, RNA-dependent post-transcriptional gene regulation has recently been recognized as an important layer of control. RNA thermometers are located in the 5'-untranslated region of temperature-controlled mRNAs, typically encoding heat shock or virulence genes. They prevent translation at low temperatures by trapping the Shine-Dalgarno (SD) sequence in a complex structure that prevents ribosome access. Partial melting of the structure by a temperature increase to 37 °C (virulence genes) or higher (heat shock genes) liberates the SD sequence thus permitting access of the 30S ribosome and translation initiation. Since melting and base pairing are reversible processes, translation is shut off upon return to lower temperatures.

1 RNA as Sensory Element

The importance of proteins in the sensing of environmental cues and subsequent signal transduction is well established. Messenger RNAs, on the other hand, have long been considered solely as passive carriers of genetic information required for protein synthesis. The protein-centric view of gene regulation has changed dramatically during the last decade (Narberhaus and Vogel, 2009; Waters and Storz, 2009). Apart from numerous small non-coding RNAs that play important roles in post-transcriptional gene regulation (Storz *et al.*, 2005), it has been discovered that many mRNAs are able to sense and respond to chemical and physical signals on their own (Serganov and Patel, 2007). Such sensory and regulatory mRNAs contain an in-built control element, in bacteria usually in the 5'-untranslated region (5'-UTR),

* Chair of Biology of Microorganisms, Ruhr-Universität Bochum, Universitätsstrasse 150, NDEF 06/783, 44780 Bochum, Germany

that folds into a complex structure. Signal-induced structural changes dictate the fate of the transcript.

Two different types of sensory mRNAs can be distinguished: riboswitches and RNA thermometers. Both exploit the potential of RNA to fold into complex secondary and tertiary structures. Despite many similarities there are a few fundamental differences between these regulatory elements (Narberhaus, 2010). Riboswitches bind intracellular metabolites or ions with high specificity and affinity (Winkler and Breaker, 2005, see also chapter 11). They control biosynthetic pathways or transport mechanisms directly related to the bound ligand, e.g. vitamin precursors, purine nucleotides or amino acids. While riboswitches are able to control diverse cellular processes, such as transcription termination, translation initiation, mRNA processing and, in eukaryotes, mRNA splicing, all known RNA thermometers control translation initiation. Most genes under the control of RNA thermometers are bacterial heat shock or virulence genes (Narberhaus *et al.*, 2006). Here, the structured mRNA blocks access of the 30S ribosome to the Shine-Dalgarno (SD) sequence at low temperatures. Melting of the heat-labile structure upon a temperature upshift permits ribosome binding followed by translation initiation. A return to lower temperatures promotes formation of the inhibitory RNA structure and reduces expression of the downstream gene (Fig. 1).

Since melting and base pairing of the thermosensor are gradual processes, they produce a graded response directly correlated with the temperature. A riboswitch, in contrast, does not act gradually but exists in two alternative conformations depending on whether its cognate ligand is bound or not. The majority of riboswitches known to date operate as off-switches that control gene expression in the metabolite-bound conformation through negative feedback.

The modular architecture of riboswitches is comprised of a conserved ligand-binding aptamer domain and an output domain, the so-called expression platform (Winkler and Breaker, 2003). The discovery of certain aptamer domains in diverse organisms, including eukaryotes, suggests an early evolutionary origin of riboswitches (Vitreschak *et al.*, 2004, for details see chapter 11). The temperature-controlled base-pair formation and melting of RNA thermometers does not necessarily rely on a specific sequence context. Although thermosensory RNAs have been

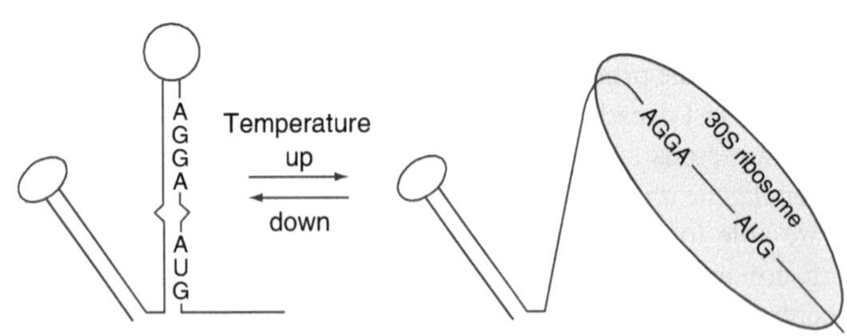

Fig. 1. The reversible temperature-dependent mode of action of RNA thermometers that control bacterial heat shock or virulence genes. At low temperatures (left), the Shine-Dalgarno sequence (AGGA) and in some cases the AUG start codon, are occluded in a hairpin structure. The structure melts at high temperatures (right) and gives access to the 30S ribosome

discovered in various distantly related Gram-negative and Gram-positive bacteria (Narberhaus *et al.*, 2006; Narberhaus, 2010), the degree of sequence conservation between them is very low.

2 RNA Measures Temperature Directly

Temperature is one of the many physico-chemical parameters that free-living microorganisms constantly monitor because a sudden temperature upshift or downshift can severely interfere with essential cellular processes. To counteract temperature-induced damage, all organisms are able to rapidly mount a protective response at potentially harmful temperatures, the so-called heat shock or cold shock responses at high or low temperatures, respectively (Phadtare *et al.*, 2000; Yura *et al.*, 2000). Several pathogenic bacteria possess temperature sensors to detect a temperature of 37 °C, suggestive of having successfully invaded a warm-blooded host (Konkel and Tilly, 2000).

The prerequisite for the induction of an appropriate response to temperature changes is the precise measurement of the ambient temperature. Since virtually any biomolecule is susceptible to temperature changes, numerous and quite diverse thermosensory mechanisms have been established in nature (Klinkert and Narberhaus, 2009; Schumann, 2009). For example, temperature-induced changes in DNA topology are known to control access of the RNA polymerase to some promoters. The intrinsic heat-sensitivity of some transcriptional activator or repressor proteins can be used to control gene expression in a temperature-dependent manner. The fluidity of membrane lipids largely depends on the ambient temperature and can control the activity of membrane-anchored regulators, such as sensor kinases. Not surprisingly, RNA molecules are also potent thermosensors as they often contain imperfectly paired secondary structure elements that are sensitive to temperature changes in the physiological growth range. Temperature measurement by RNA thermometers is very efficient since the sensory region is included in the transcript and thus the temperature input is directly transmitted to gene expression.

3 Control of Heat Shock Genes

Most RNA thermometers known to date control the expression of bacterial heat shock genes. The rapid induction of heat shock proteins is a universal stress response after a sudden temperature upshift. The induced chaperones and proteases prevent the misfolding and aggregation of cellular proteins (Yura *et al.*, 2000). Regulation of the heat shock response is extraordinarily complex, occurs at various levels and involves multiple factors. In *E. coli*, the alternative sigma factor Sigma32 (σ^{32}, RpoH) is the master regulator of the heat shock response. The intracellular concentration of the transcription factor is adjusted by a multi-layered network ranging from transcriptional to posttranslational events, including stability and activity control of the

RpoH protein. The chaperone-mediated negative feedback loop does not measure the ambient temperature but responds to the consequence of heat stress, namely the accumulation of misfolded proteins. The identification of an RNA thermometer in the *rpoH* gene showed that a direct thermosensor is part of the system (Morita *et al.*, 1999b). The detailed structure-function analysis of the *rpoH* thermometer was a remarkable achievement as the secondary structure of this regulatory element still belongs to the most complicated RNA thermometers known to date. Structure probing experiments revealed an extended structure, in which the temperature-responsive region is not restricted to the 5'-UTR but reaches more than 200 nucleotides in the coding region of *rpoH* (Morita *et al.*, 1999a; Morita *et al.*, 1999b). Although the SD sequence is partially single-stranded in the folded structure, base pairing of the flanking translation initiation region, including the AUG start codon, with internal regions of the *rpoH* transcript, prevent access of the ribosome. Site-directed point mutations revealed a clear inverse correlation between the calculated RNA stability and *rpoH* expression *in vivo*. Toeprinting (primer extension inhibition) experiments provided direct evidence for the temperature-controlled binding of the 30S ribosome to the SD sequence, which was efficient only at temperatures above 40 °C.

The architecture of the most common RNA thermometer is only slightly simpler than the *rpoH* thermometer. It was named ROSE (Repression Of heat Shock gene Expression) because it is frequently associated with heat shock genes and represses the synthesis of α-crystallin-type small heat shock proteins at low temperatures (Nocker *et al.*, 2001a). The responsible RNA structure is almost entirely located in the 5'-UTR and consists of several stem loops. The first identified ROSE element ($ROSE_1$), from the nitrogen-fixing soybean symbiont *Bradyrhizobium japonicum,* consists of 106 nucleotides forming a four-hairpin structure. Other rhizobial ROSE elements are equally complex and fold into three or four stem-loops (Nocker *et al.*, 2001b). The ROSE element upstream of the *E. coli ibpA* gene is 101 nucleotides long and adopts a three-hairpin structure (Waldminghaus *et al.*, 2009). Only the final hairpin, which contains the SD sequence and the start codon, is heat-labile. The heat-resistant initial two hairpins presumably serve as folding scaffold, facilitating correct base pairing of the fragile third hairpin.

Members of the ROSE family have been found in numerous α- and γ-proteobacteria (Waldminghaus *et al.*, 2005). They are characterized by a short consensus motif predicted to act as an anti-SD sequence pairing with the SD sequence, rendering it inaccessible to the 30S ribosome (Fig. 2A). The computer-predicted hairpin is destabilized by a bulged G opposite the SD sequence and at least one internal loop in close vicinity to the SD sequence. The NMR structure of this region from the $ROSE_1$ element revealed a well-ordered RNA helix that deviates from the predicted structure at several positions (Chowdhury *et al.*, 2006). Non-canonical base pairings, a triple base and a G-G pair, are the key to the heat-sensitivity of this hairpin (Fig. 2B). Destabilization of the RNA structure initiates at the triple base, giving rise to an open bubble, in which the SD is liberated as the temperature increases. UV and CD spectroscopy experiments of the entire four-hairpin structure demonstrated that melting of the RNA is possible in the appropriate temperature range without the aid of additional factors, and that the structural changes are reversible when the RNA is cooled down (Chowdhury *et al.*, 2003).

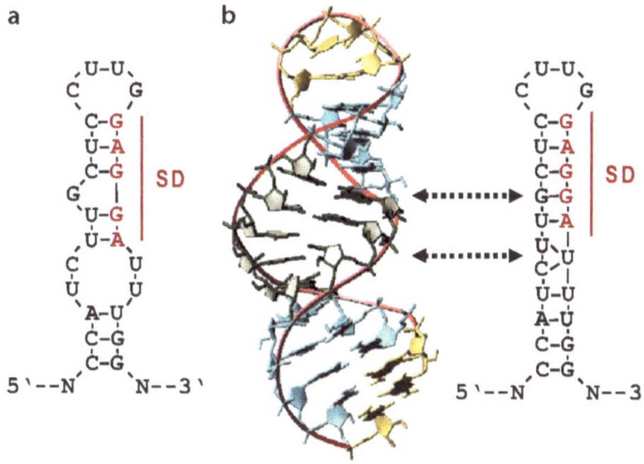

Fig. 2. The temperature-responsive stem-loop structure of the *B. japonicum* ROSE$_1$ element (A) Computer-predicted structure of ROSE$_1$ using MFOLD (Zuker, 2003). The position and sequence of the SD sequence (SD) is indicated. (B) NMR-determined structure of this region (Chowdhury *et al.*, 2006). Arrows point out the most heat-labile regions, in which the melting of the structure initiates

Another class of RNA thermometers, designated fourU, is characterized by an anti-SD sequence composed of four uridines (Waldminghaus *et al.*, 2007). The best-studied example is the *Salmonella* fourU thermometer controlling expression of the small heat shock gene *agsA*. Other heat shock genes that are potentially controlled by fourU-like structures are *groES* of *Staphylococcus aureus* and *dnaJ* of *Brucella melitensis*. Like ROSE elements, the regulatory region in the *Salmonella agsA* transcript is entirely located in the 5'-UTR and controls availability of the ribosome binding site (Fig. 3A). Unlike typical ROSE elements, however, the *agsA* thermometer is much shorter and structurally simpler. Its 57 nucleotides fold into two hairpins, of which the short first one (18 nucleotides) resists melting at heat shock temperatures. The second structure harbours the fourU element paired with the SD sequence. An internal A-G mismatch introduces instability to the hairpin. The structure is stable enough to limit access of the ribosome at 30 °C but permits efficient translation under heat shock conditions. Exchanging the internal mismatch to a stable C-G base pair stabilized the structure and prevented *agsA* expression both at low and high temperatures. Enzymatic structure-probing experiments revealed that the second hairpin of the wild-type structure is in equilibrium between open and closed structures, which shifts towards the open conformation upon exposure to high temperatures. The melting process was recently examined at atomic resolution by NMR spectroscopy. Entropy and enthalpy values for each individual nucleotide in the temperature-responsive hairpin were measured over a broad temperature range (Rinnenthal *et al.*, 2010). The results support a consecutive zipper-like mechanism in which the stacking interactions are responsible for the cooperativity of the melting process. The assumption that heat-induced liberation of the SD sequence is accompanied by translation initiation was substantiated

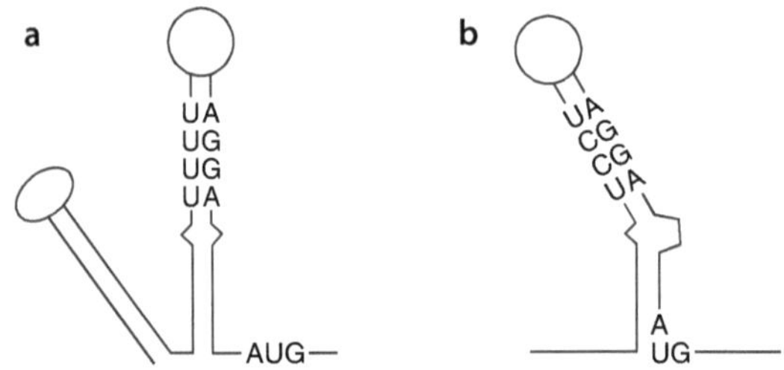

Fig. 3. Overall architecture of selected RNA thermometers.
a) The *Salmonella* fourU element consists of two stem-loop structures. The second hairpin contains the SD sequence (AGGA) pairing with the fourU motif.
b) The RNA thermometer upstream of the *Synechocystis hsp17* gene forms a single hairpin structure with an asymmetric internal loop. The SD sequence (AGGA) interacts with a complementary UCCU sequence

by toeprinting experiments, which confirmed the heat requirement for ribosome binding to the *agsA* SD sequence (Waldminghaus *et al.*, 2007).

A recently discovered novel heat shock gene thermometer controls translation of the *Synechocystis* sp. PCC 6803 heat shock gene *hsp17* (Kortmann *et al.* 2011). The exceptionally short 44-nucleotide 5'-UTR is comprised of a single hairpin. The SD sequence (AGGA) interacts with a UCCU sequence (Fig. 3B). These thermodynamically more stable (in comparison to ROSE and fourU elements), canonical Watson-Crick pairings are presumably weakened by an asymmetric internal loop. In line with temperature-mediated translational control, this region inhibits translation of the *hsp17* transcript under normal growth conditions, permits translation initiation under heat stress and shuts down Hsp17 production in the recovery phase. Point mutations that stabilized or destabilized the RNA structure deregulated reporter gene expression *in vivo* and ribosome binding *in vitro.*

The small heat shock protein Hsp17 (also known as Hsp16.6 or HspA) is a molecular chaperone required for the protection of the photosynthetic apparatus under heat shock and light stress conditions (Lee *et al.*, 1997; Török *et al.*, 2001). This provided the unique opportunity to test for the physiological importance of the thermometer by chromosomally integrating *hsp17* with the wild-type or mutated UTRs into a *Synechocystis hsp17* deletion strain. The stress-sensitive phenotype of the mutant was fully complemented by the wild-type sequence. Point mutations introducing either a closed or an open structure in the 5'-UTR of *hsp17* produced severe phenotypic defects under heat shock and high light conditions. A proper on and off function of the thermosensor was beneficial for viability of *Synechocystis*, the integrity of its photosynthetic apparatus and the oxygen evolution rate. Continuous production of Hsp17 in the derepressed thermometer mutant was detrimental in the recovery phase, when the stress declined demonstrating that shutting off heat shock gene expression in a post-stress situation is an important function of RNA thermometers.

4 Control of Virulence Genes

Multiple signals are integrated by pathogenic microorganisms to monitor whether they have reached a potential host. Several RNA-based mechanisms are involved in this process (Johansson and Cossart, 2003). Many pathogens are able to distinguish between the environmental climate and the body temperature of a warm-blooded mammal (Konkel and Tilly, 2000). First evidence for the existence of an RNA-based virulence gene thermometer was provided by Hoe and Goguen (Hoe and Goguen, 1993). They observed that the transcript of the *Yersinia pestis* virulence gene, *lcrF,* was efficiently translated only at 37 °C but not at 26 °C and concluded that the translation rate of the mRNA is thermally controlled. Although the calculated melting temperature of the stem-loop in the 5'-UTR is 52 °C, the structure was sufficiently destabilized at 37 °C to account for productive translation of the *lcrF* gene. As the *lcrF* thermosensor is of the fourU-type, this thermometer family is able to control both heat shock and virulence genes.

PrfA of *Listeria monocytogenes* is another virulence gene activator whose expression is controlled by an RNA-based thermosensor (Johansson *et al.*, 2002). The reason why virulence genes are maximally expressed at 37 °C even though the *pfrA* mRNA is present at 30 °C (Renzoni *et al.*, 1997) has long been an unsolved mystery. Thermoregulation of PrfA synthesis was retained in *E. coli* suggesting a control element in the *prfA* transcript (Johansson *et al.*, 2002). Structure-probing experiments and point mutations in the 5'-UTR provided evidence for a 127-nucleotide-long secondary structure, which makes translation at low temperatures less efficient than at 37 °C. Introduction of mismatches into the *prfA* leader sequence that allowed synthesis of the virulence gene regulator at 30 °C, increased production of virulence factors and permitted invasion of tissue-culture cells at this otherwise non-permissive temperature.

Recently, an unexpected complexity of *Listeria prfA* expression was discovered. The 5'-UTR is not only able to form intramolecular base pairs but also interacts with the *trans*-encoded SAM riboswitches SreA and SreB *in vivo* and *in vitro* (Loh *et al.*, 2009). Experiments with SreA demonstrated that the riboswitch RNA was unable to interact with the *pfrA* UTR at low temperatures when it is in a closed conformation. The riboswitch-*pfrA* interaction was independent of SAM binding. Absence of SreA in the cell increased the amount of PfrA, resulting in an elevated production of virulence factors. Overall, this study revealed that *cis*-active riboswitches can also act in trans as non-coding regulatory RNAs and suggested a RNA-controlled link between virulence and nutrient availability.

5 RNA-based Thermosensors That Do Not Act by Melting

This article would not be complete without mentioning that some RNA thermosensors act in a switch-like manner not consistent with the melting model. Alternative RNA structures were shown to determine the translation rate of the cIII gene of phage λ in response to the ambient temperature (Altuvia *et al.*, 1989). The cIII protein plays a role in the lysis-lysogeny decision of phage development. The RNA structure preferentially formed at optimal growth temperatures (37 °C) permits efficient translation of the cIII transcript because the SD sequence lies in a single-stranded region. The lysogenic pathway is favoured in the presence of high concentrations of the cIII protein. Under severe heat shock conditions (45 °C), the mRNA folds into a thermodynamically more stable conformation, in which the SD sequence is blocked. This reduces translation efficiency of cIII, allowing phage λ to enter the lytic pathway and to escape from the host.

Temperature-dependent alternative RNA structures have also been predicted to control expression of some cold shock genes at low temperatures. Like heat, low temperatures affect almost all cellular processes and the defence mechanisms counteracting cold-induced damage are manifold (Shivaji and Prakash, 2010). Several of the up-regulated genes code for so-called cold shock proteins (Csps) that serve as RNA chaperones preventing formation of stable mRNA structures that are poorly translated. The 5'-UTRs of the *E. coli* cold shock genes *cspA* and *cspE* have been postulated to control translation efficiency and differential RNA stability (Fang *et al.*, 1997; Uppal *et al.*, 2008; Yamanaka *et al.*, 1999). The switch-like mechanism of the *cspA* mRNA, was recently disclosed by extensive structure probing and ribosome binding experiments combined with mutational studies (Giuliodori *et al.*, 2010). The *cspA* gene contains a long 5'-UTR of 160 nucleotides. At an optimal growth temperature (37 °C), the UTR recruits more than 60 nucleotides of the coding region into a complex structure that stably sequesters the SD sequence and the AUG start codon and thus impedes ribosome binding. At a cold shock temperature (10 °C), the RNA undergoes a major structural rearrangement that exposes the SD sequence in a loop and positions the start codon in a labile stem. This "cold shock structure" facilitates binding of the 30S ribosome. In addition, it might be more resistant to ribonucleolytic attack than the structure at a higher temperature.

6 Are There More RNA Thermometers?

The structural and functional diversity of the RNA-based thermosensors described above suggests that there are many more temperature-responsive RNA elements yet to be discovered. In contrast to metabolite-binding riboswitches, which require a certain sequence context for ligand recognition, the evolutionary pressure to conserve temperature-responsive stem-loop structures is rather low. Therefore, the known RNA thermometers have little in common except for thermodynamically unstable hairpins that are sensitive to temperature changes in the mesophilic temperature range. Only ROSE and fourU elements share a few consensus nucleotides as a hallmark of these thermometer families. It is very conceivable that many other 5'-UTRs are used to modulate the expression of temperature-responsive genes. In support of this assumption, it has been fairly straight-forward to design simple synthetic RNA thermometers that respond to a temperature upshift as well as natural thermosensors (Neupert *et al.*, 2008; Waldminghaus *et al.*, 2008). The sequences and structures of these artificial elements are unrelated to known bacterial thermosensors. Apparently, temperature-dependent control of translation initiation is possible in a large variety of sequence and structure contexts. It might be rewarding to systematically search for new RNA thermometers, in particular in bacteria with a growth optimum at low (psychrophiles) or high (thermophiles or even hyperthermophiles) temperatures.

Another interesting aspect is whether RNA thermosensors exist that control processes other than translation initiation. It is striking that, in contrast to the more versatile riboswitches, all known RNA thermometers exclusively control access of the ribosome to the SD sequence. There is no obvious argument against temperature-responsive RNA elements being able to control transcription termination or RNA processing. Such regulatory elements might still be hidden in the transcripts of bacterial heat shock, cold shock or virulence gene transcripts.

Finally, it is possible that there are many more complex RNA thermometers that depend on additional factors yet to be discovered although several lines of evidence suggest that known temperature sensors act alone. (i) Various ROSE elements, the *Salmonella* fourU thermometer and 5'-UTRs of the *Synechocystis hsp17* and the *Listeria prfA* gene are functional in reporter gene studies conducted in the heterologous host *E. coli* (Johansson *et al.*, 2002; Nocker *et al.*, 2001a; Waldminghaus *et al.*, 2007). However, this does not necessarily rule out the involvement of some universal factor present in all bacteria. (ii) More convincingly, *in vitro* structure probing experiments, NMR spectroscopy and toeprinting experiments have demonstrated heat-induced melting and ribosome binding of these thermometers without the aid of auxiliary factors (Chowdhury *et al.*, 2006; Johansson *et al.*, 2002; Rinnenthal *et al.*, 2010; Waldminghaus *et al.*, 2007; Waldminghaus *et al.*, 2009). Nevertheless, these observations do not strictly exclude that, in the crowded intracellular environment, either proteins or RNAs contribute to the physiological response. The trans-acting SAM riboswitches acting on the *Listeria prfA* transcript are a good example of such a scenario (Loh *et al.*, 2009). Apparently, there still is a lot to be learned about the molecular mechanism of RNA-based temperature measurement.

References

Altuvia S, Kornitzer D, Teff D, Oppenheim AB (1989) Alternative mRNA structures of the cIII gene of bacteriophage λ determine the rate of its translation initiation. J Mol Biol 210: 265–280.

Chowdhury S, Ragaz C, Kreuger E, Narberhaus F (2003) Temperature-controlled structural alterations of an RNA thermometer. J Biol Chem 278: 47915–47921.

Chowdhury S, Maris C, Allain FH, Narberhaus F (2006) Molecular basis for temperature sensing by an RNA thermometer. EMBO J 25: 2487–2497.

Fang L, Jiang W, Bae W, Inouye M (1997) Promoter-independent cold-shock induction of cspA and its derepression at 37 °C by mRNA stabilization. Mol Microbiol 23: 355–364.

Giuliodori AM, Di Pietro F, Marzi S, Masquida B, Wagner R, Romby P, Gualerzi CO, Pon CL (2010) The cspA mRNA is a thermosensor that modulates translation of the cold-shock protein CspA. Mol Cell 37: 21–33.

Hoe NP, Goguen JD (1993) Temperature sensing in Yersinia pestis: translation of the LcrF activator protein is thermally regulated. J Bacteriol 175: 7901–7909.
Johansson J, Mandin P, Renzoni A, Chiaruttini C, Springer M, Cossart P (2002) An RNA thermosensor controls expression of virulence genes in Listeria monocytogenes. Cell 110: 551–561.
Johansson J, Cossart P (2003) RNA-mediated control of virulence gene expression in bacterial pathogens. Trends Microbiol 11: 280–285.
Klinkert B, Narberhaus F (2009) Microbial thermosensors. Cell Mol Life Sci 66: 2661–2676.
Konkel ME, Tilly K (2000) Temperature-regulated expression of bacterial virulence genes. Microbes Infect 2: 157–166.
Kortmann J, Sczodrok S, Rinnenthal J, Schwalbe H, Narberhaus F (2011) Translation on demand by a simple RNA-based thermosensor. Nucleic Acids Res 39: in press.
Lee GJ, Roseman AM, Saibil HR, Vierling E (1997) A small heat shock protein stably binds heat denatured model substrates and can maintain a substrate in a folding competent state. EMBO J 16: 659–671.
Loh E, Dussurget O, Gripenland J, Vaitkevicius K, Tiensuu T, Mandin P, Repoila F, Buchrieser C, Cossart P, Johansson J (2009) A trans-acting riboswitch controls expression of the virulence regulator PrfA in Listeria monocytogenes. Cell 139: 770–779.
Morita M, Kanemori M, Yanagi H, Yura T (1999a) Heat-induced synthesis of σ^{32} in Escherichia coli: structural and functional dissection of rpoH mRNA secondary structure. J Bacteriol 181: 401–410.
Morita MT, Tanaka Y, Kodama TS, Kyogoku Y, Yanagi H, Yura T (1999b) Translational induction of heat shock transcription factor σ^{32}: evidence for a built-in RNA thermosensor. Genes Dev 13: 655–665.
Narberhaus F, Waldminghaus T, Chowdhury S (2006) RNA thermometers. FEMS Microbiol Rev 30: 3–16.
Narberhaus F, Vogel J (2009) Regulatory RNAs in prokaryotes: here, there and everywhere. Mol Microbiol 74: 261–269.
Narberhaus F (2010) Translational control of bacterial heat shock and virulence genes by temperature-sensing mRNAs. RNA biol 7: 84–89.
Neupert J, Karcher D, Bock R (2008) Design of simple synthetic RNA thermometers for temperature-controlled gene expression in Escherichia coli. Nucleic Acids Res 36: e124.
Nocker A, Hausherr T, Balsiger S, Krstulovic NP, Hennecke H, Narberhaus F (2001a) mRNA-based thermosensor controls expression of rhizobial heat shock genes. Nucleic Acids Res 29: 4800–4807.
Nocker A, Krstulovic NP, Perret X, Narberhaus F (2001b) ROSE elements occur in disparate rhizobia and are functionally interchangeable between species. Arch Microbiol 176: 44–51.
Phadtare S, Yamanaka K, Inouye M (2000) The cold shock reponse. In Bacterial stress responses. Storz, G. and Hengge-Aronis, R. (eds). Washington, D.C.: ASM Press, pp. 33–45.
Renzoni A, Klarsfeld A, Dramsi S, Cossart P (1997) Evidence that PrfA, the pleiotropic activator of virulence genes in Listeria monocytogenes, can be present but inactive. Infect Immun 65: 1515–1518.
Rinnenthal J, Klinkert B, Narberhaus F, Schwalbe H (2010) Direct observation of the temperature-induced melting process of the Salmonella fourU RNA thermometer at base-pair resolution. Nucleic Acids Res 38: 3834–3847.
Schumann W (2009) Temperature sensors of eubacteria. Adv Appl Microbiol 67: 213–256.
Serganov A, Patel DJ (2007) Ribozymes, riboswitches and beyond: regulation of gene expression without proteins. Nat Rev Genet 8: 776–790.
Shivaji S, Prakash JS (2010) How do bacteria sense and respond to low temperature? Arch Microbiol 192: 85–95.
Storz G, Altuvia S, Wassarman KM (2005) An abundance of RNA regulators. Annu Rev Biochem 74: 199–217.
Török Z, Goloubinoff P, Horváth I, Tsvetkova NM, Glatz A, Balogh G, Varvasovszki V, Los DA, Vierling E, Crowe JH, Vigh L (2001) Synechocystis HSP17 is an amphitropic protein that stabilizes heat-stressed membranes and binds denatured proteins for subsequent chaperone-mediated refolding. Proc Natl Aca. Sci USA 98: 3098–3103.

Uppal S, Akkipeddi VS, Jawali N (2008) Posttranscriptional regulation of cspE in Escherichia coli: involvement of the short 5'-untranslated region. FEMS Microbiol Lett 279: 83–91.

Vitreschak AG, Rodionov DA, Mironov AA, Gelfand, MS (2004) Riboswitches: the oldest mechanism for the regulation of gene expression? Trends Genet 20: 44–50.

Waldminghaus T, Fippinger A, Alfsmann J, Narberhaus F (2005) RNA thermometers are common in α- and γ-proteobacteria. Biol Chem 386: 1279–1286.

Waldminghaus T, Heidrich N, Brantl S, Narberhaus F (2007) FourU: a novel type of RNA thermometer in Salmonella. Mol Microbiol 65: 413–424.

Waldminghaus T, Kortmann J, Gesing S, Narberhaus F (2008) Generation of synthetic RNA-based thermosensors. Biol Chem 389: 1319–1326.

Waldminghaus T, Gaubig LC, Klinkert B, Narberhaus F (2009) The Escherichia coli ibpA thermometer is comprised of stable and unstable structural elements. RNA Biol 6: 455–463.

Waters LS, Storz G (2009) Regulatory RNAs in bacteria. Cell 136: 615–628.

Winkler WC, Breaker RR (2003) Genetic control by metabolite-binding riboswitches. Chembiochem 4: 1024–1032.

Winkler WC, Breaker RR (2005) Regulation of bacterial gene expression by riboswitches. Annu Rev Microbiol 59: 487–517.

Yamanaka K, Mitta M, Inouye M (1999) Mutation analysis of the 5' untranslated region of the cold shock cspA mRNA of Escherichia coli. J Bacteriol 181: 6284–6291.

Yura T, Kanemori M, Morita MT (2000) The heat shock response: regulation and function. In Bacterial stress responses. Storz, G. and Hengge-Aronis, R. (eds). Washington, D.C.: ASM Press, pp. 3–18.

Zuker M (2003) Mfold web server for nucleic acid folding and hybridization prediction. Nucleic Acids Res 31: 3406–3415.

Chapter 11
RNA Sensors of Intracellular Metabolites

Vineetha Zacharia*, Daniel Lafontaine**, and Wade C. Winkler*

1 Introduction. Gene Regulation in Bacteria: From Transcription Initiation to mRNA Degradation

Gene expression in bacteria can be controlled through many different possible regulatory strategies. The "expression" of a given gene can be described as a function of all of the molecular processes involved in the conversion of the gene's information first into messenger RNA (mRNA) transcripts and then into the corresponding proteins. Therefore, one approach for defining the different possible types of bacterial regulatory strategies is to consider the various stages along the information-processing pathway from DNA to protein (Fig 1).

First, transcription initiation requires that a sigma factor associate with RNA polymerase (RNAP) to provide sequence-specific recognition of the appropriate promoter region. Each bacterium encodes for a suite of sigma factors, from only a few in certain bacteria to dozens in others, therefore, certain stress or physiological conditions are likely to stimulate a competition between sigma factors for access to the RNAP apoenzyme (Mooney *et al.*, 2005; Navarro Llorens *et al.*, 2010). The process of transcription initiation can be further influenced by association of transcription initiation factors with the RNAP holoenzyme, resulting in either activation or inhibition of transcription initiatin (Browning and Busby, 2004). A given bacterium is likely to encode at least 100 different putative transcription factors; therefore, regulation of transcription initiation is clearly a universally important genetic strategy in bacteria. However, this is only one of several processes that affect the final abundance of mRNA transcripts. Once RNAP has successfully transitioned from a relatively unstable initiation complex to a much more stable transcription elongation complex (TEC), transcription elongation can also be subjected

* The University of Texas Southwestern Medical Center, Department of Biochemistry, 5323 Harry Hines Blvd., Dallas, TX USA 75390-9038

** Université de Sherbrooke, Groupe ARN/RNA Group, Département de Biologie, Sherbrooke, Québec, Canada

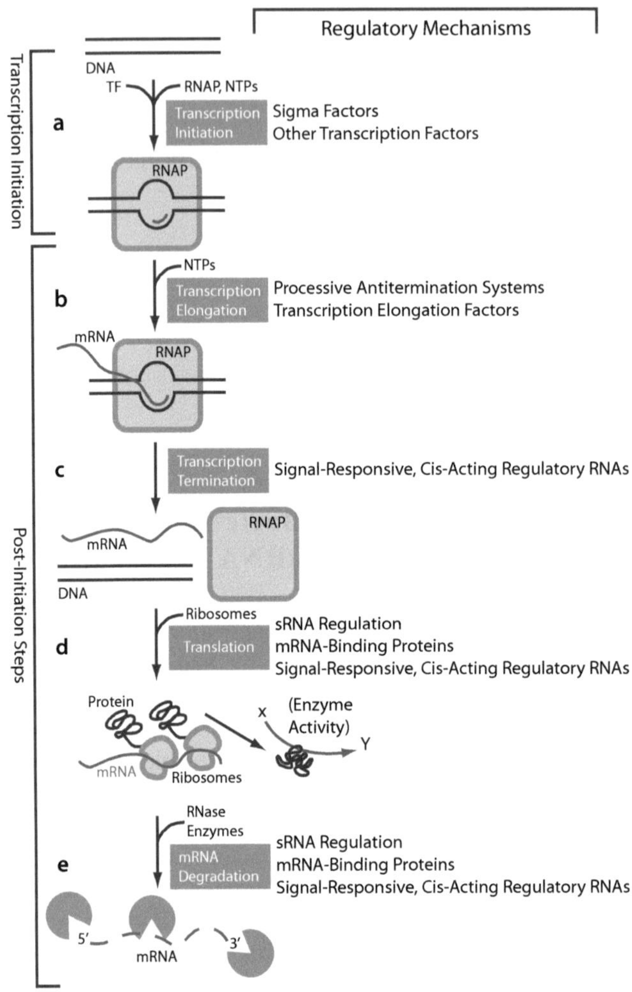

Fig. 1. Gene regulation in bacteria. A. At the level of transcription initiation, sigma factors and other transcription factors mediate the early stages of gene expression regulation (A). Post-initiation control of gene expression encompasses a variety of mechanisms including processive antitermination and other protein factors that influence transcription elongation (B), ligand-sensing cis-acting RNAs that regulate transcription termination (C), and small RNAs, RNA-binding proteins, and cis-acting regulatory RNAs that control gene expression at the level of translation and mRNA stability (D, E).

to regulatory control. Indeed, each nucleotide incorporation cycle can be viewed as a "decision" between elongation, pausing, arresting, backtracking, or termination (Greive and von Hippel, 2005; Vassylyev, 2009). Not only are molecular fates other than elongation possible, they are actually common. For example, recent data demonstrate that the TEC pauses frequently due to ubiquitous stochastic pause sites during transcription (Herbert *et al.*, 2008; Larson *et al.*, 2011). In addition to these short-duration pausing events, the TEC is also subject to influence by deterministic pause sites; specific sequence and structural elements can cause the TEC to pause for extended timescales (Landick, 2006). These sequence and structural elements may include certain sequences within the RNA:DNA hybrid of the RNAP active site as well as RNA elements that rapidly fold within the nascent transcript and associate with portions of RNAP for allosteric control of transcription. When characterized in vitro, some of these elements can lead to transcriptional pausing events with a half-life of multiple minutes (Artsimovitch and Landick, 2000). Different transcription elongation factors, which associate with the TEC, can affect the frequency and duration of different types of pausing events (Roberts *et al.*, 2008). For all of these reasons, it is unsurprising that bacteria have evolved mechanisms to exploit pause sites as key elements of certain genetic regulatory strategies (Landick, 2006). For example, a discrete pause site called the *ops* element is present within functionally related transcripts in certain bacteria. The TEC pauses upon encountering the *ops*, which allows time for association of a specialized elongation factor, RfaH. Upon escape from the *ops* pause site, the RfaH-bound TEC synthesizes the entire downstream portion of the operon. If, however, RfaH does not associate with the TEC during the lifetime of the *ops* pause, the transcriptional machinery encounters multiple, proximally located termination sites, and the downstream region is ultimately not synthesized. This process, whereby the TEC machinery is modified to allow readthrough of multiple downstream termination sites, is referred to as 'processive antitermination' (Weisberg and Gottesman, 1999). This mechanism is also employed for transcription of ribosomal RNA genes and is required for synthesis of certain long operons, including phage transcripts (Nudler and Gottesman, 2002; Roberts *et al.*, 2008) and bacterial polysaccharide operons (Irnov and Winkler, 2010).

Processive antitermination mechanisms are not the only regulatory mechanisms for control of premature transcription termination. Many cis-acting regulatory RNAs, such as metabolite-sensing riboswitches, have been discovered that exert control over transcription termination (transcription attenuation) (**Fig 2A**). However, these cis-acting regulatory elements control formation of a single termination site within the nascent transcript, as compared to the ability to readthrough many tandem termination sites as exhibited by processive antitermination complexes. For signal-responsive, cis-acting regulatory RNAs, a ligand-binding domain (the aptamer) senses the appropriate signal to harness control over the adjacent, downstream termination site. There are many different examples of these sorts of cis-acting regulatory RNAs, which individually respond to a diverse array of potential ligands, including proteins, trans-encoded RNAs, small molecules, and metals (Winkler, 2007). These types of RNA elements will be discussed further below.

In addition to their important roles in regulating transcription elongation, many different types of regulatory RNA elements can also affect the process of translation. For example, the same types of signal-responsive, cis-acting regulatory RNAs that control transcription attenuation can also be employed by bacteria to regulate the efficiency of translation initiation by controlling access to the ribosome binding site (**Fig 2B**). Another important and widespread approach for controlling translation is through the action of trans-acting regulatory RNAs (Waters and Storz, 2009; Gottesman and Storz, 2010). These small, trans-acting RNA sequences ('sRNAs') affect translation in multiple ways (**Fig 2C**). For example, sRNA regulators can associate within the 5' leader region to occlude access of the translation initiation complex. Alternatively, other sRNAs can associate within the 5' leader region in order to activate translation. More recently, certain sRNA examples have been shown to form base-pairing interactions with the coding portion of mRNA transcripts, although it

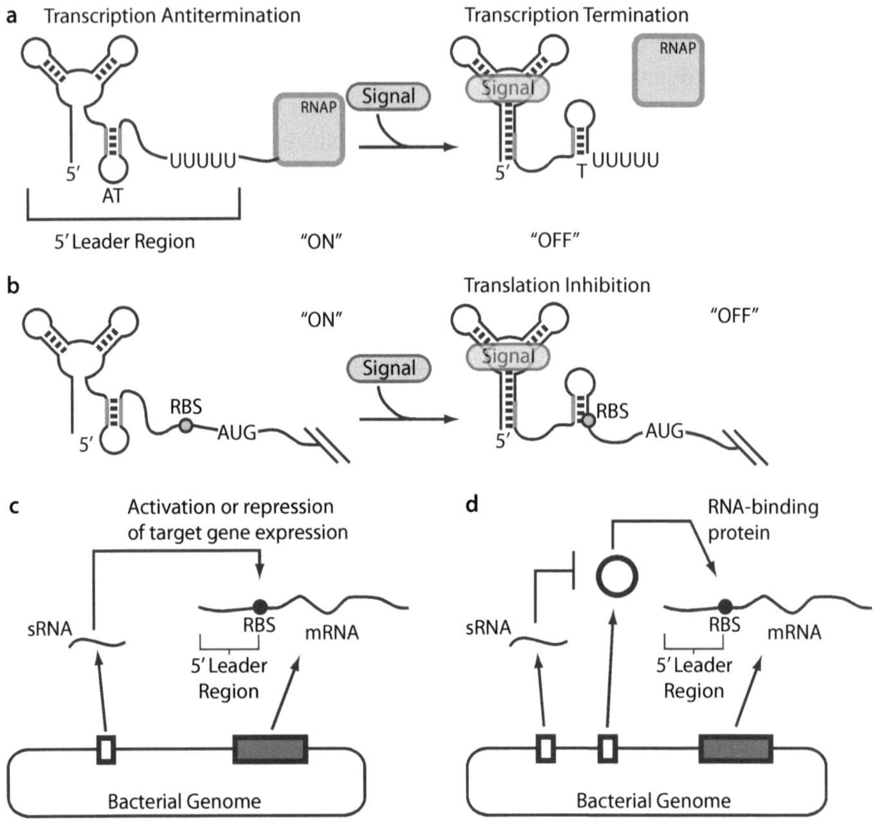

Fig. 2. Post-initiation mechanisms of cis- and trans-acting regulatory RNAs. In the absence of signal, a representative cis-acting regulatory RNA is shown in an unbound state. In this configuration, the leader region contains an anti-terminator 'AT' helix and gene expression is "ON" by permitting RNA polymerase to continue transcript synthesis (A) or by allowing ribosomes to access the ribosome binding site (B). Upon ligand binding, cis-acting regulatory RNAs change conformation to the "OFF" state which either leads to the formation of a terminator 'T' helix or a helix that sequesters the ribosome binding site, thereby terminating the transcript (A) or occluding the ribosome binding site to prevent ribosomal access, respectively (B). Trans-acting small RNAs (sRNAs) can directly interact with the target mRNA leader region to activate or repress gene expression (C). These sRNAs associate with target mRNAs to affect translation or mRNA stability. However, certain sRNAs sequester mRNA-binding proteins away from their mRNA substrates, in order to affect global expression patterns (D).

still remains to be determined whether these latter interactions are more important for regulating mRNA stability or translation (Pfeiffer *et al.*, 2009).

One of the first and most thoroughly studied examples of an sRNA is DsrA, an 85 nucleotide sequence in E. coli involved in stimulating capsule synthesis and increasing levels of stationary sigma factor, rpoS (Gottesman, 2004). Expression of the colanic acid capsule (cps) genes is induced upon increased RpoS translation and reduced levels of the histone-like nucleotide-structuring (HNS) protein, which is a silencer of capsule gene expression (Majdalani *et al.*, 1998). In the absence of DsrA only basal levels of RpoS are produced, as the result of a hairpin element within the leader region that blocks the ribosome binding site (Brown and Elliot, 1997). The DsrA secondary structure is comprised of stem-loops I, II, and III, where within loop I is a stretch of nucleotides that can form base-pairing interactions with the rpoS ribosome-sequestering hairpin. This interaction stimulates *cps* gene expression by disrupting this helix, and thereby allowing increased availability for the ribosome binding site (Majdalani *et al.*, 1998; Sledjeski and Gottesman, 1995). Conversely, a sequence stretch within stem-loop II of DsrA is predicted to interact with 13 nucleotides of the mRNA encoding HNS to inhibit translation, thereby relieving the inhibitory effect of HNS on capsule biosynthesis (Lease *et al.*, 1998). Parallel with the discovery of DsrA's role in *rpoS* regulation, it was found that Hfq, a homohexameric RNA chaperone protein, is required for intracellular stability of DsrA, and facilitates interactions between DsrA and its target mRNAs (Sledjeski *et al.*, 2001). Based on in vitro studies, Hfq binds to a single-stranded AU-rich sequence within DsrA. The interaction between Hfq and sRNAs is a commonly observed pattern since approximately one third of all sRNAs that have been identified in *E. coli* associate with Hfq in vivo, an interaction that stabilizes the sRNAs from degradation by RNase E (Møller *et al.*, 2002; Zhang *et al.*, 2003). Intriguingly, some sRNAs interact with Hfq and RNase E to facilitate degradation of target mRNAs. For example, such is the case with SgrS, a small RNA induced during sugar phosphate accumulation that forms a ribonucleoprotein complex with Hfq and RNase E to degrade ptsG mRNA, which encodes the major glucose transporter (Morita *et al.*, 2005).

In addition to sRNAs that directly affect translation and mRNA stability, some sRNAs act globally by sequestering RNA-binding proteins (**Fig. 2D**). Best characterized in this regard, *E. coli* sRNAs CsrB and CsrC act independently of Hfq and interact with an mRNA-binding protein, CsrA rather than a specific mRNA. CsrA, a post-transcriptional regulator of global gene expression, binds to mRNA and influences ribosomal binding and mRNA stability (Liu *et al.*, 1997, Liu and Romeo, 1997; Baker *et al.*, 2002). Some of the target mRNAs affected by CsrA include those involved in glycogen storage, motility, and virulence (Timmermans and Van Melderen, 2010). To down-regulate gluconeogenesis genes, CsrA binds specific sites on the target mRNA near the ribosome binding site, whereas the expression of master regulators of flagellar biosynthesis is upregulated when CsrA binds and stabilizes the 5' leader region of the target mRNA (Baker *et al.*, 2002; Wei *et al.*, 2001). To globally regulate CsrA's mRNA-binding activity, the CsrB and CsrC sRNAs influence gene expression by specifically binding CsrA and sequestering it away from its mRNA targets (reviewed in Majdalani *et al.*, 2005). The CsrB/C sRNAs contain similar repetitive sequences that bind CsrA, which allow them to outcompete mRNA targets for CsrA access. Together, these different examples of bacterial sRNAs underscore the significance of small trans-acting RNAs as regulators of global cellular processes and highlight a few of the types of RNA-based mechanisms that affect translation and mRNA stability.

Trans-acting sRNAs are not the only regulatory elements that affect mRNA stability. Although less studied overall, cis-acting regulatory RNAs can also exert an influence on mRNA stability. For example, certain metabolite-sensing regulatory RNAs ('riboswitches') are likely to regulate gene expression by affecting the RNase degradation pathway for the target mRNAs. Best characterized in this regard is the *glmS* riboswitch, which is a cis-acting regulatory RNA that directly responds to glucosamine-6-phosphate (GlcN6P) (Winkler *et al.*, 2004). Upon binding of GlcN6P, the riboswitch self-cleaves at a site-specific location near the 5' terminus of the mRNA transcript. This activity (self-cleavage) does not require accessory proteins; therefore, the *glmS* riboswitch is one of the few examples of a natural ribozyme. The 3' cleavage product of the self-cleavage reaction has a 5' hydroxyl at its 5' terminus (Winkler *et al.*, 2004; Klein and Ferré-D'Amaré, 2006). Intriguingly, the molecular features at the 5' terminus of bacterial transcripts are oftentimes a rate-limiting determinant for mRNA degradation (Bechhofer, 2009; Belasco, 2010; Condon, 2010). Indeed, in *B. subtilis*, where the *glmS* ribozyme has been best characterized, an RNase enzyme (RNase J) specifically stimulates degradation of the *glmS* transcript once the 5' terminus has been altered from a 5' triphosphate to a 5' hydroxyl, as a result of ribozyme self-cleavage (Collins *et al.*, 2007). Interestingly, the *glmS* riboswitch might not be the only riboswitch that controls mRNA stability. Several examples of riboswitches have been identified within the 3' UTR of bacterial transcripts, suggesting that they might regulate gene expression by affecting mRNA stability through uncharacterized mechanisms.

It should therefore be apparent from this brief summary that bacteria have been found to utilize a plethora of genetic regulatory strategies, from those affecting transcription initiation, transcription elongation, termination, and mRNA stability. In this chapter, we will overview an important, and growing, subclass of regulatory RNAs that sense metabolite fluctuations in vivo.

2 Sensing of Metabolites by Cis-Acting Regulatory mRNAs

2.1 Riboswitch RNAs

In the past decade, signal-responsive riboswitches have been identified that can sense amino acids, carbohydrates, coenzymes, nucleobases and metals (Serganov and Patel, 2007; Roth and Breaker, 2009). Chemical structures and molecular properties of metabolites have remained invariant throughout evolution and, correspondingly, the ligand binding domains (aptamer) of metabolite-binding riboswitches appear to be highly conserved. In contrast, the downstream expression platform is not conserved because it is involved in the regulation of gene expression. Although riboswitch classes in general are easily identified by their common aptamer structure, a few exceptions exist. The *S*-adenosylmethionine (SAM)-sensing riboswitches for example, can be categorized into various sub-classes, which differ with respect to the aptamer structure (Roth and Breaker, 2009). For most riboswitches, ligand binding induces conformational changes in the mRNA molecule, resulting in the modulation of gene expression at the levels of transcription, translation, splicing or mRNA stability (Serganov and Patel, 2009).

Extensive effort has been put forth to characterize riboswitch gene regulation mechanisms. Purine-sensing riboswitches have particularly been studied, most probably because their relatively small size makes them ideal candidates for biochemical and biophysical studies. They are involved in the recognition of adenine, guanine, cyclic di-GMP and pre-queuosine (Barrick and Breaker, 2007; Sudarsan *et al.*, 2008; Roth and Breaker, 2009). The purine riboswitch class is unique given that these riboswitches are involved in both the positive and negative regulation of gene expression. This review will focus on recent developments and applications involving adenine and guanine riboswitches.

2.2 Purine Riboswitch Gene Regulation Mechanisms

The adenine-sensing riboswitch is very similar in sequence and structure to the guanine riboswitch but exhibits very high specificity and affinity toward adenine and related ligands (Mandal and Breaker, 2004). In contrast to most riboswitches, the adenine riboswitch is a rare type of "ON" switch that activates gene expression upon ligand binding (Mandal and Breaker, 2004; Serganov *et al.*, 2004). The adenine riboswitch aptamer is organized around a three-way junction in which the core region participates in the elaboration of the ligand binding site (Mandal and Breaker, 2004; Serganov *et al.*, 2004). Although aptamer domains of related adenine riboswitches were shown to exhibit very similar ligand-binding properties, apparent discrepancies regarding the influence of the expression platform on ligand recognition were observed (Mandal and Breaker, 2004; Wickiser *et al.*, 2005a; Lemay *et al.*, 2006; Lemay and Lafontaine, 2007; Rieder *et al.*, 2007). For instance, although the presence of the expression platform of the *pbuE* riboswitch from *Bacillus subtilis* strongly inhibits ligand binding, no such effect was observed for the *add* riboswitch from *Vibrio vulnificus* (Wickiser *et al.*, 2005a; Lemay *et al.*, 2006; Rieder *et al.*, 2007). Various explanations were put forward to explain these differences such as the regulation regime that could be employed by each riboswitch (Wickiser *et al.*, 2005a; Wickiser *et al.*, 2005b; Kulshina *et al.*, 2009; Smith *et al.*, 2009; Rieder *et al.*, 2010; Lemay *et al.*, 2011). The importance of regulation regimes was first highlighted in the case of the flavin mononucleotide (FMN) riboswitch from *B. subtilis*, which was shown to operate under a kinetic regime (Wickiser *et al.*, 2005b). Under this regime, the riboswitch and the FMN metabolite do not have time to attain thermodynamic equilibrium before the RNA polymerase (RNAP) makes the genetic decision between transcription elongation or termination. Because of this situation, higher FMN concentrations are required to trigger riboswitch regulation (T_{50}) compared to the measured dissociation constant (K_D). The presence of transcriptional pause sites were also found to provide additional time for the riboswitch:FMN complex formation to occur before the transcriptional decision is made.

Because it was previously observed that the presence of the expression platform of the *pbuE* adenine riboswitch strongly inhibited ligand binding, it was proposed that the riboswitch could operate under a kinetic regime (Lemay *et al.*, 2006; Wickiser *et al.*, 2005a). It was however only recently shown that the *pbuE* riboswitch regulates gene expression using a kinetic framework (Lemay *et al.*, 2011). It was found that the *pbuE* riboswitch is strongly dependent of the transcriptional context to perform ligand binding, suggesting that ligand binding occurs in a co-transcriptional manner *in vivo* (Lemay *et al.*, 2011). Ligand binding to the RNA was found to stabilize the aptamer domain, which is important for the adoption of the ON state structure **(Fig 3A)**. *In vitro* transcription assays showed that transcription elongation is specific to adenine and related metabolites, consistent with the ligand binding specificity of the isolated aptamer (Mandal and Breaker, 2004). A strong relationship was also observed between the rate of transcription and the concentration of ligand required for riboswitch activation, as previously found for the FMN riboswitch (Wickiser *et al.*, 2005b). Interestingly, a transcriptional pause site

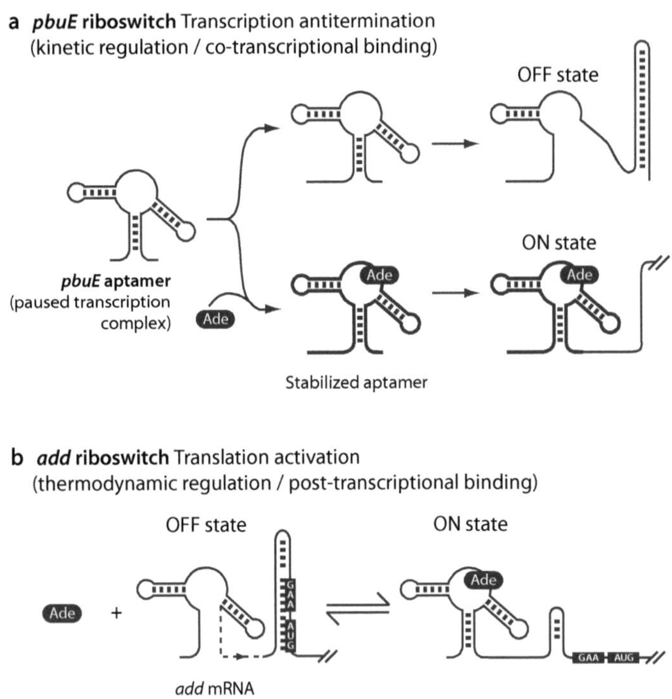

Fig. 3. Ligand-induced genetic regulation of two related adenine riboswitches. (**A**) Regulation mechanism of the *pbuE* adenine riboswitch from *B. subtilis*. In absence of sufficiently high intracellular adenine concentrations, the riboswitch adopts the OFF state which prematurely terminate transcription. However, elevated adenine concentrations favor ligand binding on paused transcription complexes, thereby stabilizing the aptamer and allowing transcription of the complete mRNA. (**B**) Regulation mechanism of the *add* adenine riboswitch from *Vibrio vulnificus*. In the OFF state, both the Shine-Dalgarno (GAA) and AUG start codons are base paired in the sequestrator stem, which does not allow ribosome binding. In the presence of adenine, the ON state is adopted thereby releasing allowing ribosome binding and gene expression. Adapted from Lemay *et al*[14].

was uncovered in the expression platform, which was initially predicted based on a U-stretch sequence (Wickiser *et al.*, 2005a). Mutation of the pause site resulted in a higher ligand concentration required to trigger riboswitch activity, consistent with the higher rate of RNAP transcription allowing less time for ligand binding. The transcription factor NusA was also observed to positively affect riboswitch regulation, most probably by reducing the transcription rate (Pan *et al.*, 1999). Taken together, these data strongly suggest that the *pbuE* adenine riboswitch regulates gene expression by employing a kinetic regime (**Fig 3A**).

In contrast to the *pbuE* riboswitch, the regulatory mechanism of the translationally-acting *add* adenine riboswitch from *Vibrio vulnificus* is consistent with a thermodynamic regime **(Fig 3B)** (Rieder *et al.*, 2007; Lemay *et al.*, 2011). In this regulatory framework, the time needed to attain an RNA-ligand equilibrium is short compared to the transcriptional time scale, and as a consequence, the K_D is determinant for the riboswitch activation. It was found that in contrast to *pbuE*, ligand binding and riboswitch regulation can occur post-transcriptionally, which are consistent with the idea that the *add* riboswitch exhibits structural reversibility as a function of ligand binding (Lemay *et al.*, 2011). This is supported by the predicted similar free energies of -23.8 kcal/mol and -23.1 kcal/mol for the OFF and ON structures, respectively (Lemay *et al.*, 2011). Moreover, because it does not require coupling between transcription and translation for translation activation, the *add* riboswitch may benefit of an extended time compared to *pbuE* to achieve genetic control given that ligand binding can occur post-transcriptionally. However, to clearly ascertain that *add* operates under a thermodynamic regime, it would require to determine that the T_{50} value approximates the K_D of the riboswitch-ligand complex *in vivo*. Recent findings about potential roles of transcription factors such as NusG indicate that additional cellular components could also be involved in riboswitch regulation (Burmann *et al.*, 2010).

2.3 Therapeutic Applications Using Purine Riboswitches

Due to their inherent capacity to bind cellular metabolites, riboswitches have been considered as promising targets for novel antimicrobial agents (Blount and Breaker, 2006). In principle, providing that a riboswitch negatively regulates the expression of a gene, or a collection of genes, that is essential for cellular growth, it could be possible to employ riboswitch ligand analogs to selectively inhibit riboswitch gene expression. For example, guanine riboswitches are negatively regulated in presence of guanine (**Fig 4A**) and appear to be excellent drug targets given their role in regulating purine metabolic pathways. As such, the disruptions of the GMP synthetase *guaA* and of the IMP dehydrogenase *guaB* were shown to lead to guanine auxotrophy and growth inhibition, respectively (Kobayashi *et al.*, 2003; Shimaoka *et al.*, 2006). Moreover, because their three-dimensional structures are highly characterized, guanine riboswitches make excellent candidates for structure-based drug design.

Crystal structures of the guanine riboswitch illustrate that the bound ligand is almost completely engulfed in the aptamer domain, where most of the functional groups of the ligand are involved in RNA interactions. The binding site also shows that regions surrounding the C2 and C6 positions of the ligand should accommodate functional group modifications without substantially disrupting binding to the aptamer. Recently, Breaker and colleagues designed a collection of C2- and C6-modified guanine analogs and characterized their ability to act as antimicrobial agents[21]. In that study, several compounds were obtained that could bind to the riboswitch *in vitro* with affinities comparable to that of the natural ligand. Further analysis shown that several of these candidates could also inhibit *B. subtilis* growth (Kim *et al.*, 2009). Out of the three molecules exhibiting antimicrobial activity, one candidate (6-*N*-hydroxyl-aminopurine or G7) was found to repress the expression of a reporter gene under the control of a guanine riboswitch, suggesting that the other two molecules inhibited bacterial growth through an alternative riboswitch-independent action mechanism. Lastly, by allowing bacteria to develop resistance against tested antibiotics, it was found that some resistant strains carried mutations within the terminator stem of the expression platform of the *pbuE* adenine riboswitch. This is consistent with the function of PbuE which is a purine efflux pump involved in the protection against toxic purine analogs (Johansen *et al.*, 2003).

It was previously reported that aminopyrimidines could be used as ligands for guanine riboswitches, given that they recreate most of the correct network of hydrogen bonds required for complex formation (Gilbert *et al.*, 2006). By taking advantage of this situation, pyrimidine-based antibiotics were developed as a way to inhibit bacterial growth by binding to the guanine riboswitch (Mulhbacher *et al.*, 2010). Based on several criteria such as hydrogen bonding interactions and molecular planarity, a successful pyrimidine compound (PC1) was found to replicate most molecular interactions involving functional groups of guanine (**Fig 4B**). By performing antibiogram assays, it was found that PC1 showed antibacterial activity only against a subgroup of bacterial species. Upon careful examination of the genomic organization of tested bacterial strains, it was deduced that all PC1-sensitive strains make use of riboswitch regulation to control the expression of *guaA*, a

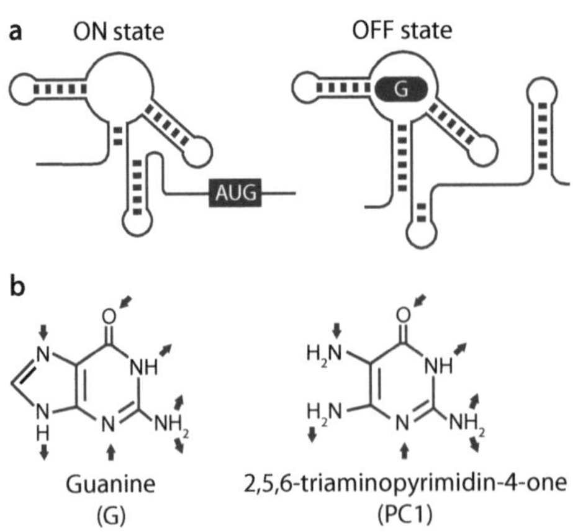

Fig. 4. The guanine riboswitch as an antimicrobial drug target. (**A**) Schematic showing the guanine riboswitch in absence (ON state) or in presence of guanine (OFF state). The formation of the OFF state is characterized by the presence of a terminator stem that promotes premature transcription termination. (**B**) Molecular recognition features for guanine (G) and predicted ones for PC1. Arrows pointing inward or outward of molecules represent hydrogen bond acceptors and donors, respectively. Adapted from Mulhbacher *et al*[23] .

GMP synthetase. Several clinical strains exhibiting multiple drug resistances were also sensitive to PC1, suggesting that a different metabolic pathway is targeted by PC1. By comparing with other known antibiotics, it was found that PC1 has an extremely rapid bactericidal activity similar to ciprofloxacin, one of the most bactericidal antibiotics. Interestingly, even after multiple serial passages of *Staphylococcus aureus* in presence of PC1, no bacterial resistance was obtained suggesting that maintaining a functional *guaA*-regulated riboswitch is a vital process. The proof of concept for the therapeutic efficacy of PC1 was performed using a murine model of *S. aureus*-induced mastitis. A strong therapeutic effect was observed for PC1 when compared to known antibiotics, such as amoxicillin. Because antibacterial activity is only observed in strains where *guaA* is under a riboswitch control, it suggests that analog binding to riboswitches is not the only determinant to achieve selective and efficient bacterial effects.

Although studies discussed above provide clear optimism for novel approaches to validate new classes of antibacterial drugs, a significant bottleneck resides at the beginning of the process in the identification of potential drug candidates. Indeed, *de novo* design of antimicrobial compounds is not straightforward as it requires to take into account multiple parameters including hydrogen bonding interactions and planarity of ligand analogs. Recent advances in this field have been obtained by using molecular docking as a virtual screen to investigate riboswitch-ligand interactions (Ling *et al.*, 2009; Daldrop *et al.*, 2011). These new approaches are very promising to explore on a large-scale basis RNA structures for antimicrobial analogs.

3 Indirect Sensing of Metabolites by Cis-Acting Regulatory RNAs

3.1 Sensing of Amino Acids via tRNA Charging Ratios

There are many different post-initiation regulatory strategies that are used by bacteria for responding to metabolic fluctuations, in addition to metabolite- and metal-responsive riboswitches described above. Certain post-initiation regulatory strategies involve RNA structural elements that regulate gene expression by indirectly sensing small organic metabolites through accessory factors. A variety of macromolecules are used for these accessory purposes, including tRNAs, ribosomes, and specialized proteins. We will briefly overview two general themes on this subject: (1) the indirect sensing of amino acids through tRNA charging ratios, and (2) the use of signal-responsive RNA-binding proteins for controlling gene expression (**Fig 5**).

Bacteria must continually monitor the production of their amino acids. Amino acids are energetically expensive to produce and toxic when they accrue to inappropriately high levels. Therefore, the genes encoding for biosynthesis and transport of amino acids are generally subjected to tight regulatory control. As discussed above, in certain instances, elaborate riboswitch structures directly bind to certain amino acids and regulate downstream gene expression. However, only a few amino acid classes are sensed in this way, and such riboswitch mechanisms are typically restricted to specific bacterial lineages. Therefore, it can be presumed that other post-initiation regulatory strategies are also likely to be used for the control of amino acid homeostasis. Several such genetic regulatory strategies rely on the monitoring of tRNA charging ratios rather than sensing amino acids directly. By following intracellular fluctuations in the pool of charged tRNAs, cells can closely monitor their overall amino acid requirements based upon the availability of aminoacylated tRNAs. Interestingly, certain operons are subject to overlapping regulatory control by an amino acid-responsive transcription factor and an additional mechanism that responds to tRNA charging ratios. This suggests that bacterial cells may, in certain instances, respond to multiple amino acid-related signals that differ slightly from one another and that are important under differing physiological reasons. Post-initiation regulatory strategies therefore allow the bacterium the opportunity to potentially "layer" its various genetic strategies.

---

next page:
►**Fig. 5.** Post-initiation regulatory mechanisms that sense amino acids. Representative examples of cis-acting regulatory RNAs that sense amino acid concentrations are shown in (A) and (B), while representative examples of regulatory RNAs that sense tRNA charging ratios are shown in (C) and (D). (A) Under tryptophan limiting conditions in *B. subtilis*, the *trp* RNA-binding attenuation protein (TRAP) remains inactive and an anti-terminator helix 'AT' in the *trpE* leader region forms, as depicted by the 2:3 base-pairing element, allowing transcriptional readthrough. High levels of tryptophan (white 'W's) bind and activate TRAP to associate with a tandem series of (G/U)AG triplet sequences located within the leader RNA (shown as grey circles). This prevents AT formation and allows the 3:4 base-pairing to form the terminator helix 'T'. (B) Upon lysine accumulation, the lysine riboswitch directly binds lysine, thereby allowing base-pairing between 3:4 (terminator), rather than between sequences 2:3 (anti-terminator) to terminate transcription (C) Ribosome-mediated attenuation governs *trp* operon regulation in *E. coli*. Under low levels of charged $tRNA^{Trp}$, the ribosome stalls allowing the default anti-terminator helix in the leader region to form, which permits transcription of the *trp* operon. High levels of tryptophan and sufficient $tRNA^{Trp}$ levels allow ribosomes to fully translate the 14-residue leader peptide, which prevents anti-terminator helix formation to instead allow formation of the anti-antiterminator 'A-AT' and transcription terminator helices. (D) T-box attenuation mechanisms regulate aminoacyl-tRNA synthetase genes by directly sensing of the cognate uncharged tRNA by the leader region. The specifier loop located at the base of 'helix I' of the leader RNA base pairs with the anti-codon of cognate tRNA to form an important determinant for tRNA specificity. The T-box leader region also interacts with the uncharged tRNA through the terminal nucleotides at the tRNA acceptor end. Specifically, these nucleotides pair with a conserved side-bulge of the antiterminator helix, thereby stabilizing it over the mutually exclusive terminator hairpin.

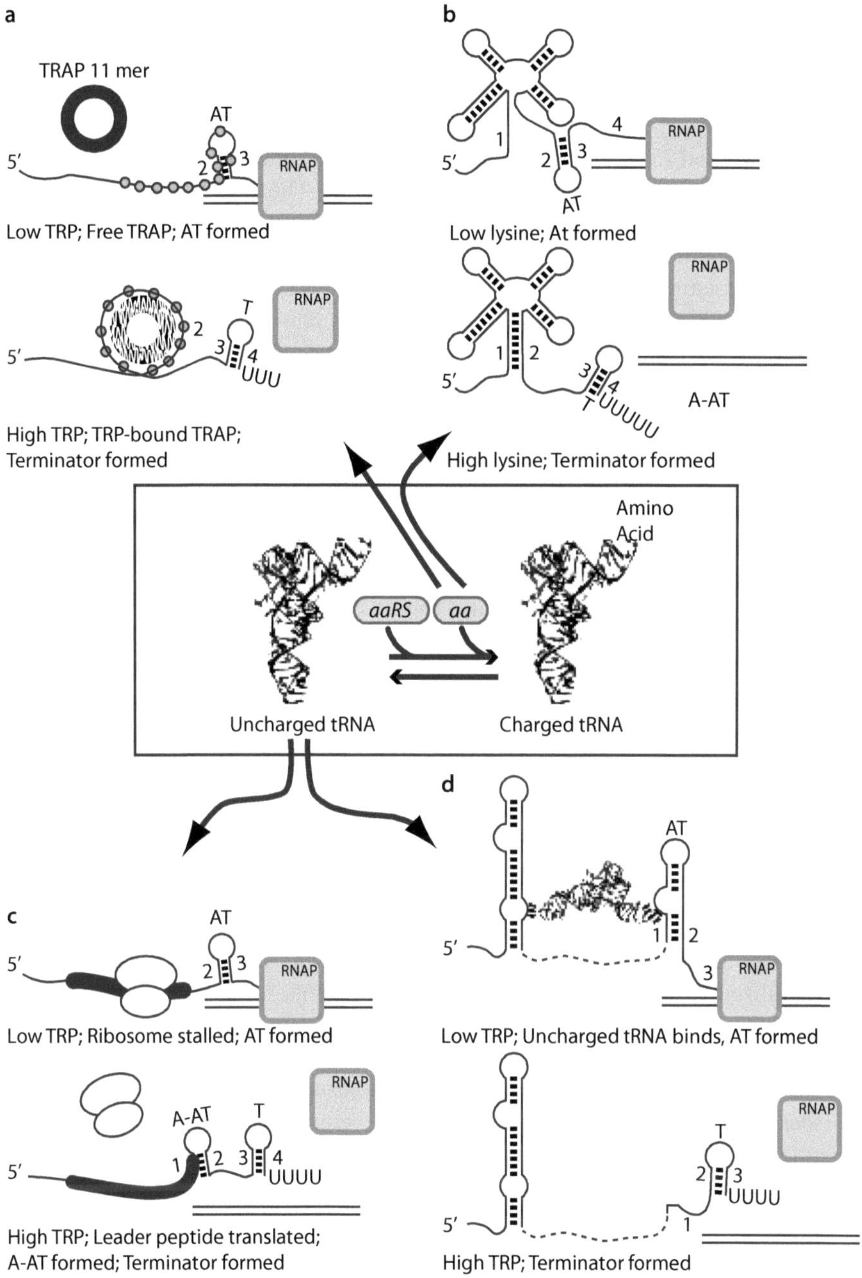
a
TRAP 11 mer
AT
5′
2
3
RNAP
Low TRP; Free TRAP; AT formed
T
RNAP
2
3
4
UUU
High TRP; TRP-bound TRAP;
Terminator formed
b
1
4
RNAP
2
3
AT
Low lysine; At formed
RNAP
1
2
3
4
T
UUUUU
A-AT
High lysine; Terminator formed
Amino
Acid
aaRS
aa
Uncharged tRNA
Charged tRNA
d
AT
5′
1
2
3
RNAP
Low TRP; Uncharged tRNA binds, AT formed
RNAP
T
2
3
UUUU
1
High TRP; Terminator formed
c
AT
5′
2
3
RNAP
Low TRP; Ribosome stalled; AT formed
RNAP
A-AT
T
1
2
3
4
UUUU
High TRP; Leader peptide translated;
A-AT formed; Terminator formed

3.2 Sensing of Amino Acids via tRNA Charging Ratios: Ribosome-Mediated Attenuation

Starting with post-initiation regulatory mechanisms that control expression of *Salmonella his* and *Escherichia coli trp* biosynthetic pathways, several tRNA-sensing mechanisms have been discovered (Yanofsky, 2000). The latter example will be briefly summarized herein. Five genes that encode for synthesis of tryptophan (*trpEDCBA*) from chorismate are co-expressed within a single *E. coli* operon (Landick *et al.*, 1996). Regulation of this *trp* operon involves detection of changes in tryptophan pools via multiple regulatory mechanisms. When intracellular levels of tryptophan become elevated, the homodimeric tryptophan repressor (TrpR) is activated to inhibit transcription initiation of the *trp* operon (Arvidsen *et al.*, 1986; Zhang *et al.*, 1994). Each subunit of the *trp* repressor is comprised of six alpha helices, of which two have been shown to bind DNA via a helix-turn-helix motif (Zhang *et al.*, 1987). Upon binding of two tryptophan molecules, a conformational change in TrpR is induced that allows the helix-turn-helix element to bind two adjacent grooves within the operator region, thereby preventing RNAP from initiating transcription (Arvidsen *et al.*, 1986). However, in addition to this initiation-based regulatory mechanism, the availability of charged $tRNA^{Trp}$ is also sensed by a post-initiation mechanism. Specifically, levels of charged $tRNA^{Trp}$ dictate whether transcription terminates within the 5' leader region of the *trp* operon, or whether transcription is allowed to proceed past the premature termination site for synthesis of the full transcript (reviewed in Landick *et al.*, 1996; Yanofsky, 2000; Gollnick and Babitzke, 2002). The 5' leader region contains the following structural elements: a small hairpin that acts as a functionally relevant transcriptional pause site, an intrinsic terminator hairpin, and a mutually exclusive antiterminator helix. Under initial conditions of tryptophan limitation, TrpR-mediated repression of the *trp* operon is relieved, yet transcription still prematurely terminates within the 5' leader region at the intrinsic terminator site due to the sufficient presence of charged $tRNA^{Trp}$. However, as tryptophan starvation continues, and as uncharged $tRNA^{Trp}$ accumulates, the antiterminator configuration is stabilized in the 5' leader region, which prevents formation of the premature termination site and allows synthesis of full-length *trp*. The $tRNA^{Trp}$-charging ratio therefore controls transcription attenuation for this regulatory system (**Fig 5C**). This is mechanistically achieved by coordinating transcription attenuation within the 5' leader region with translation of a short open reading frame. This 14-residue leader peptide includes two tandem tryptophan codons, which are critical for coupling the tRNA charging ratio to attenuation. Under conditions of replete tryptophan, ORF-associated ribosomes fully translate the leader peptide. Upon reaching the terminal codon of the leader peptide, the ribosomes prevent formation of the antiterminator helix and thereby allow formation of the downstream terminator hairpin. In contrast, under conditions of diminished tryptophan, the ORF-associated ribosomes stall at the tryptophan codons. The physical placement of the stalled ribosome on the nascent transcript is such that the default antiterminator helix is allowed to form, instead of the mutually exclusive terminator hairpin, thereby allowing synthesis of *trp* in an amino acid-dependent manner.

After discovery of the initial *trp* and *his* attenuation systems, remarkably similar regulatory systems were identified for control of other amino acid biosynthetic operons in Gram-negative bacteria (Landick *et al.*, 1996). Each of these different examples relies on ribosome stalling at specific codons as a means for controlling RNA folding pathways and preventing terminator formation within the nascent transcript. Interestingly, a variation of the ribosome-mediated regulatory mechanism has also been discovered for regulation of pyrimidine biosynthetic operons (*pyr*) in enteric bacteria (Turnbough and Switzer, 2008). For example, the *E. coli pyr* operon is preceded by a 158 bp leader region, which contains elements that resemble the *trp* regulatory mechanism, such as an intrinsic terminator, an alternative antiterminator hairpin, and a short open-reading frame. Kinetic analysis of *pyr* transcription revealed a pause site that was induced under conditions of low UTP levels. While the RNAP complex remains paused at this site, due to low UTP, ribosomes are allowed sufficient time to associate with the nascent transcript and for translation of the leader peptide. These ribosomes translate up to the stalled RNAP complex and prevent formation of the downstream terminator hairpin after the polymerase has escaped the pause site. However, under conditions of high UTP, the RNAP complex synthesizes the leader region without pausing and reaches the termination site before ribosomes can successfully translate the leader peptide, which thereby remain uncoupled from the transcription complex. Under these conditions, the RNAP complex terminates transcription. Therefore, ribosome-mediated attenuation mechanisms are not only utilized for sampling tRNA charging ratios during genetic control, although the full extent of these regulatory mechanisms remains to be determined.

3.3 Sensing of Amino Acids via tRNA Charging Ratios: Direct Sensing of Uncharged tRNAs

Another commonly used transcription attenuation mechanism that monitors the tRNA charging ratio is oftentimes referred to as T box antitermination (reviewed by Green *et al.*, 2010). These regulatory systems are named for a characteristic oligonucleotide sequence called the T box sequence, which is required for formation of an antiterminator hairpin element within 5' leader sequences or within intercistronic regions. As with most other transcription attenuation regulatory systems, this antiterminator element competes with a mutually exclusive adjacent terminator hairpin in order to exert control over downstream gene expression. The T box attenuation mechanism is widespread in Gram-positive bacteria and is primarily employed for regulation of aminoacyl-tRNA synthetase genes, although it is also oftentimes utilized for control of amino acid transporters and biosynthetic gene clusters (Vitreschak *et al.*, 2008). Indeed, the majority of the aminoacyl-tRNA synthetase genes in *B. subtilis* has been shown to be regulated by T box attenuation (reviewed in Grundy and Henkin, 2003; Green *et al.*, 2010). In this mechanism, the charging ratio of the cognate tRNA is sensed not through an indirect means, such as the ribosome-mediated mechanism exemplified by *E. coli trp*, but rather through direct association of the cognate uncharged tRNA with the T box-containing leader region (**Fig 5D**). Therefore, these cis-acting regulatory RNAs function effectively as tRNA sensors, which promote an antiterminator element only upon association with the cognate, uncharged tRNA. Importantly, accessory protein factors are not required for tRNA-directed attenuation; tRNA-mediated readthrough of transcription termination has been demonstrated in the absence of protein factors other than the RNA polymerase complex (Grundy *et al.*, 2002; Putzer *et al.*, 2002). Therefore, the tRNA-sensing leader region is likely to contain all of the necessary information content for tRNA selectivity and antitermination activity. Many examples of these tRNA-sensing regulatory RNAs have been identified (Vitreschak *et al.*, 2008; Gardner *et al.*, 2009). Presumably, the primary sequence and secondary structural elements that are conserved amongst these different examples are required for direct recognition of the correct uncharged tRNAs.

The conserved features of tRNA-sensing regulatory RNAs include several helical domains with characteristic bulges and loops, a terminator helix and the highly conserved T box sequence, which comprises part of the antiterminator element (Grundy and Henkin, 1993; Rollins *et al.*, 1997; Luo *et al.*, 1998; Vitreschak *et al.*, 2008). The first domain, 'helix I', oftentimes contains a kink-turn structural element (Klein *et al.*, 2001) at its base (Winkler *et al.*, 2001). These secondary structural motifs typically introduce a sharp angle between two adjacent helices, and oftentimes also serve as protein recognition determinants (Klein *et al.*, 2001; Schroeder *et al.*, 2010). The molecular function of this structural element is not yet known for tRNA-sensing attenuation by T box regulatory RNAs. Mutations in the kink-turn motif disrupt antitermination in vivo, but for reasons not completely understood, have a lesser effect on antitermination in vitro (Green *et al.*, 2010). Located above the kink-turn motif is an internal loop, called the specifier loop, which contains a loop E motif and a three-nucleotide sequence called the 'specifier sequence'. The loop E element is a secondary structural motif ubiquitous in biological RNAs (Dallas and Moore, 1997; Leontis and Westhof, 1998). In the context of the specifier loop it appears to assist proper placement of the specifier sequence nucleotides (Wang and Nikonowicz, 2010) for pairing with anticodon residues of the cognate tRNA (Grundy and Henkin, 1993; Grundy *et al.*, 2002; Grundy *et al.*, 1997). Given its important role in selecting the appropriate tRNA ligand, the specifier sequence is one of the most important specificity determinants for this overall attenuation mechanism. Other, less characterized portions of the leader region are also thought to play a role in mediating tRNA specificity, since mutations in the specifier sequence do not always correspond to a corresponding change in tRNA selectivity; however, the structural basis of these interactions remain to be determined.

The second major tRNA specificity determinant is located within a side-bulge of the antiterminator helix (Green *et al.*, 2010). This side-bulge forms a highly flexible structural feature that interacts through base-pairing interactions with the four terminal nucleotides of the uncharged tRNA molecule. This overall structural element (the internal bulged region of the antiterminator helix) is highly conserved and performs three important functions. First, by associating with the terminal tRNA residues, including the discriminator base, this side-bulge assists in selection of the appropriate tRNA. Second, pairing between the appropriate tRNA and the antiterminator side-bulge stabilizes the antiterminator, thereby preventing formation of the terminator hairpin. Finally, aminoacylated tRNAs do not associate with the antiterminator side-bulge, and therefore cannot productively stabilize the antiterminator helix. Therefore, the antiterminator element is directly responsible for sensing the presence or absence of an amino acid at the 3' tRNA terminus, regardless of the actual identity of the amino acid. While metabolite-sensing riboswitches function as receptors for individual metabolite ligands, the molecular characteristics of the T box antiterminator hairpin allow it to function instead as a general platform for monitoring of intracellular amino acids. In other words, by sensing the *absence* of the appropriate amino acid, the T box antiterminator can be generally employed for regulation of any of the individual amino acid classes.

Although almost all of T box regulatory RNAs that have been identified share the basic features briefly outlined above. However, T box regulatory RNAs in Actinobacteria appear to control gene expression by altering the efficiency of translation initiation, rather than transcription attenuation (Seliverstov *et al.*, 2005; Vitreschak *et al.*, 2008). The mechanistic details of this interesting subclass have not yet been described.

3.4 Sensing of Amino Acids via tRNA Charging Ratios: mRNA-Binding tRNA Synthetases

In *E. coli*, expression of certain aminoacyl-tRNA synthetases is regulated by the cognate tRNA charging ratio, but through a mechanism that is independent of T box antitermination or ribosome-mediated attenuation. In these instances, the aminoacyl-tRNA synthetase enzyme associates with a cis-acting regulatory RNA and affects expression of downstream genes. For example, threonyl-tRNA synthetase (ThrRS), which catalyzes attachment of threonine to its cognate tRNA, can associate with a cis-acting regulatory region located within the mRNA that encodes its own expression (Butler *et al.*, 1986). Association of ThrRS to the *E. coli thrS* mRNA leader region results in decreased translation. The leader region of the *thrS* mRNA is capable of forming four distinct helices, the first of which contains the ribosome binding site and the translation initiation codon (Moine *et al.*, 1988). Interestingly, two of these stem-loop structural elements closely resemble the anticodon arm of $tRNA^{Thr}$. Based on a combination of in vitro and in vivo data, the ThrRS protein has been demonstrated to associate directly with this anticodon-like element within the mRNA; binding of ThrRS is accompanied by a decrease in translation efficiency. Therefore, the mRNA leader region competes through molecular mimicry against tRNA pools for binding to the ThrRS enzyme, again coupling tRNA charging ratios to regulation of gene expression.

3.5 Sensing of Metabolites by RNA-Binding Proteins: Amino Acids

In addition to the indirect regulatory mechanisms that monitor amino acid concentrations by sensing changes in tRNA charging ratios, there is an increasingly diverse catalog of RNA-binding proteins that have been found to regulate bacterial gene expression through post-initiation mechanisms. Indeed, a subset of these signal-responsive RNA-binding proteins is employed for sensing amino acids, and is likely to be functionally equivalent with the tRNA-sensing mechanisms listed above. In that regard, the *trp* biosynthetic cluster offers another excellent example of a post-initiation regulatory mechanism paradigm.

In *B. subtilis*, the *trp* operon is comprised of six of the seven genes required for tryptophan synthesis (*trpDCFBA*) and is located between aromatic amino acid and histidine biosynthesis operons. *trpG*, the seventh *trp* gene involved in generating tryptophan, is found on a separate operon encoding genes for folic acid (Slock *et al.*, 1990). This arrangement, which is unique from the single-operon *trp* arrangement observed in *E. coli,* allows for multiple levels of genetic regulation. One mode of *trp* operon regulation is through transcription attenuation, where competition between terminator and antiterminator structural elements governs gene expression. In response to varying intracellular tryptophan concentrations, an RNA-binding protein called 'TRAP' (for *trp* RNA-binding attenuation protein) influences the terminator/antiterminator 'decision' (Babitzke and Yanofsky, 1993). This protein binds to tryptophan molecules, which in turn modifies TRAP RNA-binding activity **(Fig 5A)**. Specifically, TRAP associates into a homo-oligomeric complex of 11 subunits; each of the TRAP subunits binds a single tryptophan. The tryptophan-bound TRAP complex is then activated for binding to the *trp* mRNA leader region. Each TRAP monomer binds a (G/U)AG sequence in the leader region. In total, binding of the (G/U)AG trinucleotide sequences to TRAP wraps the mRNA leader region around the doughnut-shaped TRAP complex. This prevents formation of an antiterminator helix and instead allows formation of an intrinsic terminator hairpin in the leader region. In the absence of tryptophan-bound TRAP, the antiterminator is adopted as the default conformation; therefore, amino acids are coupled to expression of the *trp* biosynthesis operon via TRAP. However, (G/U)AG trinucleotide repeats are also found as part of cis-acting regulatory RNAs for other tryptophan-related operons (*e.g., trpG, trpP, ycbK*). In these instances, tryptophan-bound TRAP associates with the appropriate mRNAs and regulates translation efficiency. For example, amino acid-bound TRAP binds the *trpE* leader region to promote formation of a ribosome binding site-sequestering helix, which inhibits ribosome binding in a TRAP-dependent manner. Similarly, amino acid-bound TRAP regulates translation of the *trpG* gene by directly competing with ribosomes for association with the ribosome binding site.

Interestingly, despite the clear differences, there is still one similarity between the regulation of *E. coli* and *B. subtilis trp* pathways. Specifically, the charging ratio of $tRNA^{Trp}$ is still also an important signal for the *B. subtilis trp* operon, as it is for *E. coli trp*, albeit through a different mechanism. The *rtpA* gene encodes a protein that antagonizes TRAP RNA-binding activity and is therefore called anti-TRAP

(AT) (Valbuzzi *et al.*, 2002; Cruz-Vera *et al.*, 2008). Interestingly, the leader region of *rtpA* includes a T box transcription attenuation regulatory system for sensing the charging ratio of $tRNA^{Trp}$. Therefore, in total, multiple tryptophan-responsive, post-initiation regulatory mechanisms corroborate to control *trp* homeostasis in *B. subtilis*.

3.6 Sensing of Metabolites by RNA-Binding Proteins: Carbohydrates and Nucleotides

In addition to amino acid-sensing RNA-binding proteins, there are many other examples of metabolite-sensing regulatory proteins. In this section, we summarize a few representative examples, which respond to sugars and nucleobases, respectively.

In Gram-positive bacteria, many sugar catabolism operons are regulated by post-initiation regulatory proteins. These RNA-binding proteins have been best characterized in *B. subtilis* (GlcT, LicT, SacY, and SacT) (Schilling *et al.*, 2004). Glucose, the preferred carbon and energy source for *B. subtilis*, is transported by the phosphoenolpyruvate:sugar phosphotransferase system (PTS). The *ptsGHI* operon encodes for PTS components and contains a leader region with a terminator helix. Similar to other transcription attenuation systems, a portion of the terminator folds into an alternative RNA structure, referred to as the RNA antiterminator (RAT) element. When glucose is abundant in *B. subtilis*, the RAT element is formed, thereby allowing expression of the *ptsGHI* operon (Stülke *et al.*, 1997; Stülke, 2002; Schilling *et al.*, 2004). Specifically, the RNA-binding protein, GlcT, is activated by the presence of glucose and associates with the RAT element to promote antitermination. GlcT belongs to a family of antiterminators that each contain PTS regulation domains (PRD) and an RNA-binding domain at the N-terminus (Stülke, 2002). During transport of glucose, a glucose-specific enzyme II PTS component (EII^{Glc}) preferentially phosphorylates the glucose molecules. However, under glucose-limiting conditions, EII^{Glc} instead transfers a phosphate group to a PRD domain of GlcT, which results in inhibition of RNA-binding activity (Amster-Choder and Wright, 1992; Stülke *et al.*, 1998; Schmalisch *et al.*, 2003). Other RAT-binding proteins, such as LicT and SacY/SacT, promote antitermination of genes involved in β-glucan utilization and sucrose catabolism, respectively (Schnetz *et al.*, 1996; Stülke, 2002). Nonetheless, in all instances, sugar availability is communicated to the RNA-binding domains by EII-mediated phosphorylation of PRD domains; the posttranslational modification of these domains ultimately governs RAT-binding activity.

In addition to detecting amino acids and sugars, certain RNA-binding proteins also sense nucleotides, amines, or nitrates, among many other potential ligands. In particular, many nucleotide biosynthesis operons are subjected to a diverse array of post-initiation regulatory control mechanisms (reviewed in Turnbough and Switzer, 2008; Switzer, 2009). For example, we briefly mentioned earlier in this chapter how certain pyrimidine biosynthesis operons are subjected to ribosome-mediated attenuation control. In contrast, other pyrimidine biosynthesis regulatory mechanisms include the use of signal-responsive, RNA-binding proteins. The best characterized example is the pyrimidine biosynthesis operon in *B. subtilis*, which is responsive to intracellular pyrimidine levels and is regulated by the PyrR protein. Interestingly, each of the first three genes of the operon (*pyrR, pyrP, pyrB*) is preceded by a characteristic signal-responsive, cis-acting regulatory RNA element. Specifically, each of these cis-acting regulatory RNAs includes an intrinsic terminator hairpin and the potential to form a mutually exclusive, characteristic antiterminator helix. A single promoter precedes the entire operon; therefore the sum of these attenuation-based regulatory systems offers several tandem checkpoints for sensing of pyrimidine levels. The PyrR regulatory protein becomes proficient to bind RNA once it associates with uridine nucleotide ligands (UMP and UTP). The uridine nucleotide-bound protein can then bind at a characteristic site just upstream of the putative antiterminator helix. This prevents antiterminator formation and instead allows folding of the intrinsic terminator element. Binding of uridine nucleotides to PyrR is antagonized by guanosine nucleotides (GMP, GDP, and GTP); therefore, elevated guanosine levels help reduce the affinity of PyrR for the *pyr* leader region, thereby allowing for formation of the default antiterminator hairpin element. The sampling of the ratio of uridine to guanosine nucleotides thereby ensures the robust metabolic regulation of the *pyr* operon. It is important to note that PyrR represents only a single example of a post-initiation regulatory strategy for control of nucleotide metabolism. In fact, many other mechanistically diverse examples have also been discovered (Turnbough and Switzer, 2008).

4 Concluding Remarks

This chapter is intended to serve as a cursory survey of some of the diverse post-initiation regulatory strategies that have been discovered in bacteria. Our brief overview of these various mechanisms is incomplete both in full coverage of the overall types of regulatory approaches, and with the molecular details of the respective mechanisms. However, it should be abundantly clear from this brief overview that post-initation regulatory strategies are frequently utilized in bacteria, and that any given bacterial operon is just as likely to be regulated by post-initiation mechanisms as it is by transcription initiation proteins. Also, many of the abovementioned regulatory strategies are mechanistically straightforward, and it is reasonable to expect that variations of these mechanisms will also be discovered in eukaryotic organisms. Therefore, the exploration of post-initiation regulatory mechanisms in bacteria is an important endeavor that will continue to reveal fundamental insights into general metabolism and genetic regulation.

References

Amster-Choder O, Wright A (1992) Modulation of the dimerization of a transcriptional antiterminator protein by phosphorylation. Science 257: 1395–1398

Artsimovitch I, Landick R (2000) Pausing by bacterial RNA polymerase is mediated by mechanistically distinct classes of signals. Proc Natl Acad Sci USA 97: 7090–7095

Arvidson DN, Bruce C, Gunsalus RP (1986) Interaction of the Trp repressor with its ligand, L-tryptophan. J Biol Chem 261: 238–243

Babitzke P, Yanofsky C (1993) Reconstitution of *Bacillus subtilis trp* attenuation in vitro with TRAP, the trp RNA-binding attenuation protein. Proc Natl Acad Sci USA 90: 133–137

Baker CS, Morozov I, Suzuki K, Romeo T, Babitzke P (2002) CsrA regulates glycogen biosynthesis by preventing translation of *glgC* in *Escherichia coli*. Mol Microbiol 44: 1599–1610

Barrick JE, Breaker RR (2007) The distributions, mechanisms, and structures of metabolite-binding riboswitches. Genome Biol 8: R239

Bechhofer DH (2009) Messenger RNA decay and maturation in *Bacillus subtilis*. Prog Mol Biol Transl Sci 85: 231–271

Belasco JG (2010) All things must pass: contrasts and commonalities in eukaryotic and bacterial mRNA decay. Nat Rev Mol Cell Biol 11: 467–478

Blount KF, Breaker RR (2006) Riboswitches as antibacterial drug targets. Nat Biotechnol 24: 1558–1564

Brown L, Elliott T (1997) Mutations that increase expression of the *rpoS* gene and decrease its dependence on *hfq* function in *Salmonella typhimurium*. J Bacteriol 179: 656–662

Browning DR, Busby SJ (2004) The regulation of bacterial transcription initiation. Nat Rev Microbiol 2: 57–65

Burmann BM, Schweimer K, Luo X, Wahl MC, Stitt BL, Gottesman ME, Rösch P (2010) A NusE:NusG complex links transcription and translation. Science 328: 501–504

Butler JS, Springer M, Dondon J, Grunberg-Manago M (1986) Posttranscriptional autoregulation of *Escherichia coli* threonyl tRNA synthetase expression in vivo. J Bacteriol 165:198–203

Collins JA, Irnov I, Baker S, Winkler WC (2007) Mechanism of mRNA destabilization by the *glmS* ribozyme. Genes Dev 21: 3356–3368

Condon C (2010) What is the role of RNase J in mRNA turnover? RNA Biol 7: 316–321

Cruz-Vera LR, Gong M, Yanofsky C (2008) Physiological effects of anti-TRAP protein activity and tRNA(Trp) charging on trp operon expression in *Bacillus subtilis*. J Bacteriol 190: 1937–1945

Daldrop P, Reyes FE, Robinson DA, Hammond CM, Lilley DM, Batey RT, Brenk R (2011) Novel ligands for a purine riboswitch discovered by RNA-ligand docking. Chem Biol 18: 324–335

Dallas A, Moore PB. (1997) The loop E-loop D region of *Escherichia coli* 5S rRNA: the solution structure reveals an unusual loop that may be important for binding ribosomal proteins. Structure 5: 1639–1653

Gardner PP, Daub J, Tate JG, Nawrocki EP, Kolbe DL, Lindgreen S, Wilkinson AC, Finn RD, Griffiths-Jones S, Eddy SR, Bateman A (2009) Rfam: updates to the RNA families database. Nucleic Acids Res 37: D136–140

Gilbert SD, Mediatore SJ, Batey RT (2006) Modified pyrimidines specifically bind the purine riboswitch. J Am Chem Soc 128: 14214–14215

Gollnick P, Babitzke P (2002) Transcription attenuation. Biochem Biophys Acta 1577: 240–250

Gottesman S (2004) The small RNA regulators of *Escherichia coli*: roles and mechanisms. Annu Rev Microbiol 58: 303–328

Gottesman S, Storz G (2010) Bacterial small RNA regulators: Versatile roles and rapidly evolving variations. Cold Spring Harb Perspect Biol 1: 1–16

Green NJ, Grundy FJ, Henkin TM (2010) The T box mechanism: tRNA as a regulatory molecule. FEBS Lett 584: 318–324

Greive SJ, von Hippel PH (2005) Thinking quantitatively about transcriptional regulation. Nat Rev Mol Cell Biol 6: 221–232

Grundy FJ, Henkin TM (2003) The T box and S box transcription termination control systems. Front Biosci 8: d20–31

Grundy FJ, Henkin TM (1993) tRNA as a positive regulator of transcription antitermination in *B. subtilis*. Cell 74: 475–482

Grundy FJ, Hodil SE, Rollins SM, Henkin TM (1997) Specificity of tRNA-mRNA interactions in *Bacillus subtilis tyrS* antitermination. J Bacteriol 179: 2587–2594

Grundy FJ, Winkler WC, Henkin TM (2002) tRNA-mediated transcription antitermination in vitro: Codon–anticodon pairing independent of the ribosome. Proc Natl Acad Sci USA 99: 11121–11126.

Herbert KM, Greenleaf WJ, Block SM (2008) Single-molecule studies of RNA polymerase: motoring along. Annu Rev Biochem 77: 148–176

Irnov I, Winkler WC (2010) A regulatory RNA required for antitermination of biofilm and capsular polysaccharide operons in Bacillales. Mol Microbiol 76: 559–575

Johansen LE, Nygaard P, Lassen C, Agerso Y, Saxild HH (2003) Definition of a second *Bacillus subtilis* pur regulon comprising the pur and xpt-pbuX operons plus pbuG, nupG (yxjA), and pbuE (ydhL). J Bacteriol 185: 5200–5209

Kim JN, Blount KF, Lim J, Link KH, Breaker R. (2009) Design and antimicrobial action of purine analogs that bind guanine riboswitches. ACS Chem Biol 4: 915–27

Klein DJ, Ferré-D'Amaré AR (2006) Structural basis of *glmS* ribozyme activation by glucosamine-6-phosphate. Science 313: 1752–1756

Klein DJ, Schmeing TM, Moore PB, Steitz TA (2001) The kink-turn: a new RNA secondary structure motif. EMBO J 20: 4214–4221

Kobayashi K, Ehrlich SD, Albertini A, Amati G, Andersen KK, Arnaud M, Asai K, Ashikaga S, Aymerich S, Bessieres P, Boland F, Brignell SC, Bron S, Bunai K, Chapuis J, Christiansen LC, Danchin A, Débarbouille M, Dervyn E, Deuerling E, Devine K, Devine SK, Dreesen O, Errington J, Fillinger S, Foster SJ, Fujita Y, Galizzi A, Gardan R, Eschevins C, Fukushima T, Haga K, Harwood CR, Hecker M, Hosoya D, Hullo MF, Kakeshita H, Karamata D, Kasahara Y, Kawamura F, Koga K, Koski P, Kuwana R, Imamura D, Ishimaru M, Ishikawa S, Ishio I, Le Coq D, Masson A, Mauël C, Meima R, Mellado RP, Moir A, Moriya S, Nagakawa E, Nanamiya H, Nakai S, Nygaard P, Ogura M, Ohanan T, O'Reilly M, O'Rourke M, Pragai Z, Pooley HM, Rapoport G, Rawlins JP, Rivas LA, Rivolta C, Sadaie A, Sadaie Y, Sarvas M, Sato T, Saxild HH, Scanlan E, Schumann W, Seegers JF, Sekiguchi J, Sekowska A, Séror SJ, Simon M, Stragier P, Studer R, Takamatsu H, Tanaka T, Takeuchi M, Thomaides HB, Vagner V, van Dijl JM, Watabe K, Wipat A, Yamamoto H, Yamamoto M, Yamamoto Y, Yamane K, Yata K, Yoshida K, Yoshikawa H, Zuber U, Ogasawara N (2003) Essential *Bacillus subtilis* genes. Proc Natl Acad Sci USA 100: 4678–4683

Kulshina N, Baird NJ, Ferré-D'Amaré AR (2009) Recognition of the bacterial second messenger cyclic diguanylate by its cognate riboswitch. Nat Struct Mol Biol 16: 1212–1217

Landick R, Turnbough CL, Yanofsky C (1996) Transcription Attenuation. In: Neidhardt FC, Curtiss III R, Ingraham JL, Lin ECC, Low KB, Magasanik B, Reznikoff WS, Riley M, Schaechter M, Umbarger HE (eds) *Escherichia* and *Salmonella* cellular and molecular biology. ASM Press, Washington, DC pp 1263–1286

Landick R (2006) The regulatory roles and mechanism of transcriptional pausing. Biochem Soc Trans 34: 1062–1066

Larson MH, Landick R, Block SM (2011) Single-molecule studies of RNA polymerase: one singular sensation, every little step it takes.

Lease RA, Cusick ME, Belfort M (1998) Riboregulation in *Escherichia coli* DsrA RNA acts by RNA:RNA interactions at multiple loci. Proc Natl Acad Sci 95: 12456–12461

Lemay JF, Desnoyers G, Blouin S, Heppell B, Bastet L, St-Pierre P, Massé E, Lafontaine DA (2011) Comparative study between transcriptionally- and translationally-acting adenine riboswitches reveals key differences in riboswitch regulatory mechanisms. PLoS Genet 7: e1001278

Lemay JF, Lafontaine DA (2007) Core requirements of the adenine riboswitch aptamer for ligand binding. RNA 13:339–350

Lemay JF, Penedo JC, Tremblay R, Lilley DMJ, Lafontaine DA (2006) Folding of the adenine riboswitch. Chem Biol 13:857ñ868
Leontis NB, Westhof E (1998) A common motif organizes the structure of multi-helix loops in 16S and 23S ribosomal RNAs. J Mol Biol 283: 571–583
Ling B, Dong L, Zhang R, Wang Z, Liu Y, Liu C (2009) Theoretical studies on the interaction of modified pyrimidines and purines with purin riboswitch. J Mol Graph Model 28: 37–45
Liu MY, Gui G, Wei B, Preston JF, Oakford L, Yuksel U, Giedroc DP, Romeo T (1997) The RNA molecule CsrB binds to the global regulatory protein CsrA and antagonizes its activity in *Escherichia coli*. J Biol Chem 272: 17502–17510
Liu MY, Romeo T (1997) The global regulator CsrA of *Escherichia coli* is a specific mRNA-binding protein. J Bacteriol 179: 4639–4642
Luo D, Condon C, Grunberg-Manago M, Putzer H (1998) In vitro and in vivo secondary structure probing of the *thrS* leader in *Bacillus subtilis*. Nucleic Acids Res 26: 5379–5387
Majdalani N, Cunning C, Sledjeski D, Elliott T, Gottesman S (1998) DsrA RNA regulates translation of RpoS message by an anti-antisense mechanism, independent of its action as an antisilencer of transcription. Proc Natl Acad Sci 95: 12462–12467
Majdalani N, Vanderpool CK, Gottesman S (2005) Bacterial small RNA regulators. Crit Rev Biochem Mol 40: 93–113
Mandal M, Breaker RR (2004) Adenine riboswitches and gene activation by disruption of a transcription terminator. Nat Struct Mol Biol 11: 29ñ35
Moine H, Romby P, Springer M, Grunberg-Manago M, Ebel JP, Ehresmann C, Ehresmann B (1988) Messenger RNA structure and gene regulation at the translational level in *Escherichia coli*: the case of threonine:tRNAThr ligase. Proc Natl Acad Sci US 85: 7892–7896
Møller T, Franch T, Højrup P, Keene DR, Bächinger HP, Brennan FG, Valentin-Hansen P (2002) Hfq: a bacterial Sm-like protein that mediates RNA–RNA interaction. Mol Cell 9: 23–30
Morita T, Maki K, Aiba H (2005) RNase E-based ribonucleoprotein complexes: mechanical basis of mRNA destabilization mediated by bacterial noncoding RNAs. Genes Dev 19: 2176–2186
Mooney RA, Darst SA, Landick R (2005) Sigma and RNA polymerase: an on-again, off-again relationship? Mol Cell 21: 335–345
Mulhbacher J, Brouillette E, Allard M, Fortier LC, Malouin F, Lafontaine DA (2010) Novel riboswitch ligand analogs as selective inhibitors of guanine-related metabolic pathways. PLos Pathog 6: e1000865
Navarro Llorens JM, Tormo A, Martinez-Garcia E (2010) Stationary phase in Gram-negative bacteria. FEMS Microbiol Rev 34: 476–495
Nudler E, Gottesman ME (2002) Transcription termination and anti-termination in *E. coli*. Genes Cells 7: 755–768
Pan T, Artsimovitch I, Fang X, Landick R, Sosnick TR (1999) Folding of a large ribozyme during transcription and the effect of the elongation factor NusA. Proc Natl Acad Sci USA 96: 9545–9550
Pfeiffer V, Papenfort K, Lucchini S, Hinton JC, Vogel J (2009) Coding sequence targeting by MicC RNA reveals bacterial mRNA silencing downstream of translational initiation. Nat Struct Mol Biol 16: 840–846
Putzer H, Condon C, Brechemier-Baey D, Brito R, Grunberg-Manago M (2002) Transfer RNA-mediated antitermination in vitro. Nucleic Acids Res 30: 3026–3033
Rieder R, Lang K, Graber D, Micura R (2007) Ligand-induced folding of the adenosinedeaminase A-riboswitch and implications on riboswitch translational control. Chembiochem 8:896–902
Rieder U, Kreutz C, Micura R (2010) Folding of a transcriptionally acting preQ1 riboswitch. Proc Natl Acad Sci U S A 107: 10804–10809
Roberts JW, Shankar S, Filter JJ (2008) RNA polymerase elongation factors. Annu Rev Microbiol 62: 211–233
Rollins SM, Grundy FJ, Henkin TM (1997) Analysis of cis-acting sequence and structural elements required for antitermination of the *Bacillus subtilis tyrS* gene. Mol Microbiol 25: 411–421
Roth A, Breaker RR (2009) The structural and functional diversity of metabolite-binding riboswitches. Annu Rev Biochem 78: 305–34

Schilling O, Langbein I, Müller M, Schmalisch MH, Stülke J (2004) A protein-dependent riboswitch controlling *ptsGHI* operon expression in *Bacillus subtilis*: RNA structure rather than sequence provides interaction specificity. Nucleic Acids Res 32: 2853–2864
Schnetz K, Stülke J, Gertz S, Krüger S, Krieg M, Hecker M, Rak B (1996) LicT, a *Bacillus subtilis* transcriptional antiterminator protein of the BglG family. J Bacteriol 178: 1971–1979
Schmalisch MH, Bachem S, Stülke J (2003) Control of the Bacillus subtilis antiterminator protein GlcT by phosphorylation. Elucidation of the phosphorylation chain leading to inactivation of GlcT. J Biol Chem 278: 51108–51115
Schroeder KT, McPhee SA, Ouellet J, Lilley DM (2010) A structural database for k-turn motifs in RNA. RNA 16: 1463–1468
Seliverstov AV, Putzer H, Gelfand MS, Lyubetsky VA (2005) Comparative analysis of RNA regulatory elements of amino acid metabolism genes in Actinobacteria. BMC Microbiol 5: 54
Serganov A, Patel DJ (2007) Ribozymes, riboswitches and beyond: regulation of gene expression without proteins. Nat Rev Genet 8:776–790
Serganov A, Yuan YR, Pikovskaya O, Polonskaia A, Malinina L, Phan AT, Hobartner C, Micura R, Breaker RR, Patel DJ (2004) Structural basis for discriminative regulation of gene expression by adenine- and guanine-sensing mRNAs. Chem Biol 11: 1729–41
Shimaoka M, Takenaka Y, Mihara Y, Kurahashi O, Kawasaki H, Matsui H (2006) Effects of xapA and guaA disruption on inosine accumulation in *Escherichia coli*. Biosci Biotechnol Biochem 70: 3069–3072
Sledjeski D, Gottesman S (1995) A small RNA acts as an antisilencer of the H-NS-silenced *rcsA* gene of *Escherechia coli*. Proc Natl Acad Sci 92: 2003–2007
Sledjeski DD, Whitman C, Zhang A (2001) Hfq is necessary for riboregulation by the untranslated RNA DsrA. J Bacteriol 183: 1997–2005
Slock J, Stahly DP, Han CY, Six EW, Crawford IP (1990) An apparent *Bacillus subtilis* folic acid biosynthetic operon containing pab, an amphibolic *trpG* gene, a third gene required for synthesis of para-aminobenzoic acid, and the dihydropteroate synthase gene. J Bacteriol 172: 7211–7226
Smith KD, Lipchock SV, Ames TD, Wang J, Breaker RR, Strobel SA (2009) Structural basis of ligand binding by a c-di-GMP riboswitch. Nat Struct Mol Biol 16: 1218–1223
Stülke J (2002) Control of transcription termination in bacteria by RNA-binding proteins that modulate RNA structures. Arch Microbiol 177: 433–440
Stülke J, Arnaud M, Rapoport G, Martin-Verstraete (1998) PRD--a protein domain involved in PTS-dependent induction and carbon catabolite repression of catabolic operons in bacteria. Mol Microbiol 28: 865–874
Stülke J, Martin-Verstraete I, Zagorec M, Rose M, Klier A, Rapoport G (1997) Induction of the *Bacillus subtilis ptsGHI* operon by glucose is controlled by a novel antiterminator, GlcT. Mol Microbiol 1997 25: 65–78
Sudarsan N, Lee ER, Weinberg Z, Moy RH, Kim JN, Link KH, Breaker RR (2008) Riboswitches in eubacteria sense the second messenger cyclic di-GMP. Science 321: 411–413
Switzer RL (2009) Discoveries in bacterial nucleotide metabolism. J Biol Chem 284: 6585–6594
Timmermans J, Van Melderen L (2010) Post-transcriptional global regulation by CsrA in bacteria. Cell Mol Life Sci 67: 2897–2908
Turnbough CL Jr, Switzer RL (2008) Regulation of pyrimidine biosynthetic gene expression in bacteria: repression without repressors. Microbiol Mol Biol Rev 72: 266–300
Valbuzzi A, Gollnick P, Babitzke P, Yanofsky C (2002) The anti-trp RNA-binding attenuation protein (Anti-TRAP), AT, recognizes the tryptophan-activated RNA binding domain of the TRAP regulatory protein. J Biol Chem 277: 10608–10613
Vassylyev DG (2009) Elongation by RNA polymerase: a race through roadblocks. Curr Opin Struct Biol 19: 691–700
Vitreschak AG, Mironov AA, Lyubetsky VA, Gelfand MS (2008) Comparative genomic analysis of T-box regulatory systems in bacteria. RNA 14: 717–735
Wang J, Nikonowicz EP (2010) Solution structure of the K-turn and Specifier Loop Domains from the *Bacillus subtilis tyrS* T-box leader RNA. J Mol Biol 408: 99–117
Waters LS, Storz G (2009) Regulatory RNAs in bacteria. Cell 136: 615–628

Wei BL, Brun-Zinkernagel AM, Simecka JW, Pruss BM, Babitzke P, Romeo T (2001) Positive regulation of motility and *flhDC* expression by the RNA-binding protein CsrA of *Escherichia coli*. Mol Microbiol 40: 245–256

Weisberg RA, Gottesman ME (1999) Processive antitermination. J Bacteriol 181: 359–367

Wickiser JK, Cheah MT, Breaker RR, Crothers DM (2005a) The kinetics of ligand binding by an adenine-sensing riboswitch. Biochemistry 44: 13404–13414

Wickiser JK, Winkler WC, Breaker RR, Crothers D (2005b) The speed of RNA transcription and metabolite binding kinetics operate an FMN riboswitch. Mol Cell 18:49–60

Winkler WC (2007) RNA-Mediated Regulation in *Bacillus subtilis*. In: Graumann P (ed) *Bacillus*: Cellular and Molecular Biology. Caister Academic Press, Friedberg pp 167–214

Winkler WC, Grundy FJ, Murphy BA, Henkin TM (2001) The GA motif: An RNA element common to bacterial antitermination systems, rRNA, and eukaryotic RNAs. RNA 7: 1165–1172

Winkler WC, Nahvi A, Roth A, Collins JA, Breaker RR (2004) Control of gene expression by a natural metabolite-responsive ribozyme. Nature 428: 281–286

Yanofsky C (2000) Transcription attenuation: once viewed as a novel regulatory strategy. J Bacteriol 182: 1–8

Zhang RG, Joachimiak A, Lawson CL, Schevitz RW, Otwinowski Z, Sigler PB (1987) The crystal structure of trp aporepressor at 1.8 A shows how binding tryptophan enhances DNA affinity. Nature 327: 591–97

Zhang A, Wassarman KM, Rosenow C, Tjaden BC, Storz G, Gottesman S (2003) Global analysis of small RNA and mRNA targets of Hfq. Mol Microbiol 50: 1111–1124

Zhang H, Zhao D, Revington M, Lee W, Jia X, Arrowsmith C, Jardetzky O (1994) The solution structures of the trp repressor-operator DNA complex. J Mol Biol 238:592–614

Chapter 12
Bioinformatics of Bacterial sRNAs and Their Targets

Rolf Backofen*

1 Computational Detection of Bacterial sRNAs

1.1 Definition of RNA Families

In analogy to proteins, non-coding RNAs (ncRNAs) can be grouped into *families* according to their evolutionary relationships. However, the sequence of ncRNAs is less conserved compared to proteins. Hence, the conservation of the secondary structure has to be considered for the definition of the family. Furthermore, it implies that standard approaches for defining protein families (such as Hidden Markov Models in the Pfam database (Bateman *et al.*, 2004)) cannot be applied. Instead, stochastic context-free grammars (SCFG) have to be used to model a common distribution over sequence *and* secondary structure. In this way, a SCFG can be used to model an RNA family. The Rfam database (Gardner *et al.*, 2009) is the largest collection of ncRNA families, where each RNA family is encoded by a SCFG of a specific form called a *covariance model*. The covariance model of a certain family is generated from a multiple alignment of RNA sequences belonging to that family (called *seed* alignment in Eddy (2002)) and represents a SCFG-based structural profile. Thus, this SCFG model of the RNA family can be used to search *and* align new family members. The full alignment listed in the Rfam database contains for each family the known members that were found this way. The tasks of building a SCFG from a multiple alignment as well as searching and aligning new members can be done using the tools cmbuild, cmsearch and cmalign from the Infernal package (Eddy, 2002; Nawrocki *et al.*, 2009).

Another method for searching members of an RNA family given by a multiple alignment is implemented ERPIN (Gautheret and Lambert, 2001; Lambert *et al.*, 2004). The input alignment is used to generate a secondary structure profile. For

* Chair of Bioinformatics, Albert-Ludwigs-Universität Freiburg

single-stranded regions, usually log-odds sequence profiles are used. For structured regions, the authors introduce helical profiles, which provide log-odds scores for every possible base pair combination.

Covariance models have also been used to directly determine RNA motifs from a set of unaligned sequences in the CMfinder approach (Yao *et al.*, 2006). To greatly oversimplify, the approach can be seen as an adaptation of the expectation maximization approaches (EM) used in sequence motif finding to the problem of finding structural motifs.

Bacterial non-coding RNAs are often called small regulatory RNAs (sRNAs). Finding members of known bacterial sRNA families is the first step when analyzing a bacterial genome. However, albeit there are already 269 bacterial sRNA families that can be found in the Rfam database (Release 10.0), it is likely that a major portion of bacterial sRNAs in a specific bacterial genome will not be found by searching the Rfam database. There are two possible reasons. First, the sensitivity of the search might be too low because of great evolutionary distances, and second there are many new bacterial non-coding RNA families not yet listed in Rfam. For the later task of finding new and unannotated non-coding RNAs in a specific genome, specialized computational ncRNA gene finders have been introduced, which will be handled in Subsection 1.1.3. The following subsection will deal with the detection of homologous RNAs, which is an important task for both finding orthologs of annotated non-coding RNAs as well as for identifying new non-coding RNAs. Again, methods successful in the analysis of protein sequences like sequence alignment are of limited use due to the low sequence conservation. Instead, sequence-structure alignment methods have to be used, which will be described in the next subsection.

1.2 Detection of Homologous Structural RNAs

Structural non-coding RNA evolves faster in sequence than in structure. This implies that we are not only searching for an alignment of the sequences, but also for a consensus structure, which is a secondary structure over the alignment columns that represents conserved base pairs. This also implies that base pairs might be conserved but not the associated bases. Such sequentially unconserved bases in base pairs are referred to as *compensatory base pair mutations*, since the mutation in the first base participating in the base pair is compensated by a mutation in the second base such that the base pair is maintained. When these compensatory mutations occur in the context of conserved structured regions, they are a strong indication for the conservation of structure. Figure 1 shows the seed alignment from Rfam (Release 10.0) of the bacterial sRNA DsrA and the associated compensatory mutations.

Most approaches used for the analysis and annotation of ncRNAs require the prediction of a secondary structure, which can be carried out using standard tools like Mfold (Zuker, 1994) and RNAfold (Hofacker *et al.*, 1994). However, the prediction of a secondary structure for a single sequence is known to be error prone.

```
# STOCKHOLM 1.0

Enterobacter.1              CCCACAUCAGAUUUCCUGGUGUAACGAAUU.UACAAGUGCUUCUUGCAUAAGCAAGUUCAUCCCGGU.CAUCCCC.AUGCCCGGGAUU
Escherichia_coli_APE.1      AACACAUCAGAUUUCCUGGUGUAACGAAUUUUUUAAGUGCUUCUUGCUUAAGCAAGUUUCAUCCCGA.CCCCCUCAGGGUCGGGAUUU
K.aerogenes.1               AACGCAUCGGAUUUCCCGGUGUAACGAAUU.UUCAAGUGCUUCUUGCAUUAGCAAGUUUGAUCCCGA.CUCCUGC.GAGUCGGGAUUU
Salmonella_enterica_.1      CUCACAUCAGAUUUCCUGGUGUAACGAAUU.UUCAAGUGCUUCUUGCAUAAGCAAGUUUGAUCCCG..ACCCGUA.GGGCCGGGAUUU
Shigella_flexneri_2a.1      AACACAUCAGAUUUCCUGGUGUAACGAAUUUUUUAAGUGCUUCUUGCUUAAGCAAGUUUCAUCCCGACACCCUCA.GGGUCGGGAUUU
#=GC SS_cons                ...(((((((.....)))))))...(((((.((((((((.....)))))))).)).))).(((((((..((.....)).)))))))..
#=GC RF                     aacaCauCgGAuuucCcGguGuAAcgaauu.uucauGugcuucuugCaugagcaaGuuuGauCCCGa.acCcuua.GgGuCGGGauuu
```

Fig. 1. Alignment of the bacterial sRNA DsrA. Conserved stems are indicated in light grey. Compensatory base pair mutation are shown in dark grey

For example, the average accuracy of secondary structure prediction, measured as the number of base pairs correctly predicted, was estimated to be about 67% in Mathews *et al.* (2004). Albeit the use of machine learning techniques to infer optimized parameters from known secondary structures (mostly from ribosomal RNAs) has improved the situation (Do *et al.*, 2006; Andronescu *et al.*, 2007), the prediction quality is greatly enhanced if structural information of several related ncRNAs is combined. Thus, using an alignment of several sequences from one RNA family to infer a conserved consensus structure for that family with associated compensatory base pair mutations is an important phylogenetic approach for improving the quality of secondary structure prediction.

Using a classification by Gardner and Giegerich (2004), there are three different main approaches for finding an alignment together with its associated consensus structure (see Figure 2). The first approach is shown in the left branch of Fig. 2 and has been very successfully used in the past. It consists of generating a multiple sequence alignment and predicting a consensus structure from this sequence

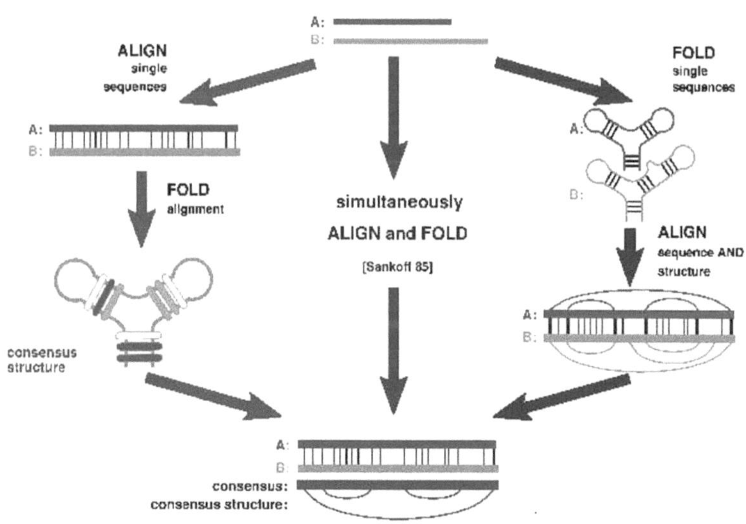

Fig. 2. Three main approaches to sequence-structure alignment according to Gardner and Giegerich (2004)

alignment using thermodynamic and/or covariation information (i. e., information about compensatory base pair mutations). The consensus structure prediction can be done by a modification of the usual secondary structure folding that works on alignment columns instead of single sequence positions. Pfold (Knudsen and Hein, 1999, 2003) is based on a SCFG model that uses mainly phylogenetic information. This class of SCFGs is commonly referred to as *phylo-SCFGs* and combines the ability of SCFGs to model a distribution on RNA structures with the possibility of modelling an observed substitution process in alignment columns. In contrast, RNAalifold (Bernhart *et al.*, 2008) calculates a consensus structure, for which the energy is the average of the energies in the single sequences, using covariation as additional information. Petfold (Seemann *et al.*, 2008) combines phylogenetic and thermodynamic information by means of an expected accuracy scoring approach, showing that the combination improves the consensus structure prediction.

If the initial multiple sequence alignment is of good quality, then the first approach is very successful in finding the correct consensus structure. However, due to the low sequence conservation, it is often impossible to find a good multiple sequence alignment, which is crucial for the correct consensus structure prediction. In fact, it has been shown that sequence alignments of structured RNAs already fail at pairwise sequence identities below 60% (Gardner *et al.*, 2005).

Hence, two other types of approaches have been introduced that use secondary structure information already for the construction of the alignment. The approach indicated by the right branch in Figure 2 is based on given secondary structures. The structure for each sequence can either be derived from experimental data, which is still a rare case, or predicted by tools like Mfold (Zuker, 1994) and RNAfold (Hofacker *et al.*, 1994). Then, the secondary structures for all sequences are used to build an alignment that shows conservation on sequence level in the single-stranded parts and conservation on base pairs in the structured parts of the single sequence foldings. To this end, specific scoring rules that evaluate base pair matches and mismatches (i. e., compensatory base pair mutations) and the different types of base pair insertions and deletions (i. e., either substituting a pair of bases that can base pair by a non-base pair, or deleting one or both ends of the base pair) were introduced. The sequence-structure alignment of a pair of sequences is then defined to be the alignment that is optimal under this scoring system. In a seminal work, Jiang *et al.* (2002) introduced such a flexible scoring scheme for the first time. Previous approaches (e. g. see Shapiro and Zhang (1990) or Evans (1999b,a)) used a different representation of RNA structures like trees or arc-annotated sequences to define similarity of RNA structures. However, these approaches were not well suited to handling deletions or insertions of base pairs, in particular.

The problem of comparing RNA structures becomes computationaly feasible if only nested structures are used as an input, and the scoring system satisfies some restrictions on the scoring of base pair deletion and insertion (Jiang *et al.*, 2002; Blin *et al.*, 2007). One major problem is that these approaches work on pairwise sequencestructure alignment, and it is not always possible to extend the pairwise methods to multiple alignment using a progressive approach. The reason is that when considering deleted or inserted base pairs in addition to matched base pairs, the result is similar to a non-nested, i. e. pseudoknotted, structure. Thus, the pairwise method cannot usually be applied to align the outputs of its own application, which makes a progressive alignment impossible. This problem was first solved in MARNA (Siebert and Backofen, 2005) and later in Lara (Bauer *et al.*, 2007) by using the multiple sequence alignment approach of T-Coffee (Notredame *et al.*, 2000) to combine the pairwise sequence-structure alignments. RNAforester (Höchsmann *et al.*, 2003) is the only exception that uses directly a progressive method to build a multiple alignment. RNAforester aligns RNA structures, which can also be represented as trees, by tree alignment. Tree alignment is a restricted form of sequence-structure alignment, where the resulting alignments are again trees. Thus, one can directly use a progressive method to build a multiple alignment.

All these methods provide good alignments when the initial secondary structures are of reasonable quality. In practice, the limited accuracy of structure prediction is a major obstacle; in particular, predicting structures for unaligned, single sequences neglects their relation to the remaining sequences of the alignment. For this reason, the gold standard is to simultaneously predict an alignment *and* a consensus structure for two or more unaligned sequences. This approach was initially proposed by Sankoff (1985) and constitutes the third branch in Figure 2. However, in its full form, the algorithm requires $O(n^6)$ CPU time and $O(n^4)$ memory, where n is the length of the given RNA sequences. For this reason, many variants of this approach have been introduced using various simplifications to be applicable to real world data.

There are three main variants of the Sankoff algorithms. The first class uses a full energy model for the folding of consensus RNA structures and is implemented in FoldAlign (Gorodkin *et al.*, 1997; Havgaard *et al.*, 2005) and dynalign (Mathews and Turner, 2002). The energy of the consensus structure can be interpreted as an average of the energies for the associated structure in each sequence and is evaluated during the alignment computation. The second class of approaches is based on SCFGs, which are used to model simultaneously a distribution on alignment and consensus structures. An implementation is provided by, e. g., Stemloc (Bradley *et al.*, 2008).

The third class finally uses a precomputation step to infer structural features, which are in a second step evaluated during the alignment process. The first representatives of this class were PMcomp (Hofacker *et al.*, 2004) and its successor LocARNA (Will *et al.*, 2007). Later, this idea was picked up by foldalignM (Torarinsson *et al.*, 2007), PARTS (Harmanci *et al.*, 2008) and RAF (Do *et al.*, 2008). They use a fully featured energy model in their precomputation step to compute a matrix of base pair probabilities, which can be done by using McCaskill's algorithm (McCaskill, 1990) separately for each input sequence. During the alignment process, these base pair probabilities are used for assessing the similarity of the secondary structures. This strategy saves time during the process of pairwise alignment. Since this is the major time-critical process for the calculation of a multiple alignment, the overall performance is greatly improved. Nevertheless, it guides the simultaneous alignment and folding precisely in accordance with the RNA energy model. LocARNA uses the sparsity of this base pair probability matrix to drastically improve speed and it is currently one of the fastest methods for sequence-structure alignment. RAF combined sparsity on the structure and sequence level with a light-weight scoring scheme, which significantly improves its efficiency over other Sankoff-style methods. Furthermore, more efficient approaches that avoid the computation of unnecessary subproblems (using a technique called sparsificiation) have led to further improvements in speed and space (Havgaard *et al.*, 2007; Ziv-Ukelson *et al.*, 2008; Backofen *et al.*, 2009).

All aforementioned approaches are primarily used to either detect homology between two or more known non-coding RNAs or to search for orthologs of a known ncRNA in other genomes.

1.3 ncRNA Gene Finders

Finding structural ncRNA genes is the problem of the *ab-initio* detection of non-coding RNAs in a complete genome. The main idea of general approaches for ncRNA gene finding is the detection of conserved RNA secondary structures. This is achieved by analyzing genome-wide multiple sequence alignments, which are produced by standard techniques like MULTIZ (Schwartz *et al.*, 2003; Blanchette *et al.*, 2004). QRNA (Rivas and Eddy, 2001), as one of the first representatives of a general ncRNA finder, uses three different stochastic models for assessing the likelihood of the given sequence alignment. The three stochastic models are based on the pair Hidden Markov model, a general method for modelling distributions for sequence alignment. The models are named RNA, COD and IND and evaluate the different constraints imposed on the distribution of sequence alignment by deviating evolutionary requirements of non-coding RNA, coding region or normal non-coding DNA. The probabilities of the sequence alignment under the different models is then used to classify a sequence alignment (and hence the sequences of the alignment) as ncRNA, coding region or non-coding DNA.

The second approach using the comparison of different probabilistic models is Evofold (Pedersen *et al.*, 2006). In contrast to QRNA, it is not based on HMMs but uses a variant of the phylo-SCFG as introduced by Knudsen and Hein (1999, 2003) in Pfold. In contrast to Pfold, two phylo-SCFGs are used: one models an RNA structure (called *structural model*), and the other (called *non-structural model*) does not introduce any base pair. Evofold outputs both a consensus structure (similar to Pfold) and a score (called folding potential score (FPS)), which is the log-odd for the probability of the alignment under the structural and non-structural model, respectively.

The third method, RNAz (Washietl *et al.*, 2005), uses a completely different approach and is based on thermodynamics. For a given input alignment, a consensus structure (using RNAalifold) and the minimum free energy (MFE) structures for all individual sequences (using RNAfold) are predicted. Then, different features of the secondary structure(s) and the alignment are classified using a support vector machine (SVM). The alignment-related features are the number of sequences in the alignment and the mean pairwise identity. The first structural feature is the *structural conservation index* (SCI), which is the ratio of the consensus structure energy and the averaged MFE structure energies. If this ratio is close to 1, then the consensus structure has the same energy as a typical MFE structure of the individual sequences, which indicates that a large portion of the consensus structure is conserved. The second feature is the averaged z-score for the MFE structure energies. It evaluates how much more stable than random the MFE structures in each sequence are. To save computation time, the z-score calculation is not done using sampling but using SVM regression.

The above mentioned methods are generic, comparative ncRNA gene finders and thus not optimized for bacterial genomes, albeit one of the first applications of QRNA was the detection of sRNAs in *E.coli* (Rivas *et al.*, 2001). However, there are additional non-comparative signals in bacteria that can be exploited for screening sRNAs in bacteria. One of the first examples was the *E.coli* ncRNA screen by Argaman *et al.* (2001), where only intergenic regions were considered and the prediction of promoters and rho-independent terminators were additionally used to select candidates. There are several screens on bacteria using different selection criteria (for current reviews, see e.g. Pichon and Felden (2008) or Backofen and Hess (2010)). Of course, it is possible to combine the non-comparative approach based on genome-specific features with generic comparative ncRNA gene finders. A nice example is Axmann *et al.* (2005), where alignments of intergenic regions were scored using ALIFOLDZ (Washietl and Hofacker, 2004). The sRNApredict2 tool (Livny *et al.*, 2005; Livny and Waldor, 2007; Livny *et al.*, 2006) is an example of a complex pipeline using both comparative and non-comparative information.

2 Computational Target Prediction

In the first subchapter we considered the problem of predicting sRNAs, many of which are awaiting functional characterization. The critical bottleneck then really is the identification of the targets of these sRNAs. Experimental approaches for the detection of sRNA targets include standard genetic screens, gene knockouts and overexpression of the sRNA of interest, followed by proteomics and microarray analysis, co-immunoprecipitation of direct interaction partners and the characterization of relevant ribonucleoprotein particles. However, all of these methods require a large amount of effort and are very time-consuming. Therefore, a highly sensitive biocomputational prediction followed by a focused experimental analysis is highly desirable. For reviews on these subjects see e.g. Pichon and Felden (2008), Vogel and Wagner (2007) and Sharma and Vogel (2009).

2.1 Search for Complementary Regions

The basis for computational target prediction is the search for complementary regions both in mRNA and sRNA. However, the "strength" of complementarity of these regions is measured in many different ways. In contrast to microRNAs, there are currently few additional steps such as conservation of specific regions or the enforcement of a seed region that are used on a regular basis. Currently, it looks like sRNA-mRNA interactions are even more flexible than miRNA-mRNA interactions, which makes it hard to determine single significant features for sRNA-mRNA interactions comparable to the 6–8nt long seed region found to be important for the prediction of miRNA regulation (Baek *et al.*, 2008; Selbach *et al.*, 2008). For the evaluation of the complementarity, pure sequence-based methods like BLAST (Altschul *et al.*, 1990) can be used to search for long stretches of complementarity. However, it is important to consider also the non-Watson Crick G-U pairs, which can be done using GUUGle (Gerlach and Giegerich, 2006). The individual base pair model used by TargetRNA (Tjaden *et al.*, 2006) can also be considered as a pure sequential approach. Here, A–U and G–C base pairs are given the same score. For this reason, Mandin *et al.* (2007) introduced a similar model where the scoring of individual base pairs is inspired by the strength of the base pair, which is especially important for genomes with low GC-content such as *Listeria* (Glaser *et al.*, 2001). The main advantage of these approaches is their simplicity; the computational costs usually grow at most geometrically with the input length. Another advantage is that one can easily calculate the significance of the matches, which will be discussed later.

2.2 Duplex Evaluation

The next step in complexity are approaches that do not score the base pairs of the interaction independently, but use a scoring system that is used in RNA folding, namely the scoring of stacked base pairs and internal loops. This leads to the thermodynamic scoring of mixed duplexes consisting of mRNA and sRNA sequences, and can be considered as a restricted and specialized version of full RNA secondary structure folding (like in Mfold (Zuker, 1994) and RNAfold (Hofacker *et al.*, 1994)). The first approach in this direction was RNAhybrid (Rehmsmeier *et al.*, 2004) (also implemented as RNAduplex in the *Vienna RNA Package* (Gruber *et al.*, 2008)), whose main application is the prediction of microRNA targets. Here, the scoring of a base pair (i,k), where i is a position in the mRNA, and k is a position in the sRNA, depends on the immediately following base pair (i',k') with $i' > i$ and $k < k'$ (assuming that both mRNA and sRNA is annotated $5' \rightarrow 3'$, and the interaction is anti-parallel as usual). If $i' = i+1$ and $k' = k-1$, then the two base pairs form a stack, which is usually energetically favourable. Otherwise, the two base pairs close an internal loop or bulge. The energy parameters for this scoring are the same as in RNA secondary structure folding, and represent free energies (in kcal/mol) that were derived from experimental data using the nearest neighbor model by Mathews *et al.* (1999). Later, a similar approach was used in TargetRNA for the prediction of sRNA targets. Both above mentioned approaches use a restriction on the length of internal loops in the mixed duplexes, since long internal loops are energetically not favorable and in addition increase the computational complexity (where the maximal loop length L contributes quadratically to the run time). RNAplex (Tafer and Hofacker, 2008) has an energy model similar to RNAduplex and RNAhybrid with the only exception of internal loops. Whereas explicit energy tables are used for small internal loops, big internal loops are usually evaluated using a logarithmic length term and an asymmetry penalty. In RNAplex, the length dependent term is replaced by an affine gap penalty, which allows elimination of the quadratic factor introduced by the maximal loop length L. In addition, RNAplex introduces a penalty term that is proportional to the sequence length and mimics the effect of an accessibility term.

There are several advantages of this simple energy model. Firstly it is a much more realistic model of RNA-RNA interactions compared to pure sequential approaches based on complementarity, and e. g., allows temperature to be taken into account, which is an important parameter when considering the stability of duplexes. Secondly these approaches are very fast since their computational complexity is comparable to simple local sequence alignment. Thirdly one can easily calculate the significance (i. e., p-values) of the hits, again due to their similarity to local sequence alignment (Rehmsmeier *et al.*, 2004). For local alignment, it is well known, due to Karlin and Altschul (1990), that the score of an optimal local alignment follows an extreme value distribution since local scores can be regarded as the maximum of a set of independent variables. Assuming an extreme value distribution with a location parameter u and a scale parameter s for the length-normalized hybridization scores, one can estimate these two parameters by fitting the extreme value distribution to an empirical distribution, generated from normalized hybridization scores for a large set of randomly generated sequences. These sequences are generated from the actual dinucleotide frequency of the mRNA space of interest. It is important to use dinucleotide instead of mononucleotide shuffling since the energy of duplexes depend on the dinucleotide frequencies due to base pair stacking. The scores have to be normalized according to length since longer putative target and sRNA sequences will tend to have more negative energies.

The main disadvantage of these approaches is that they neglect intra-molecular base pairs. This can have two effects. Firstly they could predict biologically impossible interactions where one of the interacting regions is sequestered in a stable intra-molecular structure (see Figure 3a). Secondly they tend to predict interactions that are too long, since it is usually more favourable to extend interactions if the effect of breaking intra-molecular base pairs is ignored.

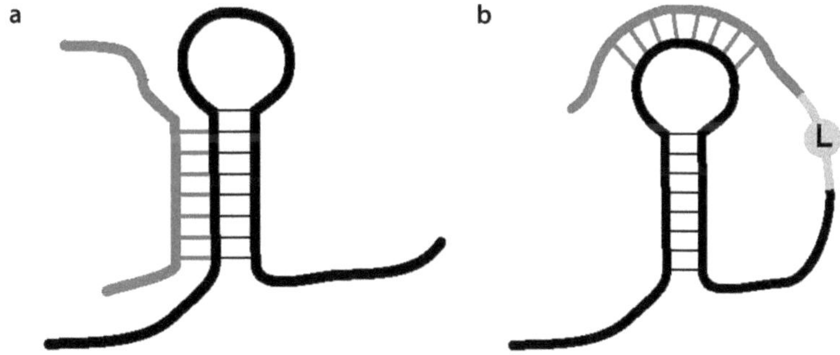

Fig. 3. a) A physiologically impossible structure that might be predicted by simple duplex scoring. **b)** A non-nested structure that cannot be predicted by concatenation approaches

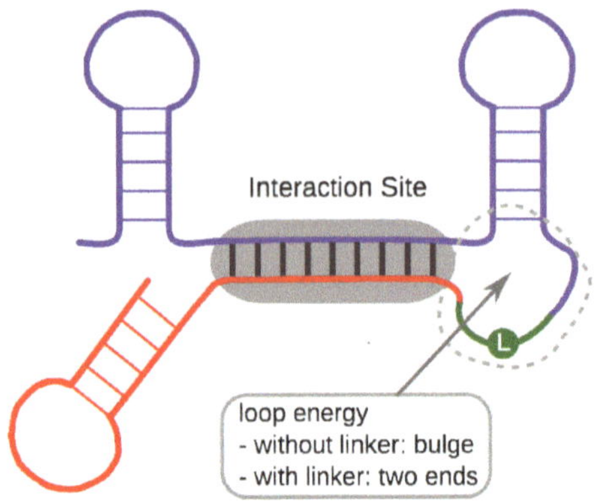

Fig. 4. Joint structure of two RNAs using a concatenation with linker element (in green). Without a special treatment of the linker element, the associated loop would be scored as a bulge, giving rise to a high positive energy contribution.With the special treatment of the linker, there are just external bases with associated dangling end contributions

2.3 Concatenation Approaches

The drawbacks of the methods described in the previous section led to the introduction of several other approaches that incorporate the effects of the internal structures of both mRNA and sRNA. There are mainly two classes of approaches. The first class of approaches, with pairfold (Andronescu *et al.*, 2005), RNAcofold (Bernhart *et al.*, 2006b) and the method presented by Dirks *et al.* as part of the NUpack package (Dirks *et al.*, 2007) being their first representatives, consider joint structures of mRNA and sRNA that are generated by the concatenation of the two sequences using a special linker character. Then, a modified version of the usual RNA folding (like in Mfold (Zuker, 1994) and RNAfold (Hofacker *et al.*, 1994)) is applied. Basically, the recursive structure is the same but there is a special treatment of loops that contain the linker symbol. The reason is just that an internal loop containing the linker element is in fact not an internal loop, but consists only of external bases (see Figure 4). As a result, these approaches predict joint structures that are nested in the sequence of two concatenated input sequences, since this is the restriction crucial for the recursive calculation of the joint structure. In the following, we will thus refer to this class of approaches as *concatenation approaches*. Figure 3b) shows an example of a possible joint structure that cannot be predicted using these approaches.

An advantage of the concatenation approach is that all techniques regularly used in RNA secondary structure prediction can be transfered to the cofolding approach. Hence, it is also possible to calculate the partition function of all joint structures as well as base pair probabilities (intra-molecular as well as base pairs between the two sequences) using a variant of McCaskill's approach (McCaskill, 1990). The partition function *ZS* for a sequence *S* (which might be composed by two sequences using a linker symbol) is the sum of all Boltzmann-weighted energies of all structures *R* that sequence *S* can take, i. e.,

$$Z(S) = \sum_{R \text{ structure of } S} e^{-\frac{E(S)}{RT}}.$$

After calculation of the partition function as in the case of RNA folding, then the $E(S)$ Boltzmann probability of a specific structure can be calculated by $\frac{e^{-\frac{E(S)}{RT}}}{Z}$. Even more importantly the probability of a base pair (i, j) (where i, j are positions in either of the sequences) can be calculated using a modification of the partition function computation to sum up the Boltzmann-weighted energies for all structures that contain the base pair, i. e., to calculate

$$Z^{(i,j)} = \sum_{R \text{ contains } (i,j)} e^{-\frac{E(S)}{RT}}.$$

The probability of the base pair (i, j) is then given by $Z^{(i,j)}/Z$. Furthermore, the concatenation approach allows for two interacting molecules A and B the calculation of the partion function for the single molecules A and B, for the homo-dimers AA and BB and for the hetero-dimer AB. With these partition functions, it is possible to calculate concentration dependent melting temperatures, which can then be compared to experimental data.

An interesting application of the concatenation approach is presented in Cao *et al.* (2009); Zhao *et al.* (2008). It addresses the problem that most applications of interaction prediction restrict the search to a region around the start codon. However, many different settings concerning the length and position of this region are used. Here, a classification method was used to solve this problem. First, multiple overlapping regions were considered, and the joint minimum free energy structure for all the overlapping regions was calculated using a simple concatenation approach (without a special treatment of the linker sequence). The minimum free energy structure of a region is then used to extract different features such as percent composition of bases in interior loops, bulge loops etc. In addition, sequence features like the percentage of A + U bases (since Hfq is supposed to bind to AU-rich regions) were used. In total, 10 features for 1000 overlapping sequences in the region +30 to –30 around the start codon were computed, giving rise to a *secondary structure profile*. These features were then used to train two classifiers (based on Naive-Bayes and SVM) on a data set of 46 positive samples and 86 negative samples.

2.4 Accessibility-based Approaches

Since concatenation approaches predict nested joint structures, i. e., without pseudoknots, they cannot handle important structures like kissing hairpins. For this reason, another class of approaches has been introduced that can handle this class of interactions. The basic idea is not to predict a single joint structure (or an ensemble of joint structures), but to investigate first the ensemble properties of the single sequences that are important for a putative interaction. Basically, an interaction site must be accessible (i. e., not covered by intra-molecular base pairs) since its positions will be bound to the interaction partner. Thus, for any two positions $a < b$ in a sequence, one computes the energy that is required to make the sequence stretch between a and b free of intra-molecular base pairs. Then, one calculates the partition function $Z^{sg(a,b)}$ for the ensemble of structures that leave the putative interaction site ranging from a to b single-stranded (see Figure 5). Now one calculates the *ensemble energy* by the formula $E^{sg(a,b)} = -RT \ln(Z^{sg(a,b)})$. Defining the energy of the ensemble of all structures by Eall $= -RT \ln(Z)$ where Z is the total partition function, we get the energy $ED(sg(a, b))$ that is required to make the interaction site accessible as

$$ED(a,b) = E^{sg(a,b)} - E_{all}.$$

Note that the above term is positive and thus can be considered as a penalty. Now approaches like RNAup (Muckstein *et al.*, 2006) and IntaRNA (Busch *et al.*, 2008) use precalculated ED-values for all possible interaction regions to calculate a combined energy consisting of the ED-values and the duplex energy. Thus, an interaction of two regions $i..i'$ of the first sequence and $k..k'$ of the second sequence is evaluated as shown in Figure 6. The ED-values for all regions in one sequence can be precalculated in basically the same complexity as the calculation of base pair probabilities in normal RNA folding using the RNAplfold approach (Bernhart *et al.*, 2006a; Bompfunewerer *et al.*, 2008). For the combined energy E, the recursion is basically similar to pure hybridization approaches like RNAhybrid, with the exception that all possible right end points have to be considered separately to determine the optimal region for the ED-values. This approach, which is used in RNAup, leads to a quadratic overhead depending on the maximal length of inter-

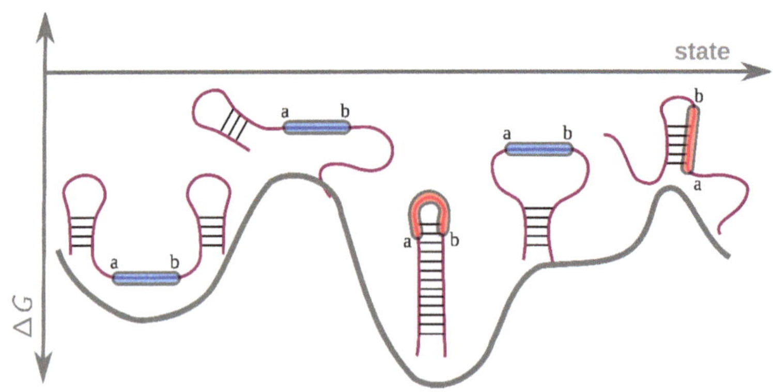

Fig. 5. Energy Landscape and accessibility. Given a putative interaction site between positions a and b, there are several structures where this site is single-stranded (denoted by a blue oval), whereas others cover the interaction site. The later ones cannot adopt a joint structure using $a..b$ as an interaction site. The partition function $Z^{sg(a,b)}$ for the ensemble of structures with the subsequence between a and b being single-stranded is the sum of all Boltzmann-weighted energies for the structures with blue ovals

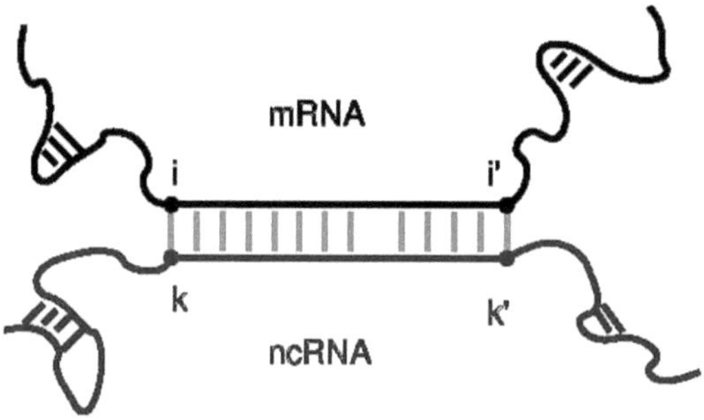

$$E = E^{hybrid} + ED^{mRNA}(i, i') + ED^{sRNA}(k, k')$$

Fig. 6. Evaluation of an interaction in RNAup and IntaRNA. The ED-values are precalculated for all possible regions in both sequences

action site considered. In IntaRNA, this overhead is avoided by using a heuristic approach. Nevertheless, the prediction quality of RNAup and IntaRNA are basically equal since IntaRNA uses a seed condition in addition.

2.5 Full Joint Structure Prediction

When comparing the concatenation approaches (RNAcofold, pairFold and NUpack) and the approaches working with accessibility (RNAup and IntaRNA), then it is apparent that both restrict the set of joint structures that are taken into account. The concatenation approaches can only predict joint structures where the interaction arcs (i.e., the base pairs between two RNAs) are not covered by intra-molecular base pairs. An equivalent condition is that the interaction arcs may only occur at external positions. A position k is external in a structure if there is no base pair (i, j) in the structure that covers k (i.e., where $i < k < j$). Given a joint structure, then the external positions are those that are external in the two substructures which are generated by restricting the joint structure to the single sequences. The approaches using accessibility, on the other hand, assume a single interaction site, which may not contain any base participating in an intra-molecular base pair. The restriction is due to the fact that the unrestricted problem (i.e., finding the best joint structure of two interacting RNAs without any restriction on the type of structures) is computationally a very hard problem. Recently, it was shown in Alkan *et al.* (2006) that the general problem is NP-complete, which means in practice that an exact algorithm would require exponential time[1].

---

[1] The precise definition of NP-complete is more complex. NP is a class of problems, which are currently believed to be different from the class P of problems that can be solved in polynomial time. Unless NP=P (which is believed to be very unlikely), there cannot be an algorithm that exactly solves the general interaction problem in polynomial time for *all* instances. However, there might be algorithms that solve the problem in reasonable time for most practical instances.

However, there are known interactions like OxyS with *fhlA* that have more than one kissing hairpin interaction (see Figure 7). Such interactions can neither be predicted by the concatenation approaches nor by the approaches using accessibility of one single interaction site. For this reason, new methods have been introduced that extend the class of allowed joint structures. The IRIS tool (Pervouchine, 2004) introduced a new recursive scheme that allowed several kissing hairpins to be considered for the first time. It uses an energy model that maximizes the number of base pairs. Then Alkan *et al.* (2006) considered the extension to a more realistic energy model, which is inspired by the standard nearest neighbor energy model of single RNA sequence folding. Furthermore, they gave a precise definition of the class of structures treated by their approach.

Both approaches can handle the OxyS-*fhlA* interaction, and both approaches predict a single structure with the minimum free energy (the structure with maximal number of base pairs in the case of IRIS). However, as already observed in folding of a single RNA, the MFE structures is often wrong. The standard way to overcome this problem is to use a partition function variant as already described above for the concatenation approaches. Since one has to calculate the sum over all possible joint structures, it is necessary to reformulate the recursion equations such that every joint structure is decomposed in a unique way. This problem was solved independently by Chitsaz *et al.* (2009b) and Huang *et al.* (2009). Thus, both approaches allow the calculation of important quantities like melting temperatures and base pair probabilities. As shown in e. g. (Chitsaz *et al.*, 2009b), the melting temperatures calculated by the algorithm are in good agreement with the experimentally measured ones, as exemplified for the wild type and three mutated constructs of the OxyS-*fhlA* interaction.

All above mentioned tools for the prediction of a joint structure have still a very high computational complexity (the computation time is in the order of O(*n*6),

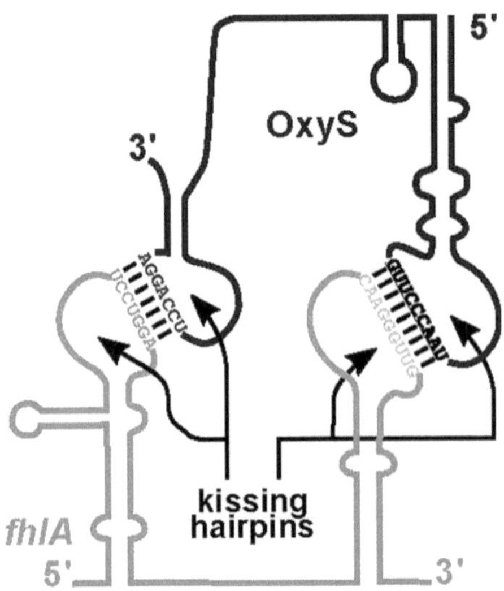

Fig. 7. Interaction components of OxyS and fhlA as presented in Argaman and Altuvia (2000)

where *n* is the length of the input sequence(s)). On the one hand, a technique called sparsification, which has already been successfully used in secondary structure prediction and alignment (Wexler *et al.*, 2006; Ziv-Ukelson *et al.*, 2008; Backofen *et al.*, 2009), greatly improved both time and space requirements for the problem of predicting an optimal joint structure (Salari *et al.*, 2010). On the other hand, there were attempts to reduce this complexity by considering approximations to the original problem. A very intuitive way is to use accessibility (as in RNAup/IntaRNA), but to allow more than one interaction site. The ED value for measuring the energy required to make a site accessible can be calculated from the probability that this site is single-stranded. Now it is immediately clear that these probabilities are not independent for different interaction sites. Thus, conditional probabilities have to be used instead. Albeit that seemed to be too complex to be calculated, a Bayesian approximation of these conditional probabilities was introduced by Chitsaz *et al.* (2009a) and Salari *et al.* (2009). This allowed a fast calculation of these conditional probabilities and resulted in a fast heuristic method for predicting the specific (multiple) binding sites of two interacting RNAs.

References

Alkan C, Karakoc E, Nadeau JH, Sahinalp SC, Zhang K (2006) RNA-RNA interaction prediction and antisense RNA target search. Journal of Computational Biology 13(2):267–82

Altschul SF, Gish W, Miller W, Myers EW, Lipman DJ (1990) Basic local alignment search tool. Journal of Molecular Biology 215(3):403–10

Andronescu M, Zhang ZC, Condon A (2005) Secondary structure prediction of interacting RNA molecules. Journal of Molecular Biology 345(5):987–1001, DOI 10.1016/j.jmb.2004.10.082

Andronescu M, Condon A, Hoos HH, Mathews DH, Murphy KP (2007) Efficient parameter estimation for rna secondary structure prediction. Bioinformatics 23(13)

Argaman L, Altuvia S (2000) fhla repression by OxyS RNA: kissing complex formation at two sites results in a stable antisense-target RNA complex. Journal of Molecular Biology 300(5):1101– 12

Argaman L, Hershberg R, Vogel J, Bejerano G, Wagner EG, Margalit H, Altuvia S (2001) Novel small RNA-encoding genes in the intergenic regions of Escherichia coli. Curr Biol 11(12):941– 50

Axmann IM, Kensche P, Vogel J, Kohl S, Herzel H, Hess WR (2005) Identification of cyanobacterial non-coding RNAs by comparative genome analysis. Genome Biol 6(9):R73

Backofen R, Hess WR (2010) Computational prediction of sRNAs and their targets in bacteria. RNA Biol 7(1)

Backofen R, Tsur D, Zakov S, Ziv-Ukelson M (2009) Sparse RNA folding: Time and space efficient algorithms. In: Kucherov G, Ukkonen E (eds) Proc. 20th Symp. Combinatorial Pattern Matching, Springer, LNCS, vol 5577, pp 249–262, DOI 10.1007/978–3-642–02441-2 22

Baek D, Villen J, Shin C, Camargo FD, Gygi SP, Bartel DP (2008) The impact of microRNAs on protein output. Nature 455(7209):64–71, DOI 10.1038/nature07242

Bateman A, Coin L, Durbin R, Finn RD, Hollich V, Griffiths-Jones S, Khanna A, Marshall M, Moxon S, Sonnhammer ELL, Studholme DJ, Yeats C, Eddy SR (2004) The Pfam protein families database. Nucleic Acids Research 32 Database issue:D138–41

Bauer M, Klau GW, Reinert K (2007) Accurate multiple sequence-structure alignment of RNA sequences using combinatorial optimization. BMC Bioinformatics 8:271, DOI 10.1186/ 1471–2105-8–271

Bernhart SH, Hofacker IL, Stadler PF (2006a) Local RNA base pairing probabilities in large sequences. Bioinformatics 22(5):614–5

Bernhart SH, Tafer H, Muckstein U, Flamm C, Stadler PF, Hofacker IL (2006b) Partition function and base pairing probabilities of RNA heterodimers. Algorithms Mol Biol 1(1):3

Bernhart SH, Hofacker IL, Will S, Gruber AR, Stadler PF (2008) RNAalifold: improved consensus structure prediction for RNA alignments. BMC Bioinformatics 9:474, DOI 10.1186/1471–2105-9–474

Blanchette M, Kent WJ, Riemer C, Elnitski L, Smit AFA, Roskin KM, Baertsch R, Rosenbloom K, Clawson H, Green ED, Haussler D, Miller W (2004) Aligning multiple genomic sequences with the threaded blocks*et al.*igner. Genome Res 14(4):708–15

Blin G, Fertin G, Rusu I, Sinoquet C (2007) Extending the hardness of RNA secondary structure comparison. In: Combinatorics, Algorithms, Probabilistic and Experimental Methodologies, First International Symposium, ESCAPE 2007, Hangzhou, China, April 7–9, 2007, Revised Selected Papers, Springer, Lecture Notes in Computer Science, vol 4614, pp 140–151

Bompfunewerer AF, Backofen R, Bernhart SH, Hertel J, Hofacker IL, Stadler PF, Will S (2008) Variations on RNA folding and alignment: lessons from Benasque. Journal of Mathematical Biology 56(1–2):129–144, DOI 10.1007/s00285–007-0107–5

Bradley RK, Pachter L, Holmes I (2008) Specific alignment of structured RNA: stochastic grammars and sequence annealing. Bioinformatics 24(23):2677–83, DOI 10.1093/bioinformatics/btn495

Busch A, Richter AS, Backofen R (2008) IntaRNA: efficient prediction of bacterial sRNA targets incorporating target site accessibility and seed regions. Bioinformatics 24(24):2849–56, DOI 10.1093/bioinformatics/btn544

Cao Y, Zhao Y, Cha L, Ying X, Wang L, Shao N, Li W (2009) sRNATarget: a web server for prediction of bacterial sRNA targets. Bioinformation 3(8):364–6

Chitsaz H, Backofen R, Sahinalp SC (2009a) biRNA: Fast RNA-RNA binding sites prediction. In: Salzberg S, Warnow T (eds) Proc. of the 9th Workshop on Algorithms in Bioinformatics (WABI), Lecture Notes in Computer Science, vol 5724, pp 25–36, DOI 10.1007/ 978–3-642–04241-6

Chitsaz H, Salari R, Sahinalp SC, Backofen R (2009b) A partition function algorithm for interacting nucleic acid strands. Bioinformatics 25(12):i365–73, DOI 10.1093/bioinformatics/btp212

Dirks RM, Bois JS, Schaeffer JM, Winfree E, Pierce NA (2007) Thermodynamic analysis of interacting nucleic acid strands. SIAM Review 49(1):65–88

Do CB, Woods DA, Batzoglou S (2006) CONTRAfold: RNA secondary structure prediction without physics-based models. Bioinformatics 22(14):e90–8

Do CB, Foo CS, Batzoglou S (2008) A max-margin model for efficient simultaneous alignment and folding of RNA sequences. Bioinformatics 24(13):i68–76, DOI 10.1093/bioinformatics/btn177

Eddy SR (2002) A memory-efficient dynamic programming algorithm for optimal alignment of a sequence to an RNA secondary structure. BMC Bioinformatics 3(1):18

Evans PA (1999a) Algorithms and complexity for annotated sequence analysis. PhD thesis, University of Alberta

Evans PA (1999b) Finding common subsequences with arcs and pseudoknots. In: CPM '99: Proceedings of the 10th Annual Symposium on Combinatorial Pattern Matching, Springer-Verlag, London, UK, pp 270–280

Gardner PP, Giegerich R (2004) A comprehensive comparison of comparative RNA structure prediction approaches. BMC Bioinformatics 5:140

Gardner PP, Wilm A, Washietl S (2005) A benchmark of multiple sequence alignment programs upon structural RNAs. Nucleic Acids Research 33(8):2433–9

Gardner PP, Daub J, Tate JG, Nawrocki EP, Kolbe DL, Lindgreen S, Wilkinson AC, Finn RD, Griffiths-Jones S, Eddy SR, Bateman A (2009) Rfam: updates to the RNA families database. Nucleic Acids Research 37(Database issue):D136–40, DOI 10.1093/nar/gkn766

Gautheret D, Lambert A (2001) Direct RNA motif definition and identification from multiple sequence alignments using secondary structure profiles. Journal of Molecular Biology 313(5):1003–11

Gerlach W, Giegerich R (2006) GUUGle: a utility for fast exact matching under RNA complementary rules including G-U base pairing. Bioinformatics 22(6):762–4, DOI 10.1093/ bioinformatics/btk041

Glaser P, Frangeul L, Buchrieser C, Rusniok C, Amend A, Baquero F, Berche P, Bloecker H, Brandt P, Chakraborty T, Charbit A, Chetouani F, Couve E, de Daruvar A, Dehoux P, Domann E, Dominguez-Bernal G, Duchaud E, Durant L, Dussurget O, Entian KD, Fsihi H, Garciadel Portillo F, Garrido P, Gautier L, Goebel W, Gomez-Lopez N, Hain T, Hauf J, Jackson D, Jones LM, Kaerst U, Kreft J, Kuhn M, Kunst F, Kurapkat G, Madueno E, Maitournam A, Vicente JM, Ng E, Nedjari H, Nordsiek G, Novella S, de Pablos B, Perez-Diaz JC, Purcell R, Remmel B, Rose M, Schlueter T, Simoes N, Tierrez A, Vazquez-Boland JA, Voss H, Wehland J, Cossart P (2001) Comparative genomics of Listeria species. Science 294(5543):849–52, DOI 10.1126/science.1063447

Gorodkin J, Heyer L, Stormo G (1997) Finding the most significant common sequence and structure motifs in a set of RNA sequences. Nucleic Acids Res 25(18):3724–32

Gruber AR, Lorenz R, Bernhart SH, Neubock R, Hofacker IL (2008) The Vienna RNA websuite. Nucleic Acids Research 36(Web Server issue):W70–4, DOI 10.1093/nar/gkn188

Harmanci AO, Sharma G, Mathews DH (2008) PARTS: probabilistic alignment for RNA joinT secondary structure prediction. Nucleic Acids Research 36(7):2406–17, DOI 10.1093/nar/ gkn043

Havgaard JH, Lyngso RB, Stormo GD, Gorodkin J (2005) Pairwise local structural alignment of RNA sequences with sequence similarity less than 40%. Bioinformatics 21(9):1815–24, DOI 10.1093/bioinformatics/bti279

Havgaard JH, Torarinsson E, Gorodkin J (2007) Fast pairwise structural RNA alignments by pruning of the dynamical programming matrix. PLoS Comput Biol 3(10):1896–908, DOI 10.1371/journal.pcbi.0030193

Höchsmann M, Töller T, Giegerich R, Kurtz S (2003) Local similarity in RNA secondary structures. In: Proceedings of Computational Systems Bioinformatics (CSB 2003), IEEE Computer Society, vol 2, pp 159–168

Hofacker IL, Fontana W, Stadler PF, Bonhoeffer S, Tacker M, Schuster P (1994) Fast folding and comparison of RNA secondary structures. Monatshefte Chemie 125:167–188

Hofacker IL, Bernhart SH, Stadler PF (2004) Alignment of RNA base pairing probability matrices. Bioinformatics 20(14):2222–7

Huang FWD, Qin J, Reidys CM, Stadler PF (2009) Partition function and base pairing probabilities for RNA-RNA interaction prediction. Bioinformatics 25(20):2646–54, DOI 10.1093/ bioinformatics/btp481

Jiang T, Lin G, Ma B, Zhang K (2002) A general edit distance between RNA structures. Journal of Computational Biology 9(2):371–88

Karlin S, Altschul SF (1990) Methods for assessing the statistical significance of molecular sequence features by using general scoring schemes. Proc Natl Acad Sci USA 87(6):2264–8

Knudsen B, Hein J (1999) RNA secondary structure prediction using stochastic context-free grammars and evolutionary history. Bioinformatics 15(6):446–54

Knudsen B, Hein J (2003) Pfold: RNA secondary structure prediction using stochastic context-free grammars. Nucleic Acids Research 31(13):3423–8

Lambert A, Fontaine JF, Legendre M, Leclerc F, Permal E, Major F, Putzer H, Delfour O, Michot B, Gautheret D (2004) The ERPIN server: an interface to profile-based RNA motif identification. Nucleic Acids Research 32(Web Server issue):W160–5

Livny J, Waldor MK (2007) Identification of small RNAs in diverse bacterial species. Curr Opin Microbiol 10(2):96–101, DOI 10.1016/j.mib.2007.03.005

Livny J, Fogel MA, Davis BM, Waldor MK (2005) sRNAPredict: an integrative computational approach to identify sRNAs in bacterial genomes. Nucleic Acids Research 33(13):4096–105, DOI 10.1093/nar/gki715

Livny J, Brencic A, Lory S, Waldor MK (2006) Identification of 17 Pseudomonas aeruginosa sRNAs and prediction of sRNA-encoding genes in 10 diverse pathogens using the bioinformatic tool sRNAPredict2. Nucleic Acids Research 34(12):3484–93, DOI 10.1093/nar/gkl453

Mandin P, Repoila F, Vergassola M, Geissmann T, Cossart P (2007) Identification of new noncoding RNAs in Listeria monocytogenes and prediction of mRNA targets. Nucleic Acids Research 35(3):962–74, DOI 10.1093/nar/gkl1096

Mathews D, Sabina J, Zuker M, Turner D (1999) Expanded sequence dependence of thermodynamic parameters improves prediction of RNA secondary structure. J Mol Biol 288(5):911–40

Mathews DH, Turner DH (2002) Dynalign: an algorithm for finding the secondary structure common to two RNA sequences. Journal of Molecular Biology 317(2):191–203, DOI 10.1006/jmbi.2001.5351

Mathews DH, Disney MD, Childs JL, Schroeder SJ, Zuker M, Turner DH (2004) Incorporating chemical modification constraints into a dynamic programming algorithm for prediction of RNA secondary structure. Proc Natl Acad Sci USA 101(19):7287–92

McCaskill JS (1990) The equilibrium partition function and base pair binding probabilities for RNA secondary structure. Biopolymers 29(6–7):1105–19

Muckstein U, Tafer H, Hackermuller J, Bernhart SH, Stadler PF, Hofacker IL (2006) Thermodynamics of RNA-RNA binding. Bioinformatics 22(10):1177–82

Nawrocki EP, Kolbe DL, Eddy SR (2009) Infernal 1.0: inference of RNA alignments. Bioinformatics 25(10):1335–7, DOI 10.1093/bioinformatics/btp157

Notredame C, Higgins DG, Heringa J (2000) T-Coffee: A novel method for fast and accurate multiple sequence alignment. Journal of Molecular Biology 302(1):205–17

Pedersen JS, Bejerano G, Siepel A, Rosenbloom K, Lindblad-Toh K, Lander ES, Kent J, Miller W, Haussler D (2006) Identification and Classification of Conserved RNA Secondary Structures in the Human Genome. PLoS Comput Biol 2(4):e33

Pervouchine DD (2004) IRIS: intermolecular RNA interaction search. Genome Inform 15(2):92–101

Pichon C, Felden B (2008) Small RNA gene identification and mRNA target predictions in Bacteria. Bioinformatics DOI 10.1093/bioinformatics/btn560

Rehmsmeier M, Steffen P, Hochsmann M, Giegerich R (2004) Fast and effective prediction of microRNA/target duplexes. RNA 10(10):1507–17

Rivas E, Eddy SR (2001) Noncoding RNA gene detection using comparative sequence analysis. BMC Bioinformatics 2(1):8

Rivas E, Klein RJ, Jones TA, Eddy SR (2001) Computational identification of noncoding RNAs in E. coli by comparative genomics. Curr Biol 11(17):1369–73

Salari R, Backofen R, Sahinalp SC (2009) Fast prediction of RNA-RNA interaction. In: Salzberg S, Warnow T (eds) Proc. of the 9th Workshop on Algorithms in Bioinformatics (WABI), Lecture Notes in Computer Science, vol 5724, pp 261–272, DOI 10.1007/978–3-642–04241-6

Salari R, Backofen R, Sahinalp SC (2010) Fast prediction of RNA-RNA interaction. Algorithms Mol Biol 5(1):5, DOI 10.1186/1748–7188-5–5

Sankoff D (1985) Simultaneous solution of the RNA folding, alignment and protosequence problems. SIAM J Appl Math 45(5):810–825

Schwartz S, Kent WJ, Smit A, Zhang Z, Baertsch R, Hardison RC, Haussler D, Miller W (2003) Human-mouse alignments with BLASTZ. Genome Res 13(1):103–7

Seemann SE, Gorodkin J, Backofen R (2008) Unifying evolutionary and thermodynamic information for RNA folding of multiple alignments. Nucleic Acids Research 36(20):6355–62, DOI 10.1093/nar/gkn544

Selbach M, Schwanhausser B, Thierfelder N, Fang Z, Khanin R, Rajewsky N (2008) Widespread changes in protein synthesis induced by microRNAs. Nature 455(7209):58–63, DOI 10.1038/nature07228

Shapiro BA, Zhang KZ (1990) Comparing multiple RNA secondary structures using tree comparisons. Comput Appl Biosci 6(4):309–18

Sharma CM, Vogel J (2009) Experimental approaches for the discovery and characterization of regulatory small RNA. Curr Opin Microbiol 12(5):536–46, DOI 10.1016/j.mib.2009.07.006

Siebert S, Backofen R (2005) MARNA: multiple alignment and consensus structure prediction of RNAs based on sequence structure comparisons. Bioinformatics 21(16):3352–9

Tafer H, Hofacker IL (2008) RNAplex: a fast tool for RNA-RNA interaction search. Bioinformatics 24(22):2657–63, DOI 10.1093/bioinformatics/btn193

Tjaden B, Goodwin SS, Opdyke JA, Guillier M, Fu DX, Gottesman S, Storz G (2006) Target prediction for small, noncoding RNAs in bacteria. Nucleic Acids Research 34(9):2791–802

Torarinsson E, Havgaard JH, Gorodkin J (2007) Multiple structural alignment and clustering of RNA sequences. Bioinformatics 23(8):926–32, DOI 10.1093/bioinformatics/btm049

Vogel J, Wagner EGH (2007) Target identification of small noncoding RNAs in bacteria. Curr Opin Microbiol 10(3):262–70, DOI 10.1016/j.mib.2007.06.001

Washietl S, Hofacker IL (2004) Consensus folding of aligned sequences as a new measure for the detection of functional RNAs by comparative genomics. Journal of Molecular Biology 342(1):19–30

Washietl S, Hofacker IL, Stadler PF (2005) Fast and reliable prediction of noncoding RNAs. Proc Natl Acad Sci USA 102(7):2454–9

Wexler Y, Zilberstein CBZ, Ziv-Ukelson M (2006) A study of accessible motifs and rna folding complexity. In: Apostolico A, Guerra C, Istrail S, Pevzner PA, Waterman MS (eds) Proc. of the Tenth Annual International Conferences on Computational Molecular Biology (RECOMB'06), Springer, Lecture Notes in Computer Science, vol 3909, pp 473–487

Will S, Reiche K, Hofacker IL, Stadler PF, Backofen R (2007) Inferring non-coding RNA families and classes by means of genome-scale structure-based clustering. PLOS Computational Biology 3(4):e65, DOI 10.1371/journal.pcbi.0030065

Yao Z, Weinberg Z, Ruzzo WL (2006) CMfinder – a covariance model based RNA motif finding algorithm. Bioinformatics 22(4):445–52

Zhao Y, Li H, Hou Y, Cha L, Cao Y, Wang L, Ying X, Li W (2008) Construction of two mathematical models for prediction of bacterial sRNA targets. Biochem Biophys Res Commun 372(2):346–50, DOI 10.1016/j.bbrc.2008.05.046

Ziv-Ukelson M, Gat-Viks I, Wexler Y, Shamir R (2008) A faster algorithm for RNA co-folding. In: Crandall KA, Lagergren J (eds) WABI 2008, Springer, Lecture Notes in Computer Science, vol 5251, pp 174–185, DOI 10.1007/978–3-540–87361-7 15, URL http://dblp. uni-trier.de/db/conf/wabi/wabi2008.html#Ziv-UkelsonGWS08

Zuker M (1994) Prediction of RNA secondary structure by energy minimization. Methods in Molecular Biology 25:267–94

Chapter 13
Computational Tools for Predicting sRNA Targets

Wuju Li, Xiaomin Ying, and Lei Cha*

Bacterial sRNAs are an emerging class of small regulatory RNAs of 40 to 500 nucleotides in length. They play a wide variety of important roles in many biological processes through binding to their mRNA or protein targets, such as expression regulation of outer membrane proteins, iron homeostasis, quorum sensing and bacterial virulence. Therefore, predicting sRNA targets plays a key role in elucidating sRNA functions. Here we intend to introduce some computational tools for predicting mRNA targets of sRNAs. Firstly, we will give an outline of some key concepts and programs associated with the prediction models developed by our center. Secondly, we will present detailed instructions on how to use the sRNATarget web server and the Perl program for both windows and the Linux system. Thirdly, we will briefly introduce the main ideas behind other tools, including IntaRNA and TargetRNA. Finally, we will present an outlook on future developments in the prediction of sRNA targets.

1 Introduction

In view of the fact that prediction of sRNA targets plays a key role in elucidating sRNA functions, some special models for the prediction of sRNA targets have been designed, including TargetRNA (Tjaden *et al.* 2006; Tjaden 2008), IntaRNA (Busch *et al.* 2008), and two others models sRNATargetNB and sRNATargetSVM, which were developed by our center using the Naïve Bayes method and support vector machines (Zhao *et al.* 2008). Since these algorithms have already been discussed in a recent review (Backofen and Hess 2010), here we shall focus only on our models. From a machine-learning point of view, three steps are involved in developing models: The first step is to construct training and test datasets; the training dataset

* Center of Computational Biology, Beijing Institute of Basic Medical Sciences, Beijing, 100850, China

is used for constructing models and the test dataset is applied to evaluate model performance. The second step is to extract features representing each sample. The third step is to determine which machine-learning methods are used. Since detailed information and processes for constructing models can be found in the paper by Zhao *et al.* (2008), here we shall focus on introducing some key concepts and programs associated with our models as follows:

1.1 Training and Test Datasets

Our models were constructed using a training dataset comprising 46 positive samples and 86 negative samples, and evaluated using a test dataset containing 22 positive samples and 1700 negative samples, which were randomly selected from a total of 4131 mRNAs in E.coli (NCBI code: NC_000913). Both the training dataset and the test dataset can be downloaded from the webpage http://ccb.bmi.ac.cn/srnatargetnb/.

1.2 RNA Secondary Structure Profile

The concept "RNA secondary structure profile" or "profile of RNA secondary structure" was first presented by our center in a paper, entitled "Construction of mathematical model for high-level expression of foreign genes in pPIC9 vector and its verification" (Wu *et al.* 2007). The key idea behind RNA secondary structure profile is that the biological functions of some short RNA fragments (namely, core region) are determined by a series of secondary structures, which are predicted from many regions around the core region. From a biological point of view, RNA secondary structure profile provides a flexible environment enabling the core region to carry out functions. For example, at the time we constructed the models for prediction of sRNA targets, we regarded the region $-30 \sim 30$ around the initial start codon as the core region because of some experimental data showing that the binding regions of target mRNAs were often located in this region. Therefore, whether an sRNA interacts with an mRNA cannot be determined by the sRNA and mRNA core region only, according to our supposition. The flanking regions around the core region are also likely to contribute to the interaction between sRNA and its targets. Nevertheless, we do not know which flanking regions are more important. It is necessary for us to consider all possible flanking regions at the same time, and then to select some more important regions for developing models by automatic processes, such as machine-learning methods. As shown in Figure 1, a total of 1000 flanking regions are considered.

To construct RNA secondary structure profile for the prediction of sRNA targets, each combination of sRNA and any one of 1000 flanking regions was considered. Then, the RNAfold program was used to predict the minimum free energy secondary structure for the linked sequences (sRNA-LLLLL- flanking region or flank-

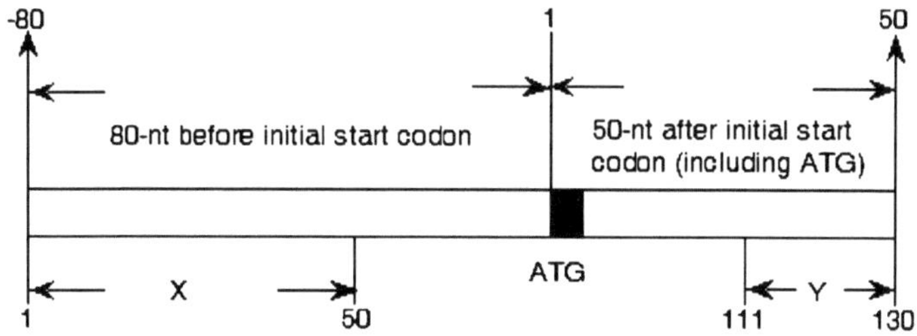

Fig. 1. Here ATG stands for the initial start codon and the 5' end region –30 ~ 30 is taken as the core region. To determine whether the core region binds to certain particular sRNAs, all flanking regions [X, Y] are considered, where $1 \le X \le 50$ and $111 \le Y \le 130$. The total number of flanking regions is 1000

ing region -LLLLL-sRNA) by default parameters (Hofacker 2003). The structure with lower free energy was used to extract the 10 features, such as percent composition of bases in interior loops, bulge loops, hairpin loops, helical regions, and multi-branch loops. Therefore, for each potential sRNA-mRNA complex, we firstly extracted 1000 flanking regions from the mRNA 5' end region. Then, for each flanking region, 10 features were extracted from the secondary structures predicted from the merged sequence. Finally, each potential sRNA-mRNA complex was represented by 10000 features. For 132 samples in the training dataset, we got a matrix with the size 10000 × 132, which is very similar to the gene expression profile produced using microarray technology. We regarded the matrix as RNA secondary structure profile of the sRNA-mRNA complex. Correspondingly, the problem of prediction of sRNA targets has been transformed into the problem of gene expression profile-based sample classification. Thus, many machine-learning methods can be used to construct prediction models.

1.3 Machine-learning Methods

To construct models for the prediction of sRNA targets, both the Tclass classification system and support vector machines (SVMs) were used. The Tclass system was originally developed for gene expression profile-based sample classification (Li and Xiong 2002), which can be downloaded from the webpage http://ccb.bmi.ac.cn/tclass/. It has also been successfully applied to construct the model for high-level expression of foreign genes in pPIC9 vector (Wu *et al.* 2007). In the Tclass system, many machine-learning methods, such as Fisher and Naive Bayes methods, were integrated with a procedure of forward feature selection. For the present task, the Naive Bayes method was used. The key idea behind the Tclass system is the presentation of the concept of "stability analysis". When we develop models for certain biological problems using machine-learning methods, for example, gene expression profile-based sample classification, we often find that there are many gene sets giving the same classification accuracy such as leave-one-out classification accuracy. In this case, we do not know which gene set is the best. To solve this problem, we presented the concept "stability analysis" in the paper (Li and Xiong 2002). For a machine-learning problem, the whole dataset is divided into a training dataset and a test dataset with some partition ratio, e.g. 75%. Then the training dataset (usually the major part) is applied to construct a model and the test dataset (usually the minor part) is used to evaluate the model performance. Ideally, the best gene set should provide the best performance, measured according to classification accuracy or other indexes, on the test dataset. However, for each time of partition, different test datasets will be generated which often give different model performances. To overcome the heterogeneity in model performances at different times of partition, we take the average value of model performances, such as classification accuracy in this case, as the ability of the gene set to distinguish different types of samples. Further studies showed that the average value (or so-called stability index) becomes stable when the number of partition times reaches a certain number, such as 1000. We call the above processes the stability analysis, through which different gene sets, even when giving the same classification accuracy on the training dataset, can be evaluated and the best gene set with the highest average value (or stability index) can be selected, which can be used to construct prediction models. Here we want to emphasize that the model constructed in this way, as a kind of ensemble classifier, is composed of 1000 classifiers rather than 1 classifier as in the normal model.

In addition to the Tclass system, we also applied support vector machines (SVMs) to build prediction models using the same training dataset (http://www.csie.ntu.edu.tw/~cjlin/libsvm). In recent years, SVMs have been successfully used to solve a variety of biological problems (Larranaga *et al.* 2006; Tarca *et al.* 2007). Here we applied it to construct models for the prediction of sRNA targets.

1.4 Construction of Prediction Models for sRNA Targets

With the training dataset, test dataset and machine-learning tools in hand, two models, sRNATargetNB and sRNATargetSVM, were constructed using the Naïve Bayes method and support vector machines. The classification accuracy, sensitivity, and specificity are 93.03%, 40.90%, and 93.71% for sRNATargetNB and 80.55%, 72.73%, and 80.65% for sRNATargetSVM, respectively. Although sRNATargetSVM gave better sensitivity, we still recommend the model sRNATargetNB for prediction of sRNA targets for the following two reasons: Firstly, for each potential sRNA–mRNA interaction, the sRNATargetNB model only needs six features. However, sRNATargetSVM requires 10,000 features, and therefore has a longer running time. Secondly, sRNATargetNB has a high specificity, meaning that a smaller number of false positive results will be generated, which can be more easily verified by experiments. Therefore, we only provide the model sRNATargetNB for users. The model can be downloaded from the webpage http://ccb.bmi.ac.cn/srnatargetnb/.

2 Program and Usage

2.1 Predicting sRNA Targets Using sRNATarget Webserver

The webserver, sRNATarget, was developed using Apache, MySql, Perl, BioPerl, PHP and Javascript. It runs on an HP server equipped with two 3.16G Intel Xeon CPUs and 4G memory (Cao *et al.* 2009). The interface is displayed in Figure 2(A). Detailed steps for using the webserver are as follows:

Step 1: Choose Bacterial Genome

To predict sRNA targets, the first thing to do is to select the bacterial genome, for which the sRNA sequence must be scanned. There are 1031 genomes provided now, which were downloaded from the NCBI bacterial genome database. We will update the genome list for sRNATarget if a user sends us information about new sequenced genomes that have not yet been included. After the genome is chosen, the NCBI code will automatically appear in the box next to the genome box. For example, if we select the genome for Escherichia coli K12 substr MG 1655, the code NC_000913 will be displayed in the right hand box (Figure 2(A)).

Step 2: Choose sRNA or Input Custom sRNA Sequence

After the genome is selected according to the procedure outlined in step 1, all known sRNA sequences will be extracted from the related genome files, and sRNA names will be displayed in the sRNA list box. If a user only wants to predict the targets for a particular, known sRNA, the appropriate sRNA should be selected from the list box by clicking the sRNA name. Then, the corresponding sRNA sequence will be displayed in the sRNA sequence box in FastA format. If a user wants to enter a new sRNA sequence, the sequence can be input directly, or pasted from the clipboard into the sequence box in FastA format. The sRNA dicF is demonstrated in Figure 2(A). Only one sequence can be processed each time.

Step 3: Set the Score and Enter Email Address

Because our model sRNATargetNB is composed of 1000 classifiers, 1000 prediction results are obtained for each potential interaction. If there are 700 classifiers, for example, to predict the interaction as the positive sample, the score will be 0.7 (i.e. 700/1000). The default score has been set to 0.5 (500/1000). In addition, an email address is required so that the prediction results can be returned to the user by email.

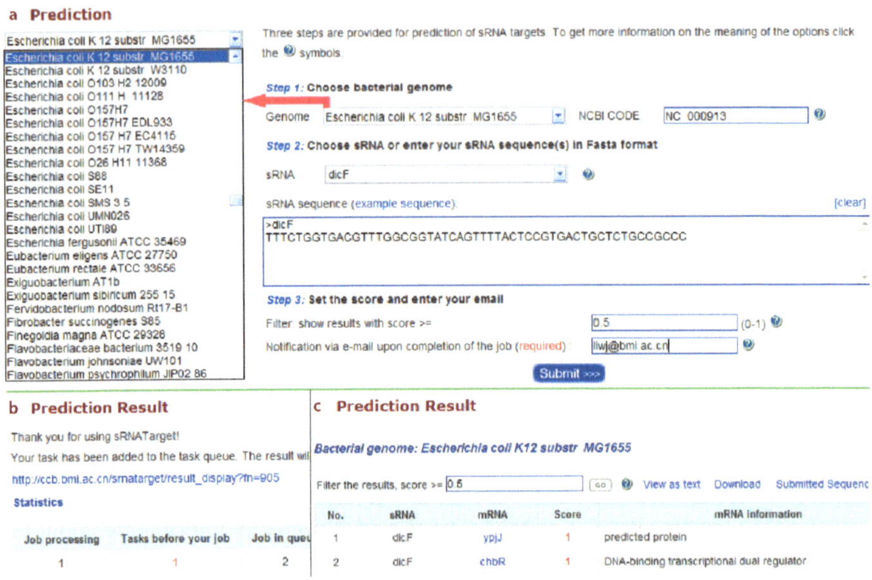

Fig. 2. The interfaces for prediction and prediction results are displayed

After steps 1–3 have been completed, task information is displayed by clicking the "submit" button (Figure 2(B)). A user can submit many tasks one by one. All the tasks are pushed into the queue according to time priority. The server will monitor the queue and process one task at a time. After computation, prediction results can be obtained in two ways, either by awaiting notification by email of the hyperlink for prediction results, or, by using the hyperlink appearing in a popup webpage. The running time for each task will be dependent on the length of the sRNA sequence and the size of the selected genome. For the sRNA DicF demonstrated here, the running time is about 30 minutes.

In the prediction results from sRNATarget (Figure 2(C)), there are five columns, representing the number of entry, sRNA name, mRNA name, score and mRNA annotation respectively. The prediction results are sorted in descending order of scores. In addition, a user can also download prediction results for all annotated mRNAs in the given genome.

2.2 Predicting sRNA Targets Using Windows System

Using the sRNATarget webserver, a user can carry out a genome-wide prediction of sRNA targets. However, before prediction results are validated using experimental protocols such as point mutation of potential mRNA targets or sRNA, it would be better to evaluate the mutated sRNA-mRNA interaction so as to ensure that the mutated mRNA target or sRNA is excluded from the interaction. This kind of analysis cannot be handled flexibly using the sRNATarget webserver. In addition, in view of concerns about confidentiality, some users prefer to use the program on their local machines. Therefore, we have provided a downloadable version of sRNATargetNB, which can be run easily on users' local machines. Detailed steps for downloading are as follows:

Step1: Install Perl and Bioperl Environments

Because program sRNATargetNB was written in Perl language and some Bioperl modules were recalled, it is necessary to install Perl and Bioperl on your local machine, which can be downloaded from http://www.activeperl.com and http://www.bioperl.org/, respectively.

Step 2: Download the Related Program

From the webpage http://ccb.bmi.ac.cn/srnatargetnb/, a user can download a compressed file named windows_version.rar, including four pieces of executable program (RNAfold.exe, b2shapiro.exe, sRNATarget.pl, and sRNATarget_genome.pl), demo files (srna.txt, srna1.txt, mrna.txt, NC_000913.fna, NC_000913.ptt) and the model file srnafun.txt. All the above files should be uncompressed in the same directory excluding any gaps. The program RNAfold.exe is downloaded from the Vienna RNA package. The program b2shapiro.exe is compiled according to the function with the same name in the Vienna RNA package. Here, Escherichia coli K-12 substr MG 1655 with NCBI code NC_000913 has been used as an example to demonstrate the usage processes of program sRNATargetNB. We also want to emphasize that model file srnafun.txt should not be changed. The detailed information for this file is as follows:

The first row: The meaning of each column has been provided. Because the model was constructed using the Tclass system, which was originally developed for gene expression profile-based sample classification, we still name each feature as gene. There are one constant term and six features in our model.

The second row: The detailed information for six features has been provided, which are the 198th, 567th, 1259th, 1839th, 4102nd and 5307th features, respectively. These features correspond to the rows in the RNA secondary structure profile, which can be downloaded from http://ccb.bmi.ac.cn/srnatargetnb/.

The third row and the 2005th row: This is just a representation of the matrix in MatLab language.

The fourth row: The first two numbers, 2 and 1000, indicate that the model is for two sample types, namely, positive and negative samples, and is composed of 1000 classifiers. There is no special meaning for the remaining six zeros, which are just used to fill in the blanks in the classification matrix.

The 5th to the 2004th row: These 2000 rows are used for the model functions, and each two consecutive rows correlate with one classifier. For example, the 5th and 6th rows are for the first classifier, 7th and 8th rows for the second classifier... and 2003rd and 2004th rows for the 1000th classifier. To determine whether a sRNA-mRNA complex is an interaction, we firstly calculate the values for the six features (the second row). Then, the six features are used to calculate two function values in each classifier. For example, we have the following two functions for the first classifier: When two values, FUN1 and FUN2, are calculated, the sRNA-mRNA complex will be classified into 'potential interaction' if FUN1>FUN2, or 'no interaction' if FUN1<FUN2.

FUN1=-67.19623+3.84889*F198+0.23376*F567+1.23650*F1259+0.06220*F1839+1.95922*F4102-0.55384*F5307

FUN2=-58.61756+3.54046*F198-0.00385*F567+1.17515*F1259+0.09715*F1839+1.53428*F4102-0.02817*F5307

The above processes are repeated for 1000 classifiers, and each sRNA-mRNA complex will have 1000 prediction results. If there are 700 prediction results, for example, to determine the sRNA-mRNA complex as the potential interaction, we call the sRNA-mRNA complex the potential interaction with probability 0.7 (700/1000).

Step 3: Prepare sRNA File and mRNA File

To evaluate whether an interaction exists between sRNA sequences and mRNA sequences, all sRNA sequences should be stored in a file in FastA format, which is a popular sequence format in bioinformatics. Each sequence is composed of two parts. The first part is an annotation line beginning with the symbol ">", and the second part consists of the sequence only. The following are the two sequences in FastA format.

Table 1. Part prediction results from sRNATargetNB program

sRNA	Gene	Score	Location	Strand	Length	PID	Gene
DicF	ftsZ	1	105305..106456	+	383	16128088	ftsZ
DsrA	hns	1	1291732..1292145	–	137	16129198	hns
GcvB	dppA	1	3704121..3705728	–	535	16131416	dppA
GcvB	oppA	1	1299206..1300837	+	543	16129204	oppA
MicA	ompA	0	1018236..1019276	–	346	16128924	ompA
MicC	ompC	1	2309668..2310771	–	367	16130152	ompC
MicF	ompF	0.967	985117..986205	–	362	16128896	ompF
OxyS	fhlA	0	2852360..2854438	+	692	16130638	fhlA
OxyS	rpoS	0	2864581..2865573	–	330	16130648	rpoS

>DicF
TTTCTGGTGACGTTTGGCGGTATCAGTTTTACTCCGTGACTGCTCTGC-CGCCC

>DsrA
AACACATCAGATTTCCTGGTGTAACGAATTTTTTAAGTGCTTCTTGCT-TAAGCAAGTTTCATCCCGACCCCCTCAGGGTCGGGATTT

The next thing is to prepare the mRNA file containing gene names only. Each line stands for a gene name. During evaluation of sRNA targets, the related sequences will be automatically extracted from the files *.fna and *.ptt, which are downloaded from the NCBI webpage. For example, the related two files for E.coli genome are NC_000913.fna and NC_000913.ptt, respectively. Here we want to emphasize that the program sRNATargetNB.pl was used to evaluate whether the nth mRNA in the mRNA file is the target of the nth sRNA sequence in the sRNA file. In addition, if there is more than one sequence with the same gene name, these sequences will be considered at the same time.

Step 4: Evaluate sRNA-mRNA Interaction

After preparing the files for sRNAs and mRNAs, we can evaluate whether there is an interaction between each sRNA sequence and a corresponding mRNA sequence using the following command in DOS window:

```
sRNATargetNB.pl srna.txt mrna.txt NC_000913.fna NC_000913.ptt result.txt
```

In the above command, srna.txt and mrna.txt stand for the sRNA sequence file and mRNA name file, respectively. E.coli genome files NC_000913.fna and NC_000913.ptt have been downloaded from the NCBI webpage. The prediction results are saved in the file result.txt, including sRNA name, gene name, interaction probability and gene annotations. Part prediction results are provided in Table 1.

Synonym	Code	COG	Product
b0095	–	COG0206D	GTP-binding tubulin-like cell division protein
b1237	–	COG2916R	global DNA-binding transcriptional dual regulator H-NS
b3544	–	COG0747E	dipeptide transporter
b1243	–	COG4166E	oligopeptide transporter subunit
b0957	–	COG2885M	outer membrane protein A (3a;II*;G;d)
b2215	–	COG3203M	outer membrane porin protein C
b0929	–	COG3203M	outer membrane porin 1a (Ia;b;F)
b2731	–	COG3604KT	DNA-binding transcriptional activator
b2741	–	COG0568K	RNA polymerase, sigma S (sigma 38) factor

Step 5: Predict sRNA-mRNA interaction on genome-scale

The command sRNATargetNB_genome.pl was developed for genome-wide prediction of sRNA targets. The command format is as follows:

sRNATargetNB_genome.pl srna1.txt NC_000913.fna NC_000913.ptt result.txt

All sRNA sequences must be saved in the file srna1.txt in FastA format. The prediction results are saved in the file result.txt.

2.3 Predicting sRNA Targets Under Linux as the Operating System

The counterparts of program sRNATargetNB.pl and sRNATargetNB_genome.pl have also been provided for the Linux operating system, which can be downloaded from the webpage http://ccb.bmi.ac.cn/srnatargetnb/. Users should put all files into the same directory. Then, the usages described in the above section can be applied for the Linux system.

3 Other Program Tools for Predicting sRNA Targets

In principal, any prediction tools for RNA-RNA interaction can be applied for the prediction of sRNA targets. In fact, to explain the excellent performance of their model IntaRNA (Busch *et al.* 2008), Busch and co-workers have applied many tools, including RNAhybrid, RNAplex, RNAup and TargetRNA, to predict the same sRNA-target interaction datasets. The results indicated that IntaRNA indeed had a better performance. Therefore, here we have only introduced two models, IntaRNA and TargetRNA, specially designed for the prediction of sRNA targets.

3.1 IntaRNA

The key idea behind the tool IntaRNA is that both binding site accessibility and seed regions are integrated into the algorithm. The related web page can be accessed at http://rna.informatik.uni-freiburg.de. From the perspective of correctly predicted base pairs on the test dataset, including 18 real sRNA-mRNA interaction data, the sensitivity and PPV are 0.783 and 0.787 respectively, which are better than the values obtained using other tools, including TargetRNA, RNAhybrid, RNAplex and RNAup. However, without the introduction of a seed feature, the model has the average sensitivity 0.699 and PPV 0.728, which are less than the values obtained using RNAup. The advantage of IntaRNA over RNAup is that IntaRNA is suitable for genome-wide target prediction because of its shorter runtime for the same sRNA sequences.

3.2 TargetRNA

The basic steps for using the TargetRNA model to predict sRNA targets are as follows (Tjaden 2008): The first step is to calculate the hybridization score of sRNA-mRNA interaction using either one of two models, basepair model or stacked basepair model. Then, extreme value distribution for the hybridization score is assumed. To this end, a large number of random mRNA sequences are generated, and the hybridization score of each of these and the sRNA sequence are calculated. Therefore, the statistical significance for each potential sRNA-mRNA interaction can be evaluated using P value. To optimize the related parameters such as the length of the seed region, a training dataset containing 12 samples was applied and eight samples were correctly predicted. The correct rate is about 66.7%. Finally, a webserver, TargetRNA, was provided, which can be accessed at http://snowwhite.wellesley.edu/targetRNA/.

4 An Example, Target Prediction for sRNA Yfr1

In the recent paper by Richter *et al.* (2010), the program IntaRNA and an experimental GFP reporter system were applied to identify targets for small RNA Yfr1, and two targets were confirmed among six potential targets experimentally tested. Here we take this sRNA, Yfr1, as an example to demonstrate the usage of our target webserver sRNATarget. See their paper for detailed sequence information for Yfl1 and two related mutation sequences.

First of all, the genome "Prochlorococcus marinus MED4" with NCBI code "NC_005072" was selected from the webserver prediction interface. Then, the small RNA Yfr1 and its two mutants were submitted as the query sequence one by one. Finally, by setting score and email and pressing "submit" button, 23 potential targets were found with scores of more than 0.5. Here we only focused on the six mRNAs that were tested in the paper by Richter *et al.* (2010). The probabilities of the six genes being potential targets of sRNA Yfr1 and its two mutants, Yfr1 M1 (CC->GG) and Yfr1 M2 (UCCU -> AAAA), are provided in Table 2.

Table 2. The probabilities of the six genes becoming potential targets using the webserver sRNATarget

	Yfr1	Yfr1 M1 (CC->GG)	Yfr1 M2 (UCCU -> AAAA)
PMM1119(som)	1	0.009	0.091
PMM0494(ppa)	1	0.155	0.477
PMM1121(som)	0	0	0
PMM1697	0	0	0
PMM0538	0	0	0
PMM0050(argJ)	0.001	0	0

The first column represents the potential targets using the model IntaRNA. The first row stands for small RNA Yfr1 and two mutants. The numbers in other cells (i,j)i=2,3,…,7; j=2,3,4 represent the interaction probability of potential target in row 'i' and sRNA column 'j', which were calculated by the model sRNATarget.

Combining the confirmed results in the paper by Richter *et al.* (2010), we can see clearly that four genes were predicted correctly, including PMM1119, PMM1697, PMM0538 and PMM0050. The correct rate is about 66.7% (4/6), which is higher than 33.3% (2/6) from the model IntaRNA. Therefore, the experimental data in their paper also gave better support to sRNATarget. In addition, we also observed that the gene PMM0494, although predicted by both IntaRNA and our webserver, was still confirmed as not being a target of Yfr1. From a statistical point of view, this gene should have a high chance of becoming targeted by the Yfr1 sRNA. Here we guess that the gene PMM0494 may become the target of Yfr1 under different conditions other than those tested in their paper. The detailed prediction results can be downloaded from the webpage http://ccb.bmi.ac.cn/srnatargetnb/.

From the above results, it can be seen that our model, as implemented in sRNA-Target, obtains better support. However, the results do not mean that our model will give better results in other cases for the following two reasons: Firstly, the number of samples tested here is still limited, which cannot be reliably applied to demonstrate performance of models. Secondly, those targets unconfirmed under one experimental condition may become targets under other experimental conditions, which leads to the difficulty of comparing different models using the data under the same experimental conditions. Thus, it is difficult to explain which model is better, IntaRNA based on given seed match and accessibility of binding sites or sRNATarget based on RNA secondary structure profile. However, the intersection set of the targets predicted from different methods may give better support for experimental verification. For example, two common genes, PMM1119 and PMM0494, were obtained when the models IntaRNA and sRNATarget were used at the same time, and one gene PMM1119 was confirmed. The correct rate was 50% (1/2), more than the present rate 33.3% (2/6). Here the model TargetRNA (Tjaden *et al.* 2006) was not used because the respective genome is not provided by their webserver.

5 Future Thinking

The main aim of this chapter has been to introduce the prediction model sRNA-TargetNB, developed by our center. Since released in March 2009, sRNATarget webserver has processed nearly 2000 tasks from more than 20 countries. However, one shortcoming of the model is that the accurate binding sites from sRNA-mRNA interactions are not provided. To provide better support for the sRNA research community, we are in the process of doing the following three things: The first thing is to construct a database for sRNA targets verified by experiments (Cao *et al.* 2010). In order to construct prediction models, it is necessary to collect as many samples as possible. The second thing is to develop more accurate models for prediction of sRNA targets. Recently we have just finished a model called sTarPicker. Compared to the sRNATargetNB model, the new model not only has higher prediction accuracy, but also provides binding sites from sRNA-mRNA interactions at the same time. The third thing is to develop sRNATarget2.0 version, providing binding sites for sRNA-mRNA interaction. We will put a hyperlink on the new information associated with sRNA target prediction on the webserver http://ccb.bmi.ac.cn/srnatargetnb/.

Acknowledgements

This work was supported by the National High Technology Development Program of China under Grant No. 2006AA02Z323, and the National Sciences Foundation of China under Grant No. 90608004 and 31071157.

References

Backofen R, Hess WR (2010) Computational prediction of sRNAs and their targets in bacteria. RNA Biol. 7:1–10.

Busch A, Richter AS, Backofen R (2008) IntaRNA: efficient prediction of bacterial sRNA targets incorporating target site accessibility and seed regions. Bioinformatics 24:2849–2856

Cao Y, Zhao Y, Cha L *et al.* (2009) sRNATarget: a web server for prediction of bacterial sRNA targets. Bioinformation 3:364–366

Cao Y, Wu J, Liu Q, Zhao Y *et al.* (2010) sRNATarBase: A comprehensive database of bacterial sRNA targets verified by experiments. RNA 16:2051–2057

Hofacker IL (2003) Vienna RNA secondary structure server. Nucleic Acids Res 31:3429–3431.

Larranaga P, Calvo B, Santana R *et al.* (2006) Machine learning in bioinformatics. Brief Bioinform 7:86–112

Li W, Xiong M (2002) Tclass: tumor classification system based on gene expression profile. Bioinformatics 18:325–326

Richter AS, Schleberger C, Backofen R *et al.* (2010) Seed-based INTARNA prediction combined with GFP-reporter system identifies mRNA targets of the small RNA Yfr1. Bioinformatics 26:1–5

Tarca AL, Carey VJ, Chen XW *et al.* (2007) Machine learning and its applications to biology. PLoS Comput Biol 3(6): e116.

Tjaden B (2008) TargetRNA: a tool for predicting targets of small RNA action in bacteria. Nucleic Acids Res 36(Web Server issue):W109-W113

Tjaden B, Goodwin SS, Opdyke JA *et al.* (2006) Target prediction for small, noncoding RNAs in bacteria. Nucleic Acids Res 34:2791–2802

Wu B, Cha L, Du Z *et al.* (2007) Construction of mathematical model for high-level expression of foreign genes in pPIC9 vector and its verification. Biochem Biophys Res Commun 354:498–504

Zhao Y, Li H, Hou Y *et al.* (2008) Construction of two mathematical models for prediction of bacterial sRNA targets. Biochem Biophys Res Commun 372:346–350

Index

MIX
Papier aus verantwortungsvollen Quellen
Paper from responsible sources
FSC® C105338

If you have any concerns about our products,
you can contact us on
ProductSafety@springernature.com

In case Publisher is established outside the EU,
the EU authorized representative is:
Springer Nature Customer Service Center GmbH
Europaplatz 3, 69115 Heidelberg, Germany

Printed by Libri Plureos GmbH
in Hamburg, Germany